D1282580

ROBOTICS AND AI

Andrew C. Staugaard, Jr.
Department of Computer Science
The School of the Ozarks

ROBOTICS AND AI:
AN INTRODUCTION TO APPLIED MACHINE INTELLIGENCE

PRENTICE HALL, INC., Englewood Cliffs, N.J. 07632

Library of Congress Cataloging-in-Publication Data

Staugaard, Andrew C.
Robotics and AI.

Bibliography: p. 337
Includes index.
1. Robotics. 2. Artificial intelligence. I. Title.
TJ211.S73 1987 006.3 86-25140
ISBN 0-13-782269-3

Editorial/production supervision and
interior design: Tom Aloisi
Cover design: 20/20 Services, Inc.
Manufacturing buyer: Carol Bystrom

To **You**, the Student: May This Book Enhance
Your Quest To Be All That You Can Be

Printed in the United States of America

10 9 8 7 6 5 4 3 2 1

ISBN: 0-13-782269-3 025

Prentice-Hall International (UK) Limited, *London*
Prentice-Hall of Australia Pty. Limited, *Sydney*
Prentice-Hall Canada Inc., *Toronto*
Prentice-Hall Hispanoamericana, S. A., *Mexico*
Prentice-Hall of India Private Limited, *New Delhi*
Prentice-Hall of Japan, Inc., *Tokyo*
Prentice-Hall of Southeast Asia Pte. Ltd., *Singapore*
Editora Prentice-Hall do Brasil, Ltda., *Rio de Janeiro*

Contents

Preface

Welcome to the fascinating world of intelligent machines! Imagine machines that might someday think, see, and communicate by voice very much the same as you and I do. Such machines will become a reality through the direct application of concepts from a branch of computer science called ***artificial intelligence,*** or ***AI***. In this book, you will learn about the basic concepts of AI and how AI is applied to produce intelligent machines.

An area that is particularly suited to applied machine intelligence is robotics. An intelligent robot must be capable of sensing its surroundings and responding to a changing environment much the same as we do. In fact, the robot that can perform humanlike intelligent tasks is the ultimate intelligent machine. For this reason, this book focuses on the applications of artificial intelligence to robotic systems. You will learn about voice communication systems, vision systems, range and navigation systems, and tactile sensing systems.

In the field of artificial intelligence, it often seems as though the author is talking to his or her colleagues rather than **you**, the student. In robotics, important operational concepts are often left out of the discussion. This book is my attempt to separate the wheat from the chaff. I have strived to write a readable text that zeros in on the important concepts of artificial intelligence and how those concepts can be applied to produce intelligent machines. At the end of this book, you will not only understand the **what's**, but more important, the **why's** and **how to's**.

One prerequisite to reading this book is a fundamental knowledge of Boolean algebra, such as you might get in a digital electronics or introductory computer science course. In addition, you should have a basic understanding of how simple electronic components, such as resistors, capacitors, inductors, diodes, and transistors, operate.

This book is meant to be used at the introductory level as a first exposure to artificial intelligence and/or intelligent machines. As a result, the book is ideal for a one semester introductory course in undergraduate technology, engineering, and computer science programs. The basic concepts learned here will prepare you for more advanced study in the areas of artificial intelligence, vision systems, voice communication systems, tactile sensing systems, and intelligent machines in general.

The first chapter lays the foundation for the rest of the text. Chapters 2 and 3 are devoted entirely to a discussion of the basic concepts of artificial intelligence and the related field of knowledge representation. Chapters 4 through 8 discuss the application areas of speech synthesis, speech recognition and understanding, vision, range finding and navigation, and tactile sensing, respectively.

Several people have made valuable suggestions and contributions during the preparation of this book. I want to thank Vince Leonard and Bob Kochersberger of Jamestown Community College, Pete Ho of the University of Missouri–Rolla, and finally, Marvin Minsky of the Massachusetts Institute of Technology for giving me insight into the tremendous potential of AI.

Andrew C. Staugaard, Jr.

1 Intelligent Robotics

The field of robotics is in a state of rapid development. Early robots were nothing more than clever mechanical devices with cams and stops that performed simple pick-and-place operations. As computer technology improved, robots became more sophisticated. Computer control allows robots to perform more precise industrial operations, such as welding, spray painting, and simple parts assembly. However, such operations do not really require the robot to "think." The robots are simply programmed to perform a series of repetitive tasks. If anything interferes with the preprogrammed task, the robot must be stopped, since it is not capable of sensing its external environment and thinking its way out of a problem. For robots to become more efficient, maintenance free, and productive, they must be capable of sensing external conditions and thinking very much like you do. Such abilities require the direct application of ***artificial intelligence*** and ***sensory perception.***

This brief chapter lays the foundation for the entire text. The chapter begins with an overview of robot technology leading to a discussion of industrial versus personal robots, the limitations of robots, and the future of robotics. Elements of intelligent robotics, such as programmability, flexibility, input/output, and sensory feedback are introduced in preparation for subsequent chapters.

1-1 ROBOT TECHNOLOGY

There has been much discussion as to what type of apparatus constitutes a robot. For instance, a pilot flying an airliner operates controls that activate various air control surfaces to allow the plane to fly a course from point A to point B. Clearly, this is not a robotic type of action, but rather the effect of direct mechanical linkages between the pilot and the control surfaces. But suppose that the airliner is equipped with a computer-based navigation control system. Such systems are capable of literally "flying" the airplane from take-off to landing without any interference from the pilot. A passenger cannot tell who is flying the airplane, the pilot or the computer. Does the plane become a robot when being flown by the computer? Many contend that a machine becomes a robot when it can perform physical tasks without human intervention. Does this make the computer-flown airplane, or for that matter your automatic dishwasher, a robot?

Webster defines a robot as "an automatic apparatus or device that performs functions ordinarily ascribed to human beings or operates with what appears to be almost human intelligence." The Robot Institute of America says, "a robot is a reprogrammable, multifunctional manipulator designed to move material, parts, tools, or specialized devices through variable programmed motions for the performance of a variety of tasks." Do either of these definitions rule out a computer-flown airliner or your automatic dishwasher from being classified as a robot?

Most people think of the devices pictured in Figure 1-1 as robots. But what makes these devices different from your automatic dishwasher? Is it the fact that they are computer controlled? Many automatic dishwashers on the market today are microprocessor based, and thus computer controlled.

As you can see, it is rather difficult to nail down a precise definition for a

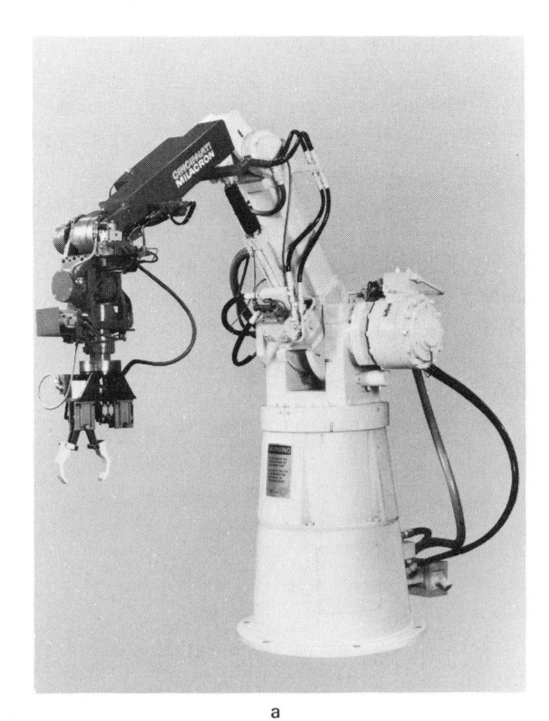

a

Figure 1-1 (a) Ideal for complex or heavy-duty materials-handling tasks, the T^3586 Robot from Cincinnati Milacron has infinitely variable six-axis positioning and a payload capacity up to 225 pounds. In addition to handling hefty loads at high speeds and performing in hazardous or difficult-to-reach places, it can load and unload machines, palletize, inspect, sort, weld, and assemble. The T^3586 features a hydraulic power system, a simple hand-held *teach pendant* for the operator to lead it through its moves, and an advanced computer control. (b) Cincinnati

Milacron's all electric, computer-controlled, $T^3$726 Industrial Robot for welding, plasma cutting, parts handling, or assembly applications. The $T^3$726 Robot features a payload capacity of 14 pounds, six axes of motion, and Milacron's unique three-roll wrist for true application flexibility. Each axis is powered by its own dc motor. (c) Cincinnati

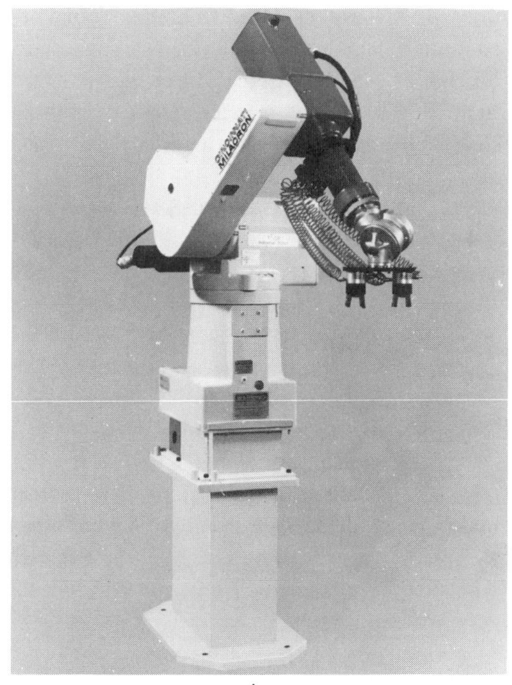

b

Milacron has introduced a low-cost new robot specifically developed for materials-handling tasks. The new $T^3$363 is the first in a series Milacron designed to bring the flexibility of robots to materials-handling applications at an affordable price. At an average cost of under $30,000, the all-electric $T^3$363 will move payloads up to 110 pounds and perform most materials-handling tasks, including palletizing, parts handling, package handling, and machine loading and unloading. It has three standard axes of motion, with the option of a fourth, and an easy-to-use microprocessor-based control capable of running a work cell. (*Courtesy Cincinnati Milacron*)

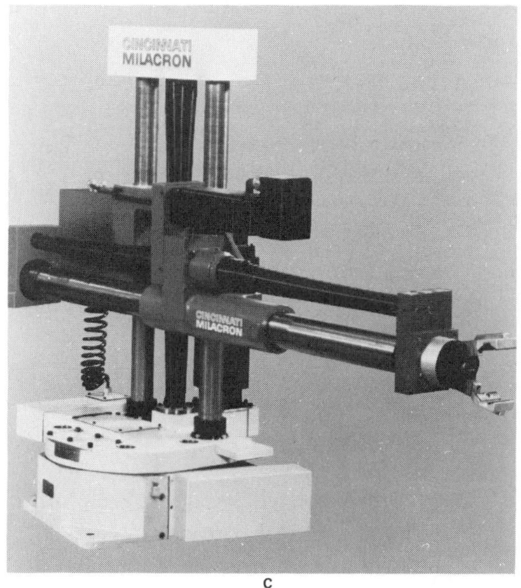

c

robot. However, most people think "robot" when they see a machine performing much like a human being, exhibiting a certain degree of "intelligence." Thus, a robot must possess a certain degree of machine, or artificial, intelligence. For now, I will leave the precise definition of a robot to *Webster* and the Robot Institute of America. However, I will attempt to define machine, or artificial, intelligence in a subsequent chapter.

Robots have developed along two paths: ***industrial*** and ***domestic.*** Industrial robots, such as those pictured in Figure 1-1, have been developed to perform a variety of manufacturing tasks, such as machine loading/unloading, welding, spray painting, and simple product assembly. However, most industrial robots have very limited sensing ability. If assembly parts are not presented to the robot in a precise, repetitive fashion, the robot cannot perform its task. If an object enters the work area of a robot, the robot and the object will likely collide, resulting in damage to both. In other words, most industrial robots are unintelligent, and cannot hear, see, or feel. Imagine trying to teach a person with all these handicaps to perform precision assembly operations. To become more efficient and productive, an industrial robot must be capable of *sensing* its surroundings and possess enough *intelligence* to respond to a changing environment much the same as we do. Thus, elements of sight, touch, and corrective action (intelligence) are essential to the evolution of industrial robotics.

Domestic or personal robots have been developed primarily for the home hobbyist market. Most of these devices are capable of voice synthesis (speaking), sensing light levels, detecting motion, and moving about with simple programmed instructions and sonar-type navigation systems. However, existing domestic robots, like their industrial counterparts, have limited intelligence, thereby limiting their applications.

Suppose you require a domestic robot that will perform a simple everyday task like cleaning off the kitchen table and washing the dishes after each meal. Imagine the sensing abilities and intelligence that such a robot must possess! First, it must be able to distinguish the dishes and eating utensils from other objects, like the morning paper, that might be on the table. Then it must have the ability to pick up and manipulate the delicate dishes, perform the washing and drying task, and place the dishes and utensils in their proper storage location. In addition, it must be capable of sweeping up and disposing of any broken dishes, as well as compensating for any change in its environment during the cleaning task. And what happens if the pet cat gets in its way during the washing operation? Of course, the robot must also be capable of receiving voice commands to warn it of any impending disaster or to alter its operation.

Now imagine the additional sensing abilities and intelligence that the same domestic robot must have to perform other household tasks, such as washing clothes, running the vacuum, cleaning the bathroom, taking out the garbage, and mowing the grass. Such an advanced robot must possess *sensing* and *intelligence* abilities similar to those of a human.

From the preceding discussion, it is clear that the two major requirements of an advanced robot (industrial or domestic) are that it must sense its surroundings and be intelligent enough to compensate for changes in its environment. In fact, you could say that ***sensory perception*** and ***intelligence*** are the common denominators of any advanced robot.

In most cases, intelligent robots will eventually incorporate elements of vision, touch, and speech recognition, all of which require the direct application of artificial intelligence. Such a robot must also possess adequate manipulators and end-effectors to perform the given task. Thus, the three key ingredients of an intelligent robot must be ***sensory perception, intelligence,*** and ***adequate end-effectors*** to perform a variety of applications tasks. Let's take a closer look at each of these three fundamental components of an intelligent robot in preparation for the material that follows in later chapters.

1-2 END-EFFECTORS

Robot end-effectors, or hands, can take on many forms. In general, end-effectors are the mechanical devices with which the robot manipulates the real world. The devices required to operate the robot's sensors must also be included. If a robot uses a TV vision system, it must be able to turn and focus the camera in the direction it wishes to see. For industrial robots, end-effectors also include the tools required to perform a given task.

Much of the research on robot end-effectors has been centered on improving the flexibility of the robot's hands and arms. Currently, most industrial end-effectors are single degree of freedom devices specialized to a given task. Examples include welding torches, paint sprayers, two-fingered grippers, and vacuum cups. Several of these specialized industrial end-effectors are illustrated in Figure 1-2. However, such designs are not well suited for general-purpose operations. As a result, many researchers are attempting to develop multifingered grippers similar to the human hand. This is not an easy task since the human hand has 22 individual joints, or axis of motion. The best attempts have only resulted in three-fingered grippers.

The design of end-effectors is a science in itself. Aside from the actual mechanical design, any associated control software must be developed. This is complicated by the fact that the end-effector must incorporate various sensing devices. The sensing devices are used to measure and feed back process variable data so that control decisions can be made. This is referred to as ***sensory feedback.*** The controller software must then coordinate any movement of the end-effector with the sensory feedback data. The mechanical design technology required for sophisticated end-effectors will not be covered in this book. Rather, my emphasis in subsequent chapters will be on the fundamental concepts of artificial intelligence and sensory feedback that must be incorporated into any sophisticated end-effector, or manipulator, design.

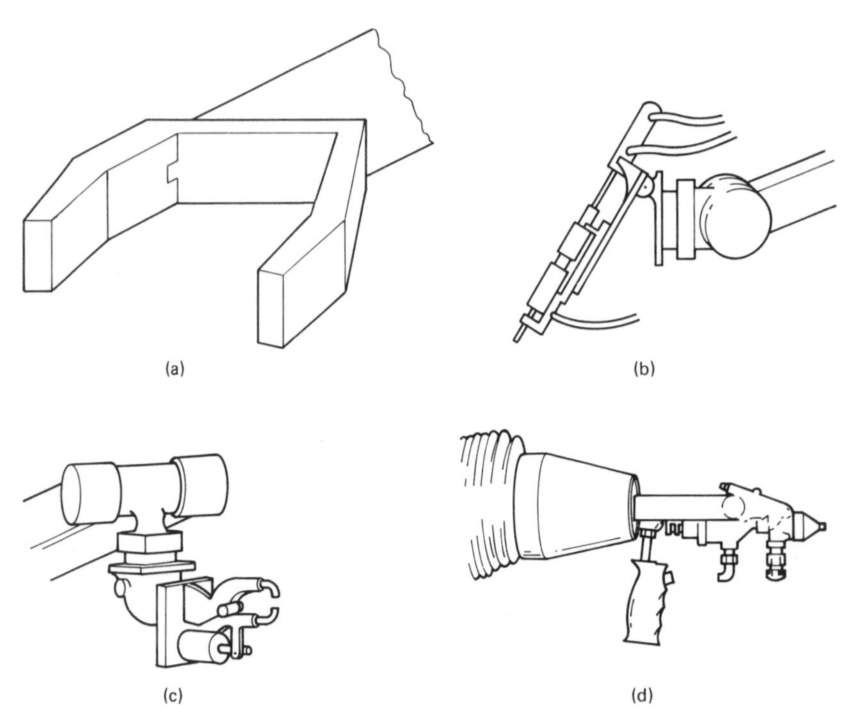

Figure 1-2 Most industrial end-effectors are single degree of freedom devices specialized to a given task: (a) standard gripper; (b) arc welder; (c) spot welder; (d) paint sprayer.

1-3 SENSORY PERCEPTION

Sensory perception, or feedback, must be incorporated into robot designs if robots are to become more than just mechanical manipulators. The categories of sensory perception include vision, tactile sensing, range finding, and voice communication. In most robot applications, the preceding order of categories represents the relative application enhancement priority. In other words, the possible range of applications of a robot is enhanced the most by vision. Tactile sensing, or touch, runs a close second, especially for industrial assembly operations. Many existing product assembly tasks employ some form of tactile sensing in lieu of vision. Range finding is required for three-dimensional (3-D) vision and also for position and proximity, or nearness, sensing. Range-finding techniques are also important for robot navigation, whether a robot is stationary and navigates its gripper or mobile and navigates its body. Finally, voice communication is less important, but makes the robot more user friendly and permits verbal person-to-robot communication for programming and control purposes. Let's take a short look at each of these sensory perception categories.

Vision

Vision is probably the single most important sensing ability that an intelligent robot can possess. An industrial robot that can "see" is capable of parts recognition, parts sorting, and precision assembly operations. Likewise, a domestic robot requires a sense of vision to perform everyday household tasks and to navigate from room to room. Any robot would be more intelligent if it could acquire information about its environment through its own vision system rather than being limited to a knowledge base provided by its programmer. Thus, a robot vision system can actually be used to build the knowledge base of the robot.

The science of computer vision is called ***visual machine perception.*** Visual perception is easy for us since once we become familiar with an object it is easily recognized. We seem to just "know" that a dog is a dog and a cat is a cat. However, visual perception is very difficult for a computer. An enormous amount of artificial intelligence is required for a computer, or robot, to distinguish between a dog and a cat.

Simple computer vision systems detect different levels of light. These systems typically utilize a light-dependent resistor (LDR) or phototransistor circuit to translate light levels to a proportional analog voltage level. The analog voltage is then converted to a digital value with an analog-to-digital (A/D) converter and read by the computer. Such a system is illustrated in Figure 1-3. A robot using this system can be readily programmed to awaken you at sunup or to detect when an object is in a given position. However, it cannot really recognize an object or perceive its surroundings. Actually, you could say that the system described is a light-detection system rather than a vision system.

True computer vision involves the transformation, analysis, and understanding of light images. As a result, the science of computer vision can be reduced to three fundamental tasks: ***image transformation, image analysis,*** and ***image understanding.***

Image transformation involves the conversion of light images into electrical signals that can be used by a computer. Many existing computer vision systems

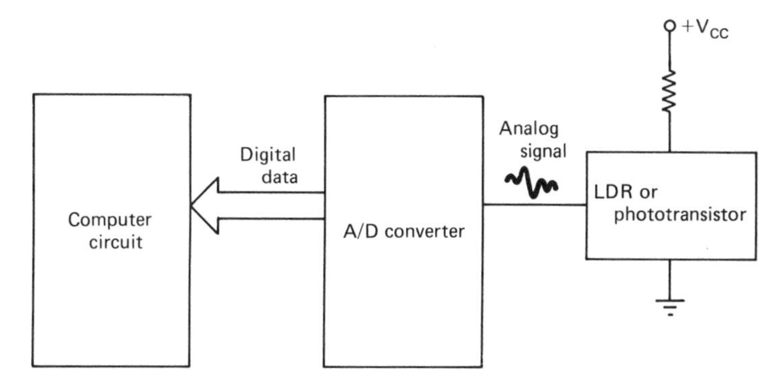

Figure 1-3 Simple robot light-detection systems only detect light levels and are not capable of vision.

utilize a TV camera, photodiode array camera, or change-coupled device (CCD) camera for the ***imaging device***, or eye, as illustrated in Figure 1-4. As the camera scans a scene, its output is converted to a digital code by an A/D converter and stored in memory as a digital image. The computer must then analyze the digital images and apply some degree of artificial intelligence to understand the scene.

Once a light image is transformed into an electronic image, it must be analyzed to extract such image information as object edges, regions, boundaries, color, and texture. In some systems, the digital images are compared to ***image templates*** to classify and recognize objects in the scene. The image templates have been previously stored in the computer memory for the comparison task. Using the template technique, the computer can recognize distinct, well-defined patterns. However, the problem is in storing enough memory templates to cover all possible scenes in a given environment. More intelligent systems use edge and region image analysis methods. In these systems, the edges and regions of objects in the scene are analyzed in order to generate information for the image-understanding process.

For a computer to truly see, it must be capable of understanding a given scene and use the knowledge gained from the scene for future problem-solving tasks. This is by far the most difficult of all computer vision tasks and requires the direct application of artificial intelligence. Image transformation, analysis, and understanding are covered in Chapter 6.

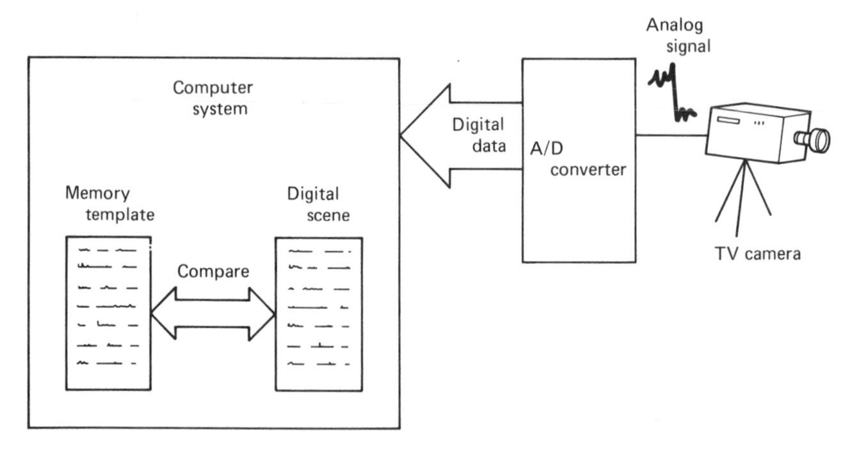

Figure 1-4 Many existing vision systems utilize a TV camera as the visual sensing device.

Tactile Sensing

Next to vision, tactile sensing, or touch, most enhances a robot's abilities. Like a blind person, a blind robot can be extremely effective in performing an assembly task using only a sense of touch. In fact, many times tactile sensing is more critical to a precision assembly operation than is vision.

During a precision assembly task, an industrial robot must be capable of sensing any problems encountered during assembly from the interaction of parts and tools. Such interference and fit problems are created by part tolerance, misalignment of parts, tool wear, and the like. For example, suppose an industrial robot is making a gear assembly. One gear is to be placed on an axle, followed by a second gear on an adjacent axle. The teeth of the second gear must mesh with the teeth of the first gear. Using its tactile sensing abilities, the robot must sense any interference between the two gears and rotate the second gear until its teeth mesh with those of the first gear. If the robot cannot see what it is doing, this would be an impossible assembly task without tactile sensing. Imagine trying to perform the same operation blindfolded and without any sense of touch–impossible!

The term ***compliance*** describes the allowed movement between mating parts for the purpose of alignment during assembly tasks. In other words, a ***compliant robot*** "complies" with external forces by modifying its motions in order to minimize those forces. A robot that could successfully perform the preceding gear assembly operation by compensating for interference between the gears would be called a compliant robot. There is both ***active*** and ***passive compliance.*** Active compliance employs sensory feedback, such as tactile sensing, whereas passive compliance does not incorporate any sensory feedback.

As you will discover in Chapter 8, simple tactile sensing can be accomplished by placing microswitches, strain gauges, pressure transducers, and optical sensors in the end-effector of the robot. Magnetic Hall-effect devices and sonar sensors are also sometimes used. However, the most important parameter that must be measured to achieve tactile sensing is force. To accomplish this, strain gauge and pressure transducers are commonly placed on the robot's arm, wrist, and fingers. In addition, artificial skin pads that sense pressure are placed on the robots gripper. These ***tactile arrays*** provide a sense of feel for the robot and can be used to determine the position and orientation of an object and to aid in the identification of unknown objects.

Information generated by these various force-sensing devices can be transmitted to the central robot computer or analyzed by ***sensing cells*** located within the manipulator. Sensing cells are single-chip microcomputers dedicated to the tactile sensing task. This subject is dealt with further in Chapter 8.

Range Finding and Navigation

Range, or distance, data must be obtained in order for a robot to create the 3-D information necessary for real-world navigation. Such information is required whether the robot is stationary and navigating its gripper or mobile and navigating its body. Vision systems can employ stereo cameras used like your eyes to determine depth through triangulation. However, vision systems are relatively sophisticated and expensive. Many simple robot navigation problems do not require such a sophisticated solution.

A simpler solution to range finding is found in time-of-flight ranging systems. These systems measure the amount of time it takes a radio signal, sound, or light to reach and return from an object. Such systems can be used independently to generate range data or in conjunction with a single-camera vision system to generate 3-D images.

Depending on the application, even simple position and proximity sensors can be used to complement or replace more sophisticated vision and time-of-flight ranging systems. Position and proximity sensing devices include mechanical limit switches, inductive sensors, capacitive sensors, magnetic sensors, and optical sensors. All these various robot range-finding, navigation, position, and proximity sensing techniques will be explored in Chapter 7.

Voice Communication

Imagine a computer, or robot, with complete voice communication abilities. Aside from being extremely user friendly, such a computer could be programmed and controlled without the use of special programming languages. And it could be easily queried as to why it performed a given operation or reached a certain conclusion. Ultimately, computers and robots will possess such communications skills.

Voice communication in robots requires the application of two separate, but related, technologies: ***speech synthesis*** and ***speech recognition.*** Speech synthesis, or speaking, is a relatively proven technology as compared to speech recognition. Many computer systems are capable of speaking electronically using voice synthesizer circuits. On the other hand, much research is taking place in the field of speech recognition. As you are about to discover, speech recognition requires the direct application of artificial intelligence.

Speech Synthesis. Two fundamental techniques are used to produce electronic speech. One technique simply stores words or phrases in read-only memory, or ROM. When the computer must speak, the given word or phrase is simply read from memory, converted from digital to an analog signal, and amplified to produce the required sound. Many "talking" appliances use this technique. However, it is not very practical in robotics. The obvious disadvantage of storing words and phrases in memory is that the entire vocabulary of the robot is fixed, and thus not very flexible. Furthermore, larger vocabularies require larger amounts of memory. For these reasons, the stored-word technique is not very widely used for speech synthesis in robots. Rather, another technique, called ***phoneme speech synthesis***, is commonly used in robotics.

In phoneme speech synthesis, the computer puts together sounds, called ***phonemes***, to produce a given word. Words are then strung together to produce phrases. Any word in the English language can be broken down into the same general set of approximately 64 sounds, or phonemes. By storing these fundamental building blocks of speech in a computer's memory, the computer is capable of an almost infinite vocabulary. Moreover, a minimum amount of memory is required to store the phonemes as compared to storing words. The computer must then be programmed to concatenate the phonemes to generate a given word or phrase. In addition, volume and inflection must be added to make the speech sound human. Speech synthesis is the topic of Chapter 4.

Speech Recognition. Speech recognition is more complicated than speech synthesis. A large amount of research and development is being applied to the speech-recognition problem. Basically, two types of systems are presently used for speech recognition: ***isolated-word recognition*** systems and ***connected-speech understanding*** systems.

Isolated-word recognition is used where the computer expects a specific input, such as *yes* or *no*. The technique requires that the computer acoustically analyze the spoken word and match the resulting digital voice pattern to a voice template stored in memory. Such systems are sometimes called ***template-matching*** word recognition systems. However, there are several problems with template matching systems. First, the system is ***speaker dependent*** and must be "trained" by the user. The training session requires that the user speak each word contained in a fixed vocabulary into the system to create the memory voice templates. This is both time consuming and may have to be performed several times to be successful. Second, larger vocabularies require large amounts of memory to store the voice templates. Finally, most template systems are single-word recognition systems and cannot recognize connected speech. This imposes speaking-rate restrictions on the user, since each word must be recognized individually.

Connected-speech understanding systems, on the other hand, are ***speaker independent*** and actually interpret the spoken phrase. In addition, these systems use a technique called ***feature analysis*** and do not require large amounts of memory for large vocabularies. Feature analysis systems do not use voice templates stored in memory. Rather, phonetic data are stored to allow the system to recognize the phonetic, or phoneme, features of a spoken word. Memory requirements are thus kept to a minimum regardless of the vocabulary size, since any language has a limited number of distinct phonetic features.

Speech understanding systems incorporate AI techniques to categorize the spoken word into broad phonetic, or sound, categories. This allows the system to generate a small set of possible word candidates. Fine phonetic distinctions are then made to determine which one of the possible word candidates was actually spoken. Further discussion of this technology is given in Chapter 5, where both the isolated-word recognition and connected-speech understanding systems will be explored.

1-4 ROBOT CONTROL AND INTELLIGENCE

A major component of any robotic system is the robot controller. In fact, the sophistication of the robot controller determines the robot's possible application range and classifies its intelligence level.

Early robot controllers used a rather sophisticated collection of mechanical levers, drums, cams, rods, gears, pins, and so on, to control the action of the robot.

The familiar music box drum shown in Figure 1-5 provides a simple example of such a controller. However, as robot operations began to require more precision and programming flexibility, digital playback controllers, such as the one shown in Figure 1-6, were required. Here a digital computer program, stored on magnetic tape or disk, is loaded into the memory of a digital computer-based type of controller. When the program is executed, the robot simply sequences through the programmed operations. However, without sensory feedback, the robot cannot be truly intelligent.

Robots that employ simple computer-based control are more than adequate to perform many common industrial operations. They are very precise and can be reprogrammed to suit a wide variety of applications. However, the controller requirements of intelligent robots are quite different from those that employ simple computer-based control. Aside from controlling its various manipulators and end-effectors, an intelligent robot must incorporate sensory perception and make decisions based on sensory feedback signals. Such sense and control operations require the direct application of artificial intelligence in a sophisticated computer

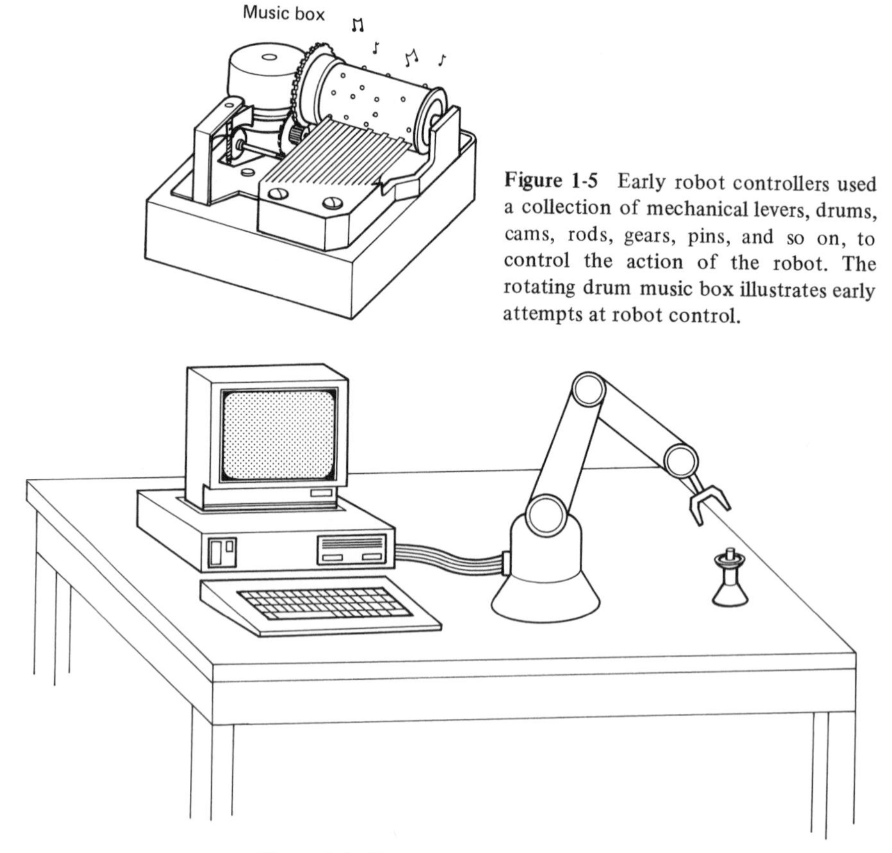

Figure 1-5 Early robot controllers used a collection of mechanical levers, drums, cams, rods, gears, pins, and so on, to control the action of the robot. The rotating drum music box illustrates early attempts at robot control.

Figure 1-6 Computer-based controller.

network. Figure 1-7 illustrates the idea that AI is the common denominator for most of the components required in an intelligent robot. For this reason, Chapter 2 is devoted entirely to the fundamental concepts of artificial intelligence.

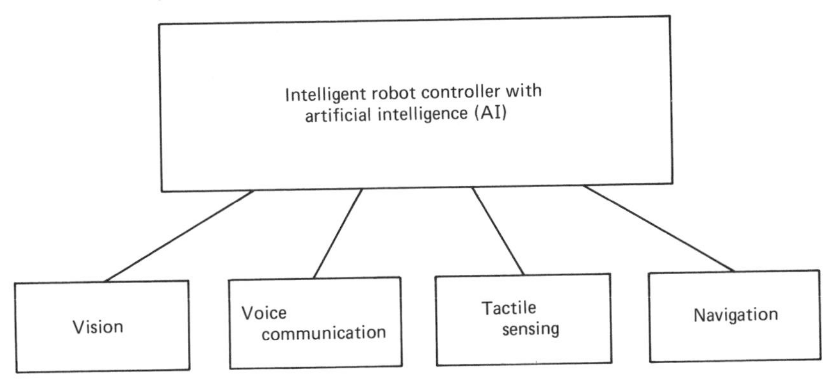

Figure 1-7 Artificial intelligence (AI) is the common denominator for most components of an advanced robot.

Knowledge Engineering

The design of artificial intelligence into a computer, or robot, involves a relatively new field of engineering called ***knowledge engineering.*** Knowledge engineering is a branch of computer engineering and thus involves two fundamental design components: hardware design and software design.

The hardware design aspect of knowledge engineering often involves the networking of a multiprocessor system, as illustrated in Figure 1-8. In a multiprocessor system, several (multi) computer processors are interconnected to form a computer network, with each processor dedicated to a specific task. A central processor

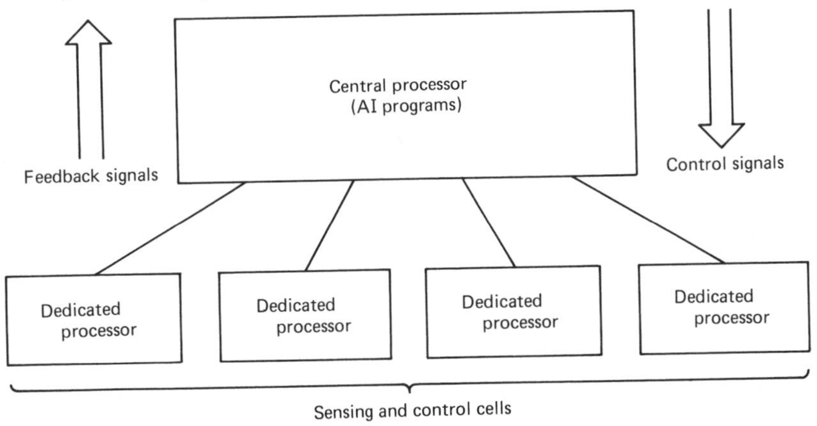

Figure 1-8 The hardware design aspect of knowledge engineering often involves the networking of a multiprocessor system.

supervises the entire system and makes the final control decisions. This is where the more sophisticated AI programs must reside. Separate processors are dedicated to the various sense and control tasks. In this text, a processor dedicated to a sensing task is referred to as a ***sensing cell.*** Likewise, a ***control cell*** is a processor dedicated to a given control task. Sensing and control cells are actually single-chip microcomputer devices made possible by very large scale integration (VLSI) technology.

Individual sensing cells are dedicated to the different sense operations, such as vision, tactile sensing, voice recognition, and navigation. These devices are used to collect, partially analyze, and transmit the sensory feedback data to the central processor. The central processor uses the sensory feedback data for overall sensory coordination and major control decisions. Simple control decisions can be made at the sensing-cell level.

Individual VLSI control cells receive control commands from the central processor and generate the control signals required to carry out the commands. Individual control cells might be applied to each individual axis of movement, depending on the sophistication of the robot.

The multiprocessor approach to robot control illustrated in Figure 1-8 requires ***hierarchical planning*** in both the hardware and software design of the robot. Hierarchical planning involves dividing problems into subproblems from both a hardware and software perspective. The central nervous system, which generates behavior in biological organisms, is a good example of a hierarchically structured system. A cross-section of the human central nervous system is shown in Figure 1-9. It has been shown that basic muscle reflexes remain if the brain stem is severed at A; the coordination of these reflexes required to stand is still possible if the brain is cut at B. Walking requires the region below C to be operable, and primitive tasks can be performed if the region below D is available. However, thinking, decision making, and more complex tasks require an intact cerebral cortex. Many believe that, if computers, or robots, are to truly think, their controllers or hardware must be hierarchically constructed similar to that of the human nervous system.

The software design aspect of knowledge engineering also involves the science of ***knowledge representation.*** If a computer is to recognize a scene, it must have prior knowledge about that scene. In other words, it must have a data base that will permit it to look, see, understand, and act. If a computer is to diagnose a disease, it must have specific knowledge as to the symptoms of that disease. Thus, knowledge is an important prerequisite for understanding and intelligence.

Knowledge representation can be divided into two categories: ***general knowledge*** representation and ***specific***, or ***expert, knowledge*** representation. Of the two, general knowledge representation provides the biggest challenge to knowledge engineers. General knowledge representation in a robot requires that the robot contain a sufficient data base to understand its surroundings, or environment, by using its various sensing abilities. Acquiring such general knowledge is the thrust of much of the research in knowledge engineering. Several researchers are studying the various levels of knowledge that have been acquired by children and exploring how such knowledge can be programmed into a computer. These researchers observe the

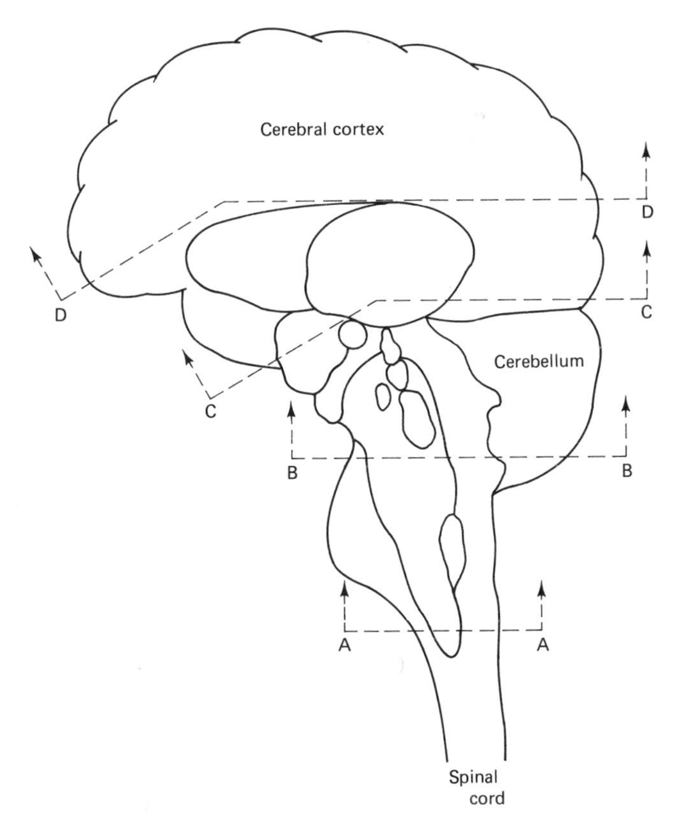

Figure 1-9 The human central nervous system is a good example of a hierarchically structured system.

ways in which children learn in an effort to develop procedures and techniques whereby computers, or robots, can learn and build on their knowledge base.

Expert knowledge, on the other hand, is extensive knowledge in a very specialized field. Several successful AI programs have been developed that incorporate the knowledge of human experts into a computer. Such expert systems must gain their knowledge base by interacting with the human expert. The knowledge engineer then acts as the interface between the system and the expert.

One of the most notable expert system programs is a disease-diagnosis program called MYCIN. The MYCIN system aids the physician in the diagnosis and selection of treatment for patients with meningitis and bacterial infections. MYCIN provides an interactive dialogue with the physician and is able to explain its line of reasoning. In this way the physician can verify that the correct line of reasoning was taken to arrive at a particular disease diagnosis and treatment. The MYCIN program also utilizes a knowledge-acquisition subsystem called TEIRESIAS. The TEIRESIAS subsystem allows the physician, with the help of the knowledge engineer, to expand and modify the knowledge base of the system. Systems using AI programs like

MYCIN are indeed expert within a very specialized field, but they are not capable of the general-purpose reasoning and common sense required for a robot.

The field of knowledge engineering and general-purpose knowledge-based systems will be explored further in Chapter 3.

SUMMARY

Robot technology has taken two evolutionary paths: industrial and domestic. To perform precision assembly and control operations, industrial robots must be capable of sensing their surroundings and possess enough intelligence to respond to a changing situation or environment. Likewise, to be truly useful, domestic robots must possess significant sensing and intelligence capabilities. Thus, sensory perception and intelligence are the common denominators of any advanced robot.

Sensory perception categories include vision, tactile sensing, range finding, navigation, and voice communication. Of these, the range of applications of a robot is enhanced most by vision, followed by tactile sensing, especially for industrial assembly applications. The science of providing vision for a robot is called visual machine perception. Most existing vision systems employ one or more cameras to act as the robot's eyes.

Tactile sensing, or touch, for a robot is accomplished by placing microswitches, strain gauges, pressure transducers, and optical sensors in the manipulator of the robot. The most important parameter that must be measured to achieve tactile sensing is force.

Simple ranging and navigation systems employ position sensing, proximity sensing, or time-of-flight devices to determine the distance to an object. However, advanced systems must employ 3-D vision to navigate within the real world.

Finally, robot voice communication requires the application of speech synthesis and speech recognition. Speech synthesis is a relatively proven technology and is generally accomplished using a technique called phoneme speech synthesis. Speech recognition, on the other hand, is more complicated and requires the application of AI. Two techniques are presently used for speech recognition: isolated-word recognition and connected-speech understanding.

Artificial intelligence, or AI, must be incorporated into intelligent robot controllers to integrate sensory information with control decisions. Thus, AI becomes the common denominator for most of the components required in an intelligent robot.

The design of AI into a robot involves knowledge engineering. Both the hardware and software design components of knowledge engineering require the networking of a multiprocessor system and hierarchical planning. The software design aspect of knowledge engineering also involves the science of knowledge representation.

Knowledge representation can be divided into two categories: general knowledge and expert knowledge representation. General knowledge representation requires a computer to contain a data base sufficient to understand its surroundings using its various sensing abilities. Expert knowledge is extensive knowledge in a very specialized field or environment. Expert system programs such as MYCIN gain their knowledge base by interacting with human experts.

PROBLEMS

1-1 What are the two common denominators of an intelligent robot?

1-2 List the five categories of sensory perception that can be incorporated into robots.

1-3 Explain how a vision system can be used to build the knowledge base of a robot.

1-4 The science of computer vision is called ______________ .

1-5 Describe a simple light-detection system.

1-6 Describe a computer vision system that uses a TV camera as the image-transformation device.

1-7 Adding a sense of touch to a robot is called ______________ .

1-8 Define compliance.

1-9 List the three fundamental tasks that must be performed by a computer vision system.

1-10 The speech synthesis technique that allows an unlimited vocabulary in any language is called ______________ .

1-11 The two techniques presently employed in speech-recognition systems are ______________ and ______________ .

1-12 Speaker-independent systems use the ______________ speech-recognition technique.

1-13 Explain why AI is the common denominator for most of the components required in an intelligent robot.

1-14 What is a sensing cell?

1-15 Explain the hardware and software aspects of knowledge engineering.

1-16 The two categories of knowledge representation are ______________ and ______________ .

1-17 The diagram in Figure 1-10 is called a ***network***. Suppose this network represents a map of roads between cities A and B. List all possible routes that could be used to get from city A to city B and indicate the total mileage for each route. Do not visit any city more than once.

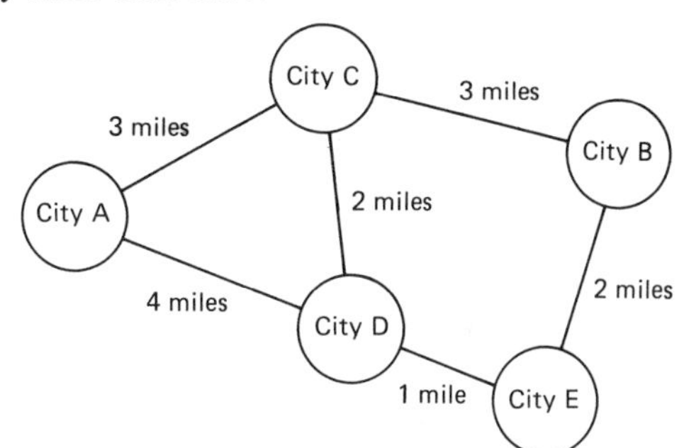

Figure 1-10 Typical network used in AI.

1-18 Using your solutions from Problem 1-17, fill in the circles in Figure 1-11 for each A to B route. The resulting diagram is called a ***search tree*** and is used for problem solving in AI.

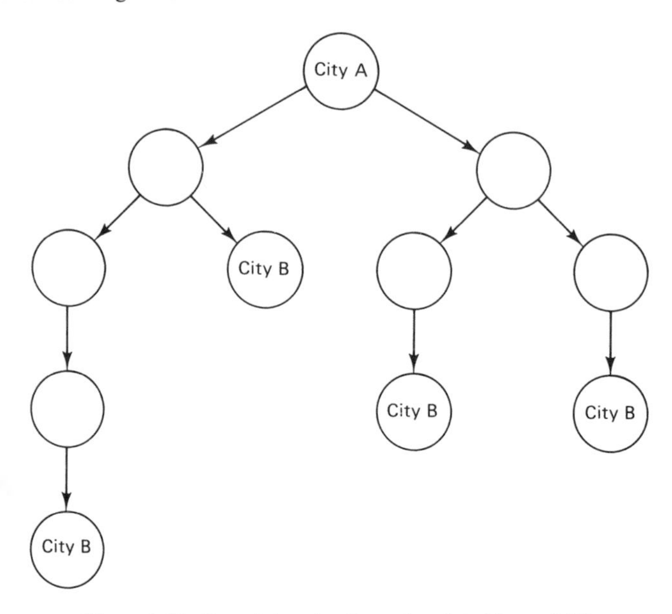

Figure 1-11 Search tree for the network in Figure 1-10.

1-19 The following is a classic problem in AI. Can you find a solution? If so, try to describe the logical thought process you used in finding the solution.

> Three missionaries and three cannibals are on the bank of a river. There is one boat and the three missionaries must get safely to the other side. The boat will carry no more than two people. However, if the cannibals outnumber the missionaries on either river bank, the cannibals will eat the outnumbered missionaries. Devise a plan to get all three missionaries safely to the other side of the river using a series of boat trips.

2 Basic Concepts of Artificial Intelligence

This chapter begins with a short philosophical discussion of artificial versus human intelligence, leading up to a working definition of artificial intelligence (AI). Once artificial intelligence is defined, you will explore the fundamental concepts of AI. The field of AI is quite sophisticated and requires an extensive background in mathematics and computer science for a complete understanding. Here my goal is to teach you the core ideas that are common to all AI programs such that you gain a fundamental knowledge of the principles required for machine, or artificial, intelligence.

2-1 INTELLIGENCE

In Chapter 1 you learned that the human brain and associated nervous system are a hierarchically structured system. Likewise, an advanced computer system capable of possessing a degree of intelligence must be a multiprocessor system that has been hierarchically planned from both hardware and software perspectives. Let's take a closer look at the human thought process in order to make a case that machines are also capable of possessing intelligence.

Digital computers consist of a network of switches. Computers perform operations and make decisions using on/off states. Thus, the basic element of any digital computer, a transistor, is simply a two-state device. Likewise, many believe that the basic element of the brain, the neuron, also exhibits two-state operation. If this is true, the basic element of the human thought process can be likened to integrated transistors within a digital computer.

The brain is composed of approximately 20 billion (2×10^{10}) nerve cells called ***neurons.*** The function of the neuron is to generate and conduct nerve impulses. From Figure 2-1 you can see that a neuron consists of a body with several fiber branches called ***axons*** and ***dendrites.*** Nerve impulses to and from the neuron are propagated along these fibers. These are electrochemical impulses of approximately 50 millivolts (mV) with a duration of about 1 millisecond (ms). The impulses create both chemical and physical changes within the neuron cell body and associated fibers. Thus, the presence or absence of nerve impulses can be viewed as an on/off, or two-state, operation.

The fibers of the neurons come together to form junction points called ***synapses***, as shown in Figure 2-2. Nerve impulses from adjacent neurons excite and

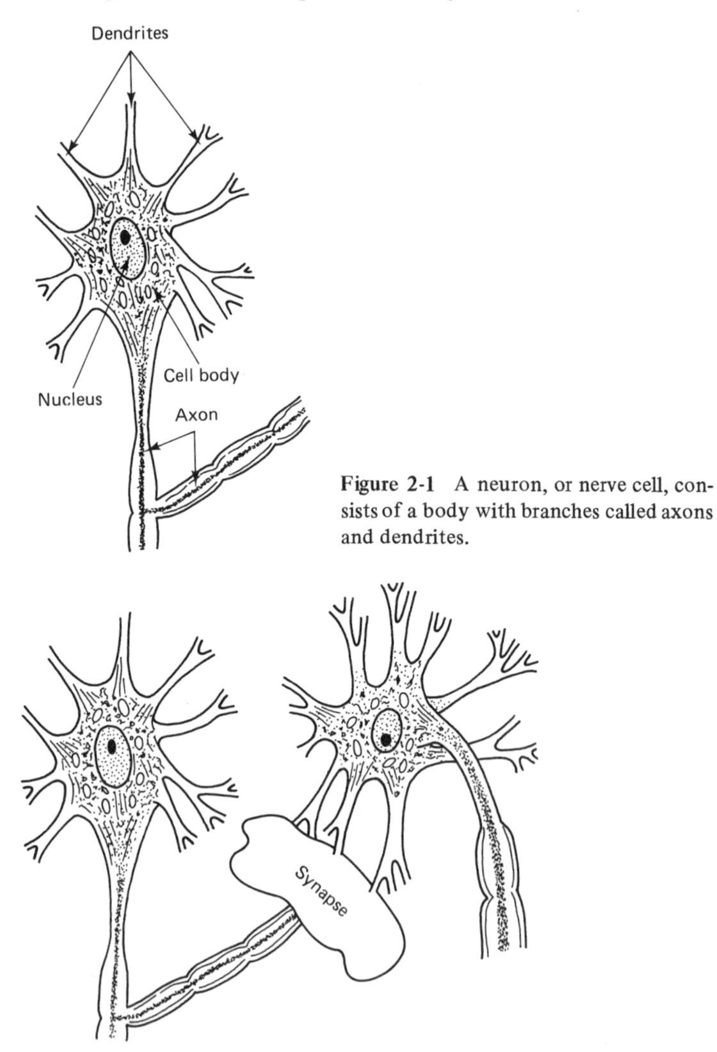

Figure 2-1 A neuron, or nerve cell, consists of a body with branches called axons and dendrites.

Figure 2-2 Neurons come together to form junction points called synapses.

bridge a synapse when the sum of several input pulses exceeds a threshold voltage level for a given synapse. The ***threshold*** level for a given synapse is the minimum stimulation level required to excite the synapse. Conversely, a synapse is said to be ***inhibited*** when the input nerve pulses do not exceed the threshold level.

This excitation and inhibition, or on/off, effect of the synapses suggests that they operate very much like simple logic gates. For example, suppose two neurons are connected by a common synapse, as shown in Figure 2-3. Also, suppose that the threshold level for the synapse is that of two nerve impulses, or approximately 100 mV. Then the synapse is excited only when both nerve impulses are present. Consequently, the synapse performs the function of a simple 2-bit AND gate. On the other hand, if the threshold level of the synapse is that of one nerve impulse, or about 50 mV, the synapse would perform the function of a 2-bit OR gate.

The conclusion to be drawn from our discussion is that the brain and associated thought process might exhibit a digital character. However, this is only one theory, which was originally proposed by John von Neumann, a famous twentieth-century mathematician and one of the original computer scientists. Other scientists disagree with this theory, contending that the nerve cells of the brain are analog in nature, operating like tiny operational amplifiers in their range of functions. One thing is certain; the sooner we figure out how the brain operates, the sooner we will be able to program a computer to emulate that operation.

So, what sets your brain apart from a digital computer? There are literally billions of neuron-synapse connections within the brain. To perform an intelligent act, you might use several thousand nerve fiber chains simultaneously. Individual nerve fiber action can be comprehended. However, the combined effect of thousands of nerve fibers performing an intelligent act is not fully understood. If it were,

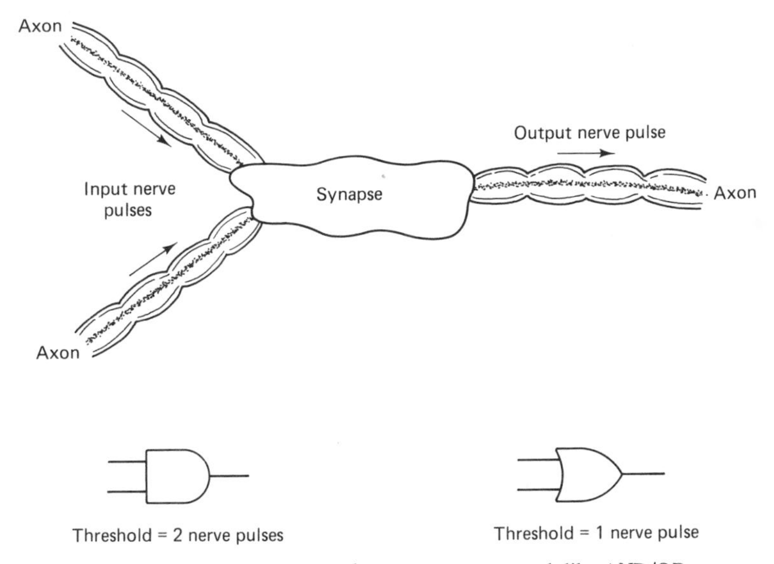

Figure 2-3 Synapses can be viewed to operate very much like AND/OR gates.

we would know how to create a truly intelligent computer. For this reason, some AI researchers study how to make computers smarter by attempting to understand the human thought process.

Let's make some other comparisons between the brain and the digital computer. The digital computer consists of five fundamental hardware sections, as shown in Figure 2-4. Recall that these are the arithmetic logic unit (ALU), control unit, memory, and input and output sections. By comparison, the brain can also be divided into essentially the same five fundamental sections. For example, the eye acts as an input device and the vocal cords act as an output device. The central nervous system then provides the computational, logic, control, and memory portions of the system. The overall structure of the brain and its sensory organs is, therefore, very similar to that of a typical digital computer.

Although the brain is structurally similar to a computer, it is much slower. A nerve impulse lasts about 1 ms (10^{-3} s), whereas computers operate in the nanosecond (10^{-9} s) range. Thus, computers are about 1 million (10^6) times faster than the brain. As a result, a computer can process a lifetime's worth of information within a few hours. But the brain makes up for its relatively slow speed by using simultaneous, or ***parallel***, processing, whereas most computers perform operations sequentially, in serial fashion.

With regard to size and density, the human brain is about 1000 times as dense as a digital computer. Its volume is about 1 liter, and it contains about 20 billion (2×10^{10}) neurons. Consequently, the size of a neuron is approximately 0.5×10^{-7} cubic centimeters (cm^3). By comparison, existing VLSI technology produces integrated circuits where the size of an integrated transistor is approxi-

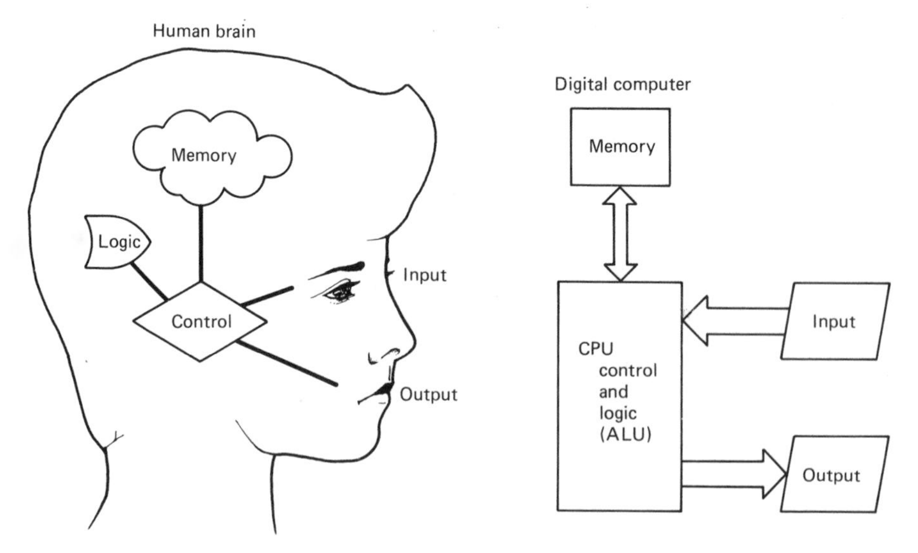

Figure 2-4 The fundamental sections of a digital computer are analogous to those of the human brain.

mately 10^{-4} cm^3. Thus, the brain is at least 1000 times more dense in terms of functional units than a digital computer of the same size.

Other brain-computer characteristics are listed in Table 2-1. Notice that characteristics of the brain include inventiveness, perception, abstract thinking, inductive reasoning, and emotion, just to mention a few. To date, computers do not exhibit these characteristics. However, computers are much faster, more accurate, and more consistent than the brain.

TABLE 2-1 Brain and Computer Characteristics

Human Brain	Digital Computer
Slow	Fast
Inductive reasoning	Deductive reasoning
Inaccurate	Accurate
Perception	Not perceptive
Forgetful	Long-term memory
Creative	Not creative
Emotion	No emotion
Learns	Must be programmed with new knowledge
Electrochemical	Electrical

In summary, the brain and computer have a similar structure. However, the brain is relatively slow but very dense, while the computer is fast and not as dense. With advances in VLSI and SLSI technology, digital computers will eventually be as dense or more dense than the brain. If digital computers will someday be as dense as the human brain and much faster, why can't computers someday be as intelligent or even more intelligent than their human counterparts? The answer to this question is the essence of all research in the field of artificial intelligence. Recall that, once man had discovered the principles that govern the flight of a bird, he was able to build a flying machine. Likewise, once we can accurately describe the human thought process, this process can possibly be duplicated in a computer to create a truly intelligent machine.

Now we are ready to define the term ***artificial intelligence***, or ***AI***. The most accepted definition is that proposed by Marvin Minsky, one of the fathers of the study of artificial intelligence. Minsky defines artificial intelligence as "the science of making machines do things that would require intelligence if done by men." The key word in this definition is ***intelligence.*** What is intelligence? Is it the ability to perform inductive reasoning? Is it the ability to learn and acquire knowledge? Or is intelligence just plain old common sense? Many researchers, including Minsky, equate intelligence with commonsense reasoning. Some researchers define AI as the mechanization, or duplication, of the human thought process. Other scientists simply refuse to define artificial intelligence, contending that it is like a biologist attempting to define life itself. In this text, I will use Minsky's definition of AI, keeping in mind that it uses the term *intelligence*, which is not clearly defined.

Now, before we discuss the fundamental principles of AI, let's dispel a common myth associated with computers.

"A computer is a stupid machine and does exactly what it is programmed to do and no more."

The above statement is sometimes referred to as the "stupid computer" myth. It has been around a long time and in many cases it is probably used to squelch many of the fears that people have about computers. In a literal sense, the stupid computer myth is true. Computers do what they are programmed to do. However, this does not mean that they are stupid. You and I also do what we are programmed to do by our genetic make-up and accumulated learning experiences. Does this make us stupid? Hardly! Actually, the fact that computers do what they are programmed to do presents the greatest challenge for the science of artificial intelligence. AI researchers must figure out how to program computers to learn, invent, create, perceive, reason, etc. These are the abilities commonly associated with human intelligence. As such programming techniques are developed, computers will indeed become smarter, and not be viewed as simply stupid machines.

It is the rigid way that computers are traditionally programmed that has created the "stupid" computer myth. So, you might be wondering, how does AI programming differ from traditional forms of programming? Consider a typical non-AI programming language like BASIC. When a problem must be solved using BASIC, you must develop a step-by-step method to solve the problem. This procedure normally requires you to flow-chart the solution steps and then translate the flowchart steps into precise BASIC language instructions that carry out the solution plan. This step-by-step type of programming is called ***algorithmic programming.*** An ***algorithm*** is a set of well-defined rules developed for the solution of a problem in finite steps.

Once you develop an algorithm to solve a given problem, the computer requires a precise amount of time to run the algorithm and find a solution. Suppose we designate the size of an algorithm with the letter S. Clearly, the time it takes to run the algorithm is proportional to its size, S. As the algorithm increases in size to, say, $2S$, $3S$, or $100S$, the amount of time to run the algorithm increases proportionally. Most computers can run such algorithms within a reasonable amount of time. Even algorithms of size S^2, S^3, and S^4 can be handled by today's computers. However, as an algorithm increases to size 10^s, S^s, or $S!$, it is increasing exponentially and is said to run in ***exponential time.*** Algorithms of this size cannot be run on even the fastest computers in a reasonable amount of time. Even the simplest everyday reasoning problems that you and I take for granted would require exponential-time algorithims. Consequently, algorithmic programming does not lend itself easily to intelligent problem solving.

Aside from the time factor, algorithmic programming is a major reason that computers are so inflexible. Such programming requires a computer to be accurate, precise, and infallible. On the other hand, humans are not so accurate, precise, and infallible; but we possess *commonsense reasoning* that permits us to solve many everyday reasoning problems that computers cannot. Many times, we simply apply commonsense rules to a problem and come up with an approximate solution. In most cases the solution is not precise, but is good enough to get the job done. If computers are to reason, they must be programmed to apply commonsense rules of thumb to problem solving much the same as we do. To accomplish this, AI programming must:

- Define and represent problems within the computer so that the application of certain rules will eventually lead to a solution if one exists.
- Provide an efficient means of directing the search for a solution to the problem utilizing existing knowledge and commonsense rules of thumb.

This is the fundamental difference between AI programming and algorithmic programming and leads us to the remaining sections of this chapter.

2-2 PROBLEM REPRESENTATION IN ARTIFICIAL INTELLIGENCE

Before a solution to a problem can be found, it must first be represented in a form that can be easily manipulated by a computer. Such a representation reduces to a ***model*** of the problem within the data structure of the computer. A problem is represented by constructing a model that is analogous to the original problem. Thus, the idea is to solve the original problem by solving its model representation.

Problem models are used as representations since they reduce the original problem to a set of states that are more easily understood and workable (within the computer) than the original problem. For example, consider the problem of explaining the operation of a diode to a nontechnical friend. You might begin with a discussion of atomic theory, leading into an explanation of the semiconductor physics associated with the diode's operation. However, do you really think your nontechnical friend could follow such a discussion, even though it might be technically correct?

A better approach would be to choose for the diode a model representation whose operation could be easily understood by a nontechnical person. For this example, the ideal model is a simple switch. The switch model gets across the idea of the diode's operation, while avoiding many of the difficult features associated with semiconductor theory. Since the operation of a switch is analogous to the operation of a diode, the switch becomes an ideal representation for the diode. The switch model could then be expanded to include a battery and resistor, thereby creating a more accurate diode model.

Problem representation in AI works the same way. The problem is represented by a model consisting of workable ***states*** within the computer's data base. These states are then manipulated using a set of ***operators***, which it is hoped will find a solution to the problem. When and where to apply the appropriate operators depend on an overall ***control strategy***. Thus, problem solving in AI reduces to states, or models, of the problem within the computer data base, operators to manipulate the states, and a control strategy that applies the operators to produce a problem solution.

To illustrate this idea, let's consider the BASIC programming language. Problem states can be represented by assigning values to variables. For example:

```
100  LET  A$ = "SUBPROBLEM 1"
110  LET   A = 10
120  LET   B = I + 1
```

and so on. Operators are then defined using IF/THEN type statements, such as:

```
130  IF  B = 5  THEN 1000
140  IF  B = A + 5 AND I = 0    THEN C = B ELSE C = A
```

The overall control strategy for problem solution in BASIC then reduces to the order in which the preceding operations are performed. Of course, this is the completed BASIC program.

In this section, you will learn two general methods for problem representation in AI programming: ***state-space representation*** and ***problem-reduction representation.*** Then, in the next section, you will explore several control strategies that are used to apply operators to these representations to find problem solutions.

State-space Representation

Imagine a robot performing a product assembly operation. The robot initially begins with a group of parts that must be assembled into a final product. To program the robot for such an operation, you would look at the problem as that of programming the robot for the series of operations required to assemble the parts into the final product. After each operation, the assembled parts take on a new form, or ***state***, until the finished product state is reached. If you view the unassembled parts as an initial state and the finished product as a goal state, the problem reduces to programming the robot for the series of operations that transform the initial state into the goal state. After each operation is performed, a new subassembly state is created until the final assembly, or goal, state is reached. This process is illustrated in Figure 2-5.

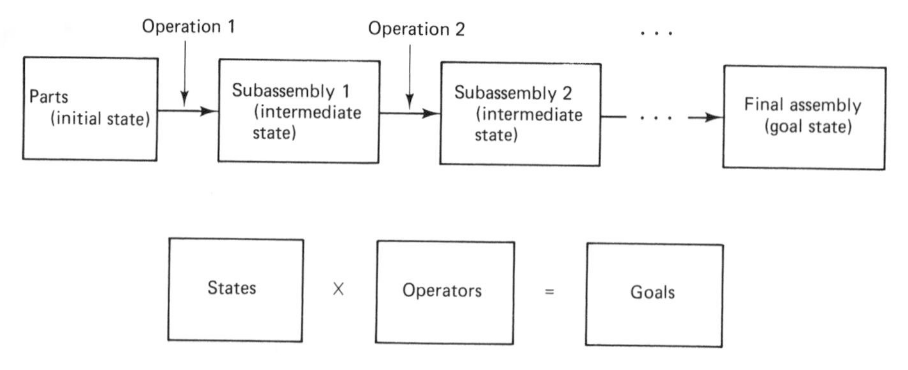

Figure 2-5 Any problem can be broken down into states, operators, and goals.

Most problems, no matter how complex, can be broken down into three key elements: ***states, operators,*** and ***goals.*** In an AI program, the states are data structures that represent the condition of the problem at a given time. The operators are a set of constraints and rules that is applied to the states to produce more desirable states. These operators consist of two parts: a ***condition*** part and an ***action*** part. The operation cannot be performed until certain conditions are met. Thus, you might say that the conditions apply *constraints* on the operation. Once all the conditions are met, the action part of the operation is a set of rules that describes all the changes that must take place to transform a given state into a new state. Finally, the goals are the final states that you are trying to produce by applying the operators. The ***solution*** to any problem, therefore, is the set of operators that must be applied to transform the initial state into the goal state.

State Graphs and Networks. Suppose you wish a robot to move the two discs from peg 1 to peg 3 in Figure 2-6. Sounds like a simple pick-and-place operation, doesn't it? However, there are some real-world constraints on the operation. First, suppose the robot's payload is not adequate to lift both discs. Second, suppose the smaller disc is made of fragile material and thus can never be placed beneath the larger disc.

Recall that any problem can be broken down into states, operators, and goals. The initial and goal states for this problem are illustrated in Figure 2-6. In addition, seven possible intermediate states are illustrated in Figure 2-7. In AI, the set of all possible states for a given problem is called a ***state space.*** An operation is required to change from any one legal state to another. Thus, the solution to the problem is for the robot to find the most efficient series of intermediate operations that will transform the initial state into the goal state. Of course, *you* could solve the problem and program the robot to perform the required operations. However, a robot with artificial intelligence should be able to duplicate your thought process and find the proper solution.

Beginning with the initial state, the robot can perform two legal operations: move disc *A* to peg 2 or move disc *A* to peg 3, as shown in Figure 2-8. With either operation, the robot transforms the initial state into a new intermediate state. When working problems such as this, it is common to designate states at different ***levels.*** The initial state is always at level 0. A state created by applying an operator to the

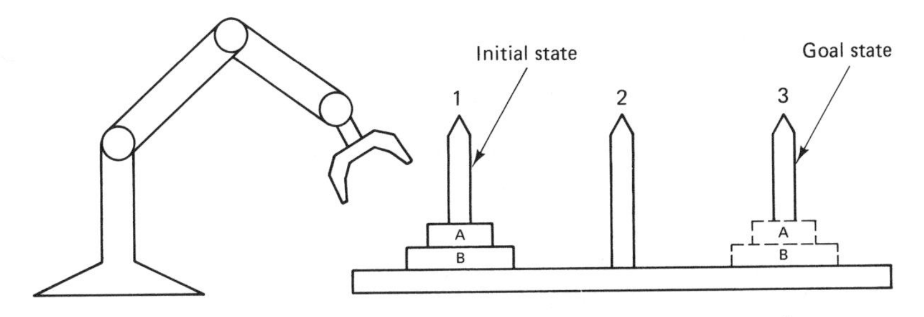

Figure 2-6 A robot must move the two discs from peg 1 to peg 3.

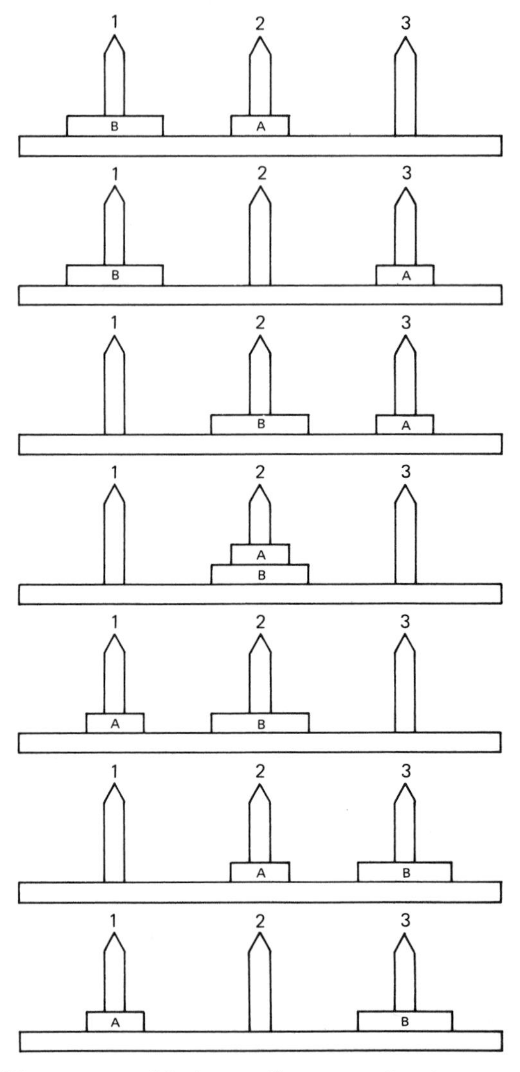

Figure 2-7 The seven possible intermediate states for the two-disc pick-and-place problem.

initial state is a level 1 state. A state created by applying an operator to a level 1 state is a level 2 state, and so on. Now, back to the problem at hand. At this point, the robot could choose to go back to the initial state (level 0) or into another intermediate state at level 2, as shown in Figure 2-9. From this level, the robot has the option of going back to level 1 or forward to one of two new states at level 3, as illustrated in Figure 2-10. Notice that the goal state appears at this level, thereby eliminating the need for any additional levels. In addition, the final graph in Figure 2-10 shows all the unique states and all possible operations for the given problem.

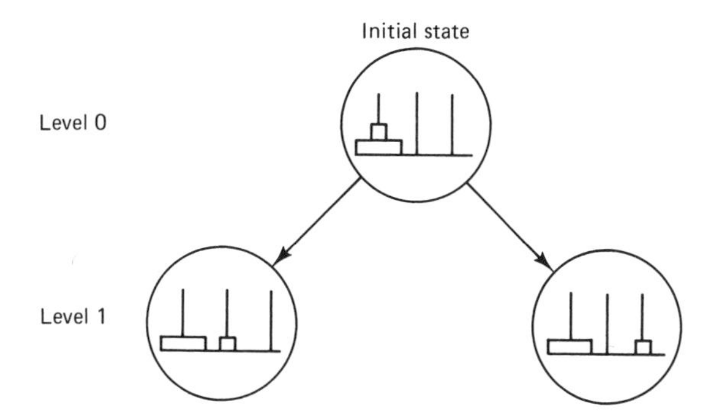

Figure 2-8 From the initial state, two intermediate states are possible.

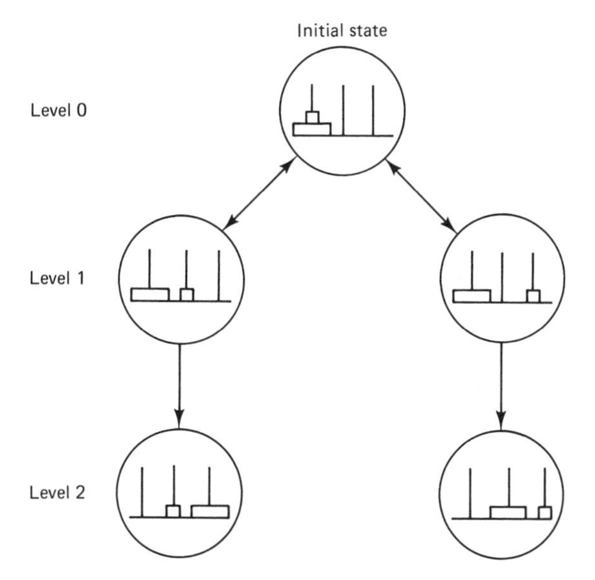

Figure 2-9 At level 1, the robot could choose to go back to level 0 or on to a new level 2 state.

The graph in Figure 2-10 is called a ***state-space graph***, or ***network.*** State-space graphs are extremely useful in AI since they illustrate all possible states, operations, and goals for a given problem. Some related terminology is appropriate at this time. The various states within a state-space graph are called ***nodes.*** The lines that connect the nodes are called ***arcs.*** These lines represent the operations that are used to transform one state, or node, into another state, or node. If the arcs have arrows on them, the state-space graph is called a ***directed graph.*** The arcs can be unidirectional (one way) or bidirectional (two way) in a directed graph. If a directed arc goes from node *A* to node *B*, then node *A* is called the ***parent***, or ***predecessor***, node and node *B* is called the ***child***, or ***successor***, node. Finally, a ***path*** is defined as a set of arcs, or

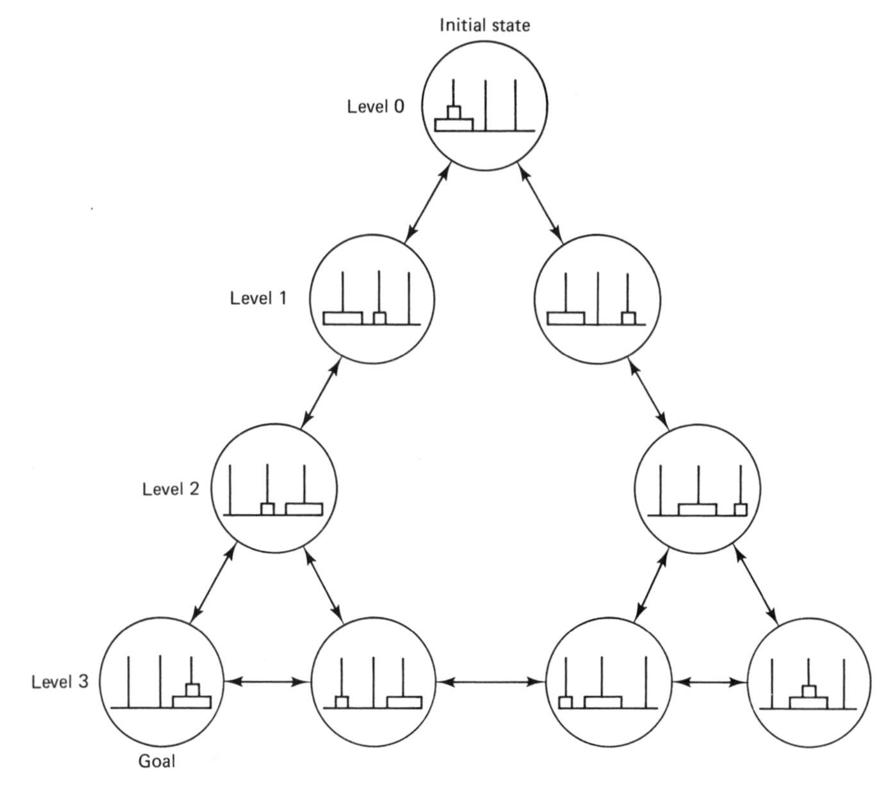

Figure 2-10 Completed state-space graph for the two-disc pick-and-place problem.

operations, that connects a series of nodes such that each node is a successor, or child, of the previous node. The general-purpose state-space graph shown in Figure 2-11 summarizes this terminology.

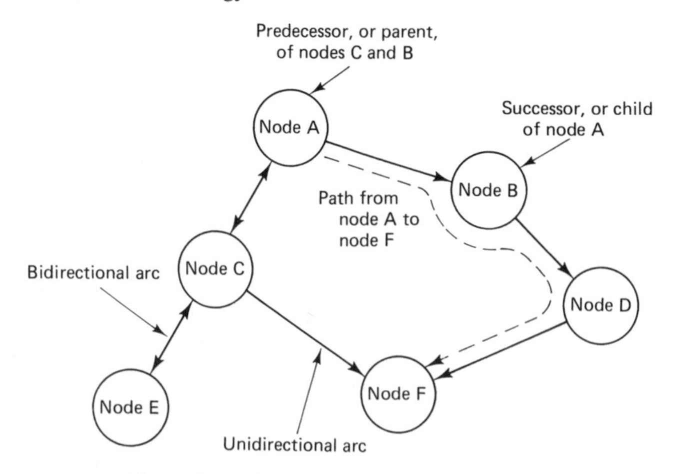

Figure 2-11 State-space graph terminology.

Search Trees. A special type of state-space graph is a ***search tree***, or simply a ***tree***. It has the following properties:

1. At level 0, there is a unique node called a ***root*** node that has no predecessors.
2. Each successor, or child, node in the tree has only one predecessor, or parent node.

These graphs are called trees since you can think of them as an upside-down tree, as illustrated in Figure 2-12. Notice that the ***root*** of the tree is at the top and has no predecessors. The ***leaves*** of the tree are the nodes at the bottom and have no successors. The arcs between the nodes within a tree are called ***branches***, for obvious reasons. Trees are always directed downward, from root to leaves. As a result, there is no need to use arrows to indicate the direction of the arcs, or branches. The level of a given node in a tree is sometimes called its ***depth***. Consequently, the root of a tree has a depth of zero and the leaves of the tree shown in Figure 2-12 have a depth of three. Notice that the overall depth of a tree is equal to the depth of its leaves. Trees are useful since they are simpler to work with than networks. There are no loops and there is at most one path from a given node to any other node.

A search tree for the two-disc robot pick-and-place problem discussed earlier is provided in Figure 2-13. If you compare this with the state graph for the same problem (Figure 2-10), you will see that the search tree in Figure 2-13 shows all the

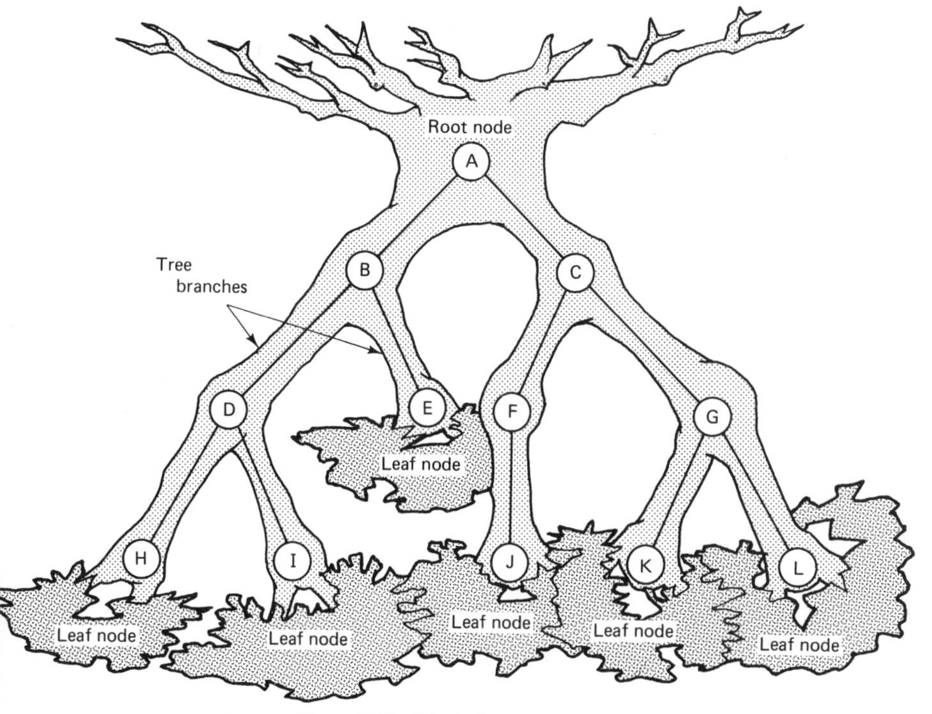

Figure 2-12 Typical state-space tree.

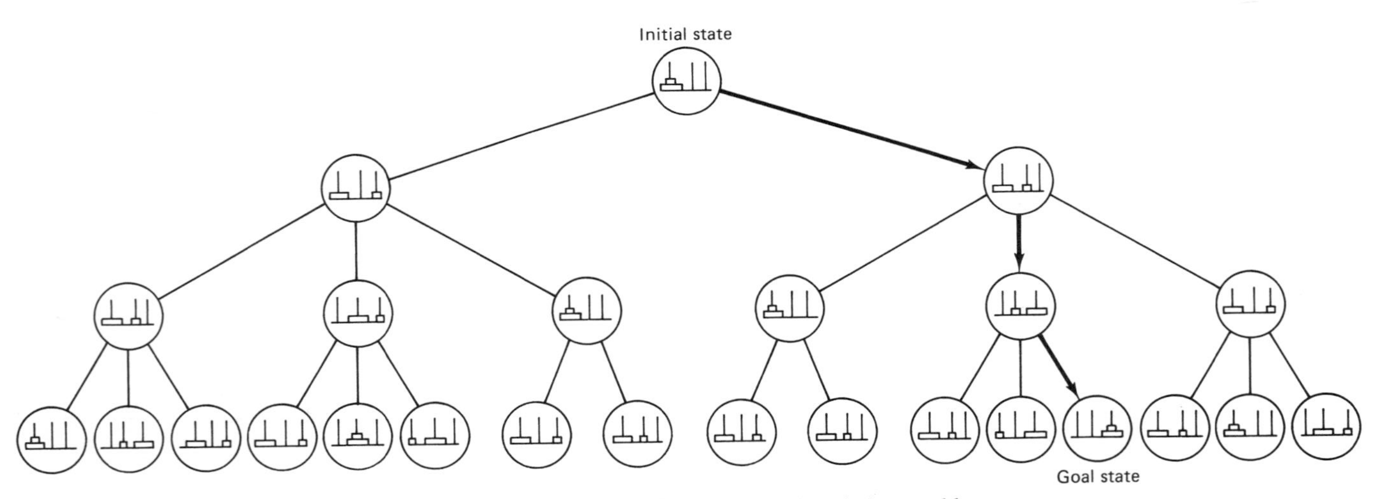

Figure 2-13 Search tree of the two-disc pick-and-place problem.

possible one-way paths that can be generated until the goal level is reached. Notice that the solution to the problem is the path from the root to the goal node. This path has been enhanced in Figure 2-13 for illustrative purposes.

The search trees shown in Figures 2-12 and 2-13 are called ***explicit*** since a diagram is used to define the tree. Explicit trees are easy for us to visualize, but difficult for a computer to work with. Even simple problems in AI have very large, almost infinite search trees, thereby requiring vast amounts of memory to store the tree. For this reason, most search trees used in AI programs are defined implicitly. An ***implicit*** definition for a search tree is a rule or mathematical relationship that is used to generate the tree.

Example 2-1

Develop a mathematical rule that could be used by a computer to generate the number tree shown in Figure 2-14.

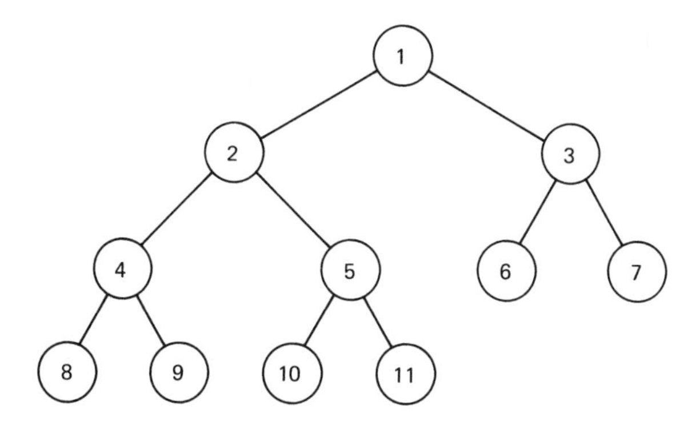

Solution

Notice that any given node, n, has two successor nodes $2n$ and $2n + 1$. Thus, the operators to be applied to any given node to generate its successor nodes are $2n$ and $2n + 1$. These are the action parts of the operator. However, recall that operators also have conditions. The conditions for the operators in Figure 2-14 are that $1 \leqslant n \leqslant 5$, and a node whose value is greater than 5 does not have any successor nodes. Thus, the implicit operators used to generate the number tree are $2n$ and $2n + 1$, where $1 \leqslant n \leqslant 5$.

Implicit definitions of search trees allow a computer to generate the tree as a path is being explored, thereby drastically reducing the memory space required for the tree. When a tree is generated from one node to another implicitly, we say that the tree is being ***expanded.*** Thus, a parent node is expanded when its children are generated. You could say that the tree "grows" by expanding its nodes. A node is said to be ***open*** if it has not yet been expanded by tree growth. Conversely, a ***closed*** node is one that has already been expanded.

Problem-reduction Representation

Up to this point, we have been discussing the ***state-space*** problem representation technique. In AI, state-space graphs and their associated search trees are typically used to solve simple problems like the two-disc pick-and-place problem discussed earlier. However, many problems are too complex to be represented by simple state graphs and their associated search trees. The search tree for a complex problem might be too large for a computer to solve within a reasonable amount of time. As a result, a more powerful problem representation technique, ***problem reduction***, is used for complex problem representation.

Simply stated, the problem-reduction technique divides a complex problem into a series of subproblems. The idea then is to solve the original problem by solving the subproblems. In most cases the individual subproblems are much easier to solve than the original problem. If you think about it, this is the way most of us go about solving a problem.

Imagine that you are about to design a robot. Do you jump right in and attempt to design the entire system at one time? Of course not! Most design engineers and technicians would break up the design task according to the functional requirements of the robot. The power supply, controller, manipulator, gripper, and so on, would all be designed separately according to the overall robot system specifications. In this way, the complex robot design problem is broken down into a series of smaller subsystem design problems. The subsystems might even be further broken down into other subsystems. Then all the subsystems are integrated together to form the complex system. As a result, a very complex problem is solved by dividing it up into a series of smaller subproblems. This is the idea behind problem-reduction representation in AI.

To see how problem reduction is used to represent problems in AI, consider the robot pick-and-place problem discussed earlier. This time suppose there are three discs (*A*, *B*, and *C*) rather than just two. The rules are the same as before: the discs must be transferred from peg 1 to peg 3, only one disc can be moved at a time, and no disc may be placed on a smaller disc. By now you might have recognized this as the classic Tower of Hanoi problem.

Recall the state graph and associated search tree that was generated earlier for the two-disc problem (Figures 2-10 and 2-13). By adding another disc, the problem becomes quite a bit more complex. With a three-disc problem there are 27 unique states, as compared to just 9 states for the two-disc problem. Now the initial problem is to move the three discs from peg 1 to peg 3. Let's divide this problem into three subproblems, as follows:

1. Move discs *A* and *B* from peg 1 to peg 2.
2. Move disc *C* from peg 1 to peg 3.
3. Move discs *A* and *B* from peg 2 to peg 3.

This series of subproblem moves is illustrated in Figure 2-15. Notice that, if all the subproblems are solved in the sequence shown, the original problem is solved.

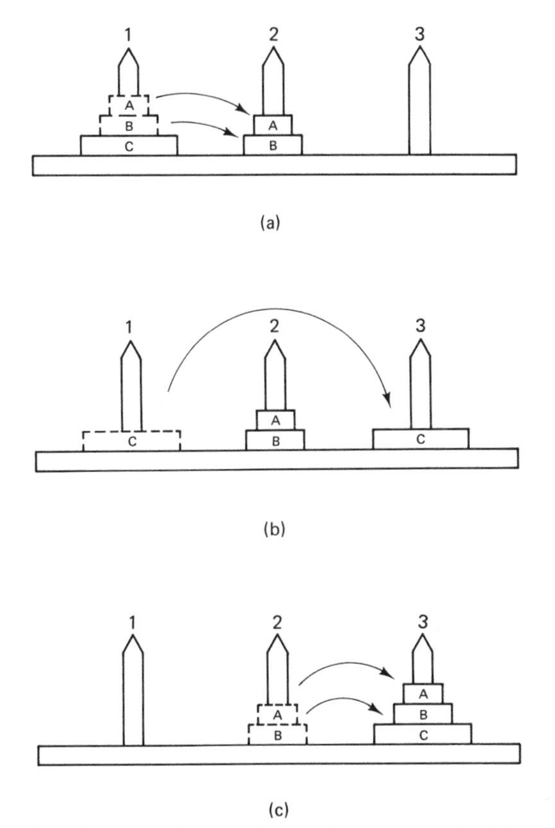

Figure 2-15 The three-disc pick-and-place problem can be divided into three subproblems: (a) subproblem 1, move discs *A* and *B* to peg 2; (b) subproblem 2, move disc *C* to peg 3; (c) subproblem 3, move discs *A* and *B* to peg 3.

Looking at the three subproblems, you can see that the only one that can be performed by the robot is subproblem 2. A subproblem such as this, which meets all the operational conditions and can be solved immediately, is called a ***primitive problem.*** Thus, the basic idea of the problem-reduction technique is to divide the initial problem into a set of primitive problems whose solutions are immediate. Since subproblems 1 and 3 are not primitive, they must each be broken down further into a set of primitive problems.

Subproblem 1 can be broken down into the set of three primitive problems shown in Figure 2-16. Notice that each primitive operation can be performed by the robot. If the three primitive operations are performed from left to right in the sequence shown, the subproblem is solved.

Likewise, subproblem 3 can be broken down into the set of three primitive problems shown in Figure 2-17. Again, the subproblem is solved by performing the three primitive operations in the order shown.

The complete problem-reduction tree is shown in Figure 2-18. Here we have

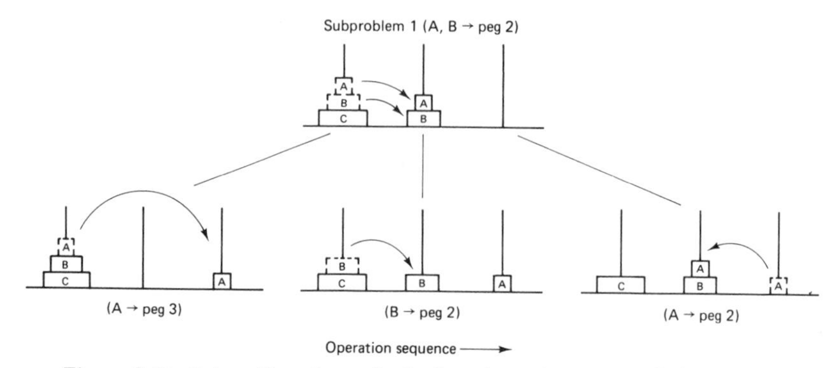

Figure 2-16 Subproblem 1 can be broken down into a set of three primitive problems.

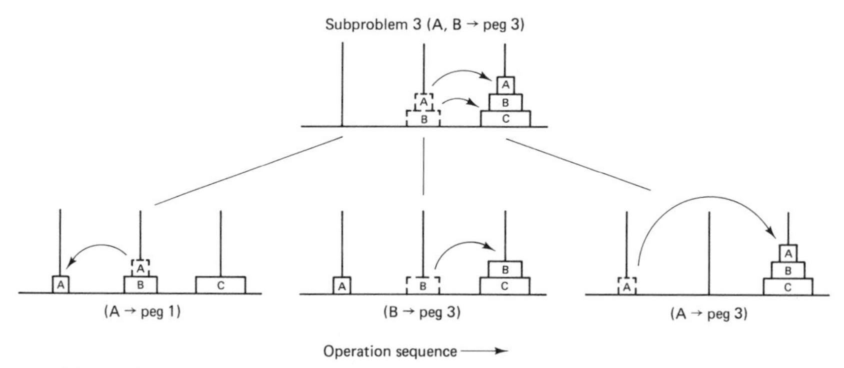

Figure 2-17 Subproblem 3 can be broken down into a set of three primitive problems.

combined the previous subproblem operations into a complete tree. Notice that the solution to the initial problem is produced by solving all the subproblems, which, in turn, are solved by solving the primitive problems. Furthermore, the final solution is dependent on a given sequence. That is, subproblem 1 must be solved first, then subproblem 2, followed by subproblem 3. For this reason, the subproblem solution sequence, sometimes called a ***solution plan***, is extremely important to the overall solution. Problem-reduction solution plans will be discussed further in Section 2-3.

Example 2-2

Suppose the number of discs in the robot pick-and-place problem discussed previously is increased to five. Using the problem-reduction technique, divide the problem into three subproblems, one of which is a primitive problem.

Solution

Label the discs A, B, C, D, and E from the smallest to the largest disc, and label the pegs 1, 2, and 3 from left to right. The object is to move the five discs from peg 1 to peg 3 using the conditions and rules given previously. This problem can be divided into three major subproblems using the problem-reduction technique, as

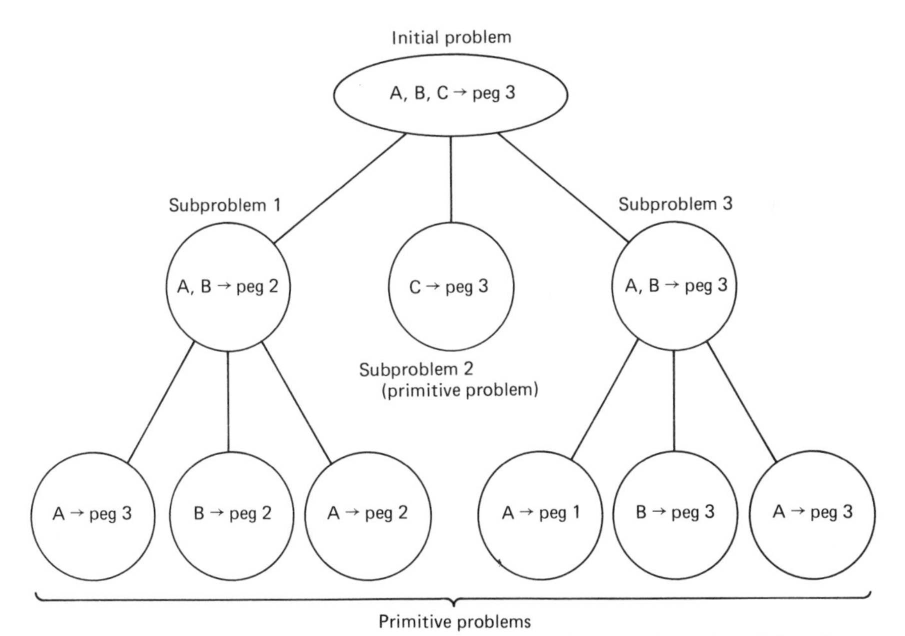

Figure 2-18 Complete problem-reduction tree for the three-disc robot pick-and-place problem.

shown in Figure 2-19. Notice, however, that only one of the subproblems is a primitive problem. The other two subproblems must be broken down further until a complete set of primitive problems is generated.

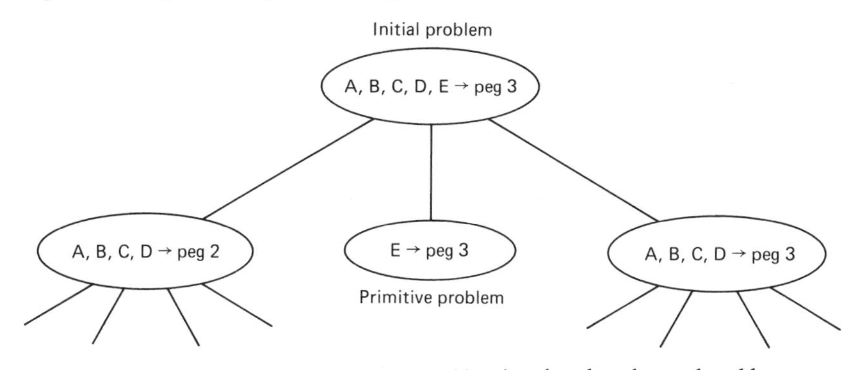

Figure 2-19 A five-disc pick-and-place problem is reduced to three subproblems.

AND/OR Trees. In the three-disc robot pick-and-place problem just discussed, the initial problem is solved only if *all* its subproblems are solved. Such a problem-reduction tree is called an *AND* tree. In other words, subproblem 1 AND subproblem 2 AND subproblem 3 must all be solved to solve the initial problem. In addition, all the primitive problems must be solved to solve the subproblems. In general, with an AND tree, the initial problem, or root node, is solved only when *all* the leaf nodes are solved.

The opposite of an AND tree is an OR tree. With an OR tree, the root node is solved if *any one* of the leaf nodes is solved. The state-space graphing technique discussed earlier typically produces OR trees, while the problem-reduction technique generates AND trees. For diagramming purposes, an arc is used between the branches of a node to designate an AND expansion, as shown in Figure 2-20(a). The absence of an arc is used to designate an OR expansion, as shown in Figure 2-20(b).

Most problems do not reduce to a pure AND tree or a pure OR tree. Rather, the problem reduces to a combination AND/OR tree, as illustrated in Figure 2-21. Consequently, an AND/OR tree is a mixed collection of pure AND expansions and pure OR expansions. The AND expansions produce a set of subproblems that must be solved to solve the initial problem. The OR expansions produce a set of *different*

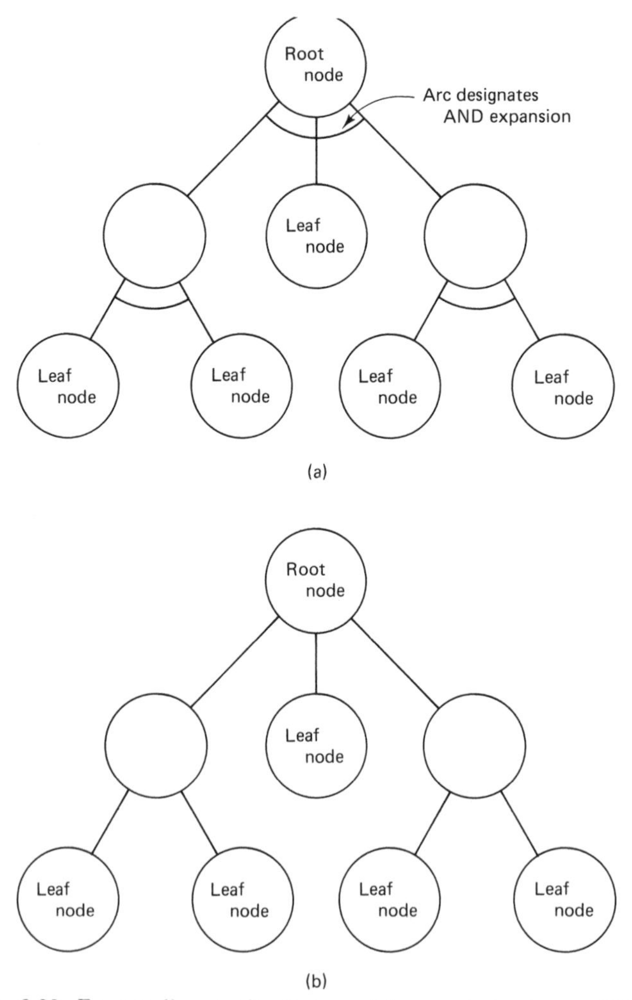

Figure 2-20 For tree diagramming purposes, (a) an arc designates an AND expansion, while (b) the absence of an arc designates an OR expansion.

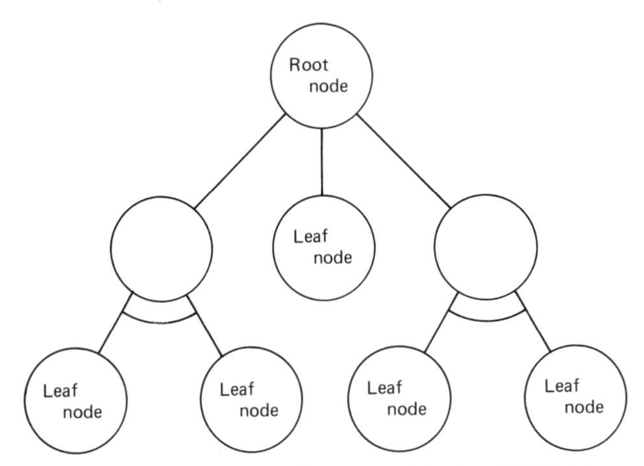

Figure 2-21 Most problems reduce to combination AND/OR trees.

plans that can be used to solve the initial problem. For instance, recall that a specific sequence of operations, or plan, was required to solve the three-disc problem. This generated a pure AND tree. However, suppose another sequence of operations existed that could also be used to solve the same problem. Such a solution plan must also generate a pure AND tree similar to the tree generated by the first plan. The two plans each produce a pure AND tree, but together they produce an AND/OR tree. This idea is illustrated by Figure 2-22. Notice that each solution plan produces a separate OR branch for the root node.

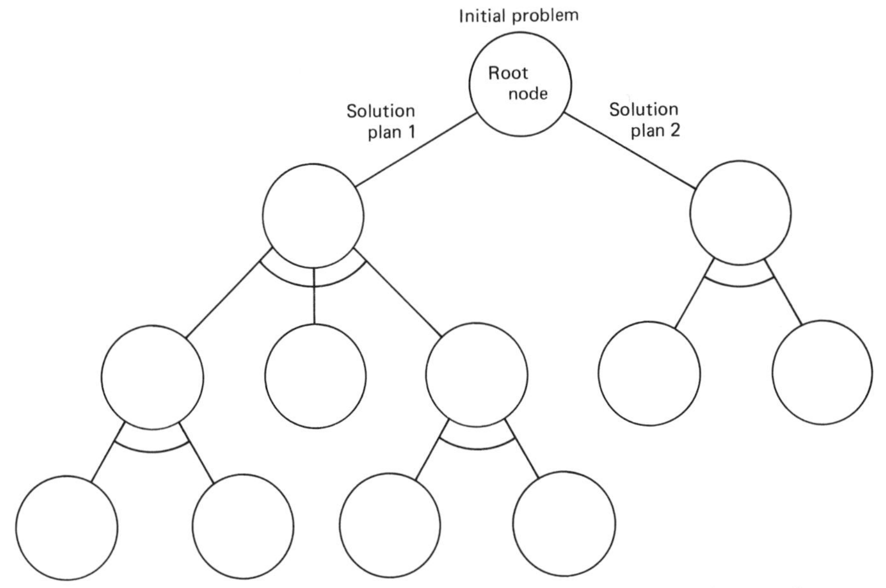

Figure 2-22 Solution plans, taken separately, produce AND trees, but together they produce an AND/OR tree.

Now that you are familiar with the two major problem-representation techniques (state-space and problem reduction), it is time to explore the various techniques that are used in AI programs to find problem solutions.

2-3 PROBLEM-SOLUTION TECHNIQUES USED IN ARTIFICIAL INTELLIGENCE

In Section 2-2, you discovered two techniques used in AI for problem representation: state-space and problem-reduction representation. With state-space representation, the problem is expressed in terms of states, operators, and goals. A solution to the problem is found by applying those operators that will transform the initial state into the goal state. In terms of a state-space tree, the solution path is from the initial, or root node, to a goal node. Thus, the idea is to take the initial state forward to the goal state, working from the root to the leaves of the tree. Such a process is referred to as ***forward reasoning.*** On the other hand, problem-reduction representation breaks down the main problem, or goal, into a set of subproblems, or subgoals. The problem is then solved by solving the simpler subproblems, thereby working from the leaves to the root of the tree to determine if the root is solvable. This type of process is called ***backward reasoning.*** This idea of forward versus backward reasoning is illustrated by Figure 2-23.

With either of these reasoning processes, a solution to the problem is found by *searching* for a solution path within the representative state-space or AND/OR tree. The tree-searching procedure should attempt to accomplish three tasks:

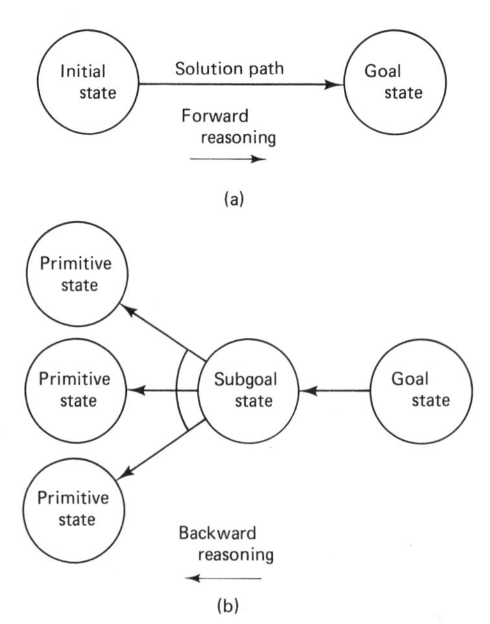

Figure 2-23 (a) State-space searches use forward reasoning, and (b) problem-reduction searches use backward reasoning.

1. Always find a solution to the problem if one exists.
2. Always find the best solution to the problem.
3. Always find the most efficient solution in terms of computer time and memory space.

In this section, you will explore various tree-searching techniques that are common in AI programs. These techniques fit into two general categories: ***blind searches*** and ***heuristic searches.*** In a blind search, the order of node expansion within the tree is more or less arbitrary. Although a general plan is followed for node expansion, blind searches do not incorporate any knowledge about the problem. Consequently, blind searches often lead to ***combinatorial explosion***, whereby the search tree grows beyond the space and time limitations of the system. Heuristic searches, on the other hand, utilize additional information, or knowledge, about the properties of a problem to reduce and thus focus the search. As a result, heuristic searches are much more efficient than blind searches.

As a general rule, blind searches are used in AI for simple problems such as the two- and three-disc pick-and-place problems discussed earlier. However, imagine the search tree required for a 16-, 24-, or 32-disc problem. Even with only 16 discs, there would be 3^{16} or over 43 million unique legal states within the tree. For complex problems such as this, the power of heuristic searching is required in order to find a solution to the problem within a reasonable amount of time and space. Let's explore each of these search techniques in more detail.

Blind State-space Searches

A state-space search begins with the initial, or root, node. If the root node is a goal node, the problem is solved. Of course, this is not usually the case. The root node must be expanded and one of its successors must be examined, and so on, until a goal node is found. When a goal node is found, the search is over and the tree shows every path that has been explored during the search. Thus, the tree grows as the search for a goal node progresses.

In blind searches, three fundamental strategies, or plans, are used for tree growth: ***breadth-first***, ***depth-first***, and ***progressive-deepening*** searches.

Breadth-first Searches. In a breadth-first search, all the nodes at a given level are examined for a goal node before the search moves on to a new level. As you can see from Figure 2-24, a breadth-first search is a level-by-level search of the state-space tree. This type of search is guaranteed to find a solution if one exists. Furthermore, it will always find the solution path of minimum length. However, the shortest solution might not be the cheapest or most efficient solution path.

A cost is associated with each operator. The cost of an operator is usually a function of the computer time and space required to apply the operator. It might be that the shortest solution requires higher-cost operations than a more lengthy solution.

Breadth-first searching is not always the most efficient search strategy that can be applied to a given problem. For example, suppose you were going on a trip

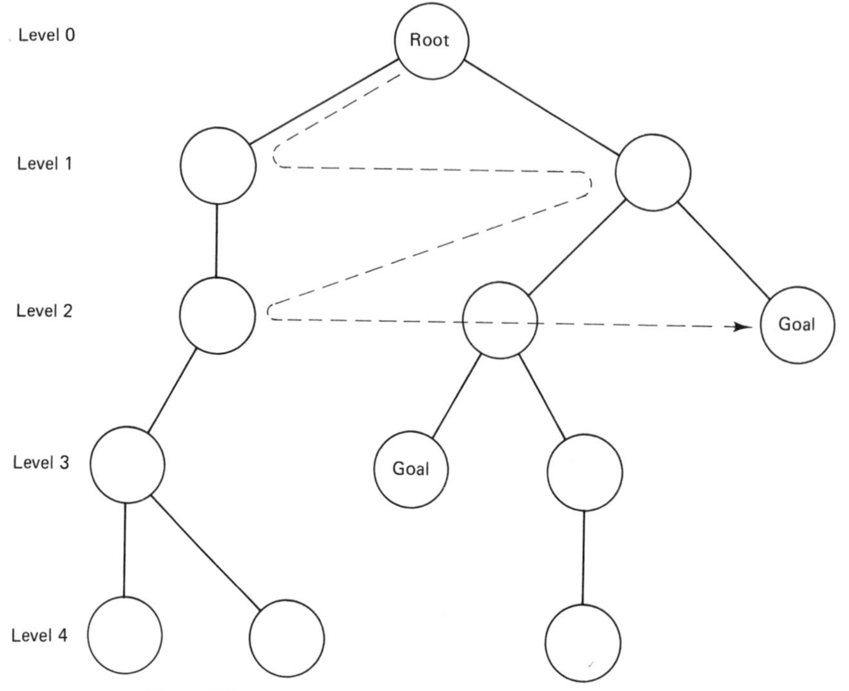

Figure 2-24 A breadth-first search is a level-by-level search.

from city A to city B, but you were not sure of the route to take. Several roads lead from city A through several small intermediate villages, as illustrated in Figure 2-25. Breadth-first searching would be analogous to leaving city A on road 1 to village 1,

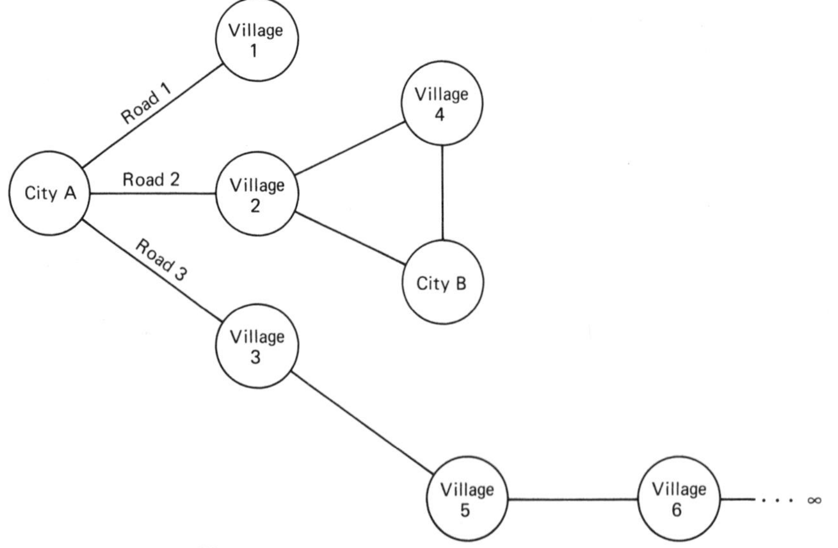

Figure 2-25 How do you get from city A to city B?

and then returning to city A on the same road. Once back in the city, you would leave on road 2 to village 2 and then return on the same road, leaving the city again on road 3. Once all the roads leaving the city are explored, all the roads leaving the first level of villages must be explored, and so on, until city B is found. Notice that this method guarantees that you will eventually get to city B if it exists. However, it might take a lot of time and energy to get there.

Computers perform breadth-first searches very much like the trip analogy just described. A data structure representing a node is examined to see if it is the goal node. If not, an implicit operator is applied to generate one of its children. The child is then examined, and if it is not a goal node, the operator is applied in reverse to get back to the parent node. Another operator is then applied to generate another child, the child is examined, and then the operator is applied in reverse to get back to the parent, and so on, until all the children of a given parent node have been examined. Once one generation of children is examined, the next generation is generated and examined in the same way. Notice that breadth-first searching requires that a given operator always be applied in reverse to get back to the parent before a new child is generated. In other words, breadth-first searching requires backing up and retracing paths until a goal node is found. Such a technique can be inefficient, especially if all paths lead to a goal node at the same level.

An alternative to this procedure would be to continue to generate children along a given path, returning to the parent only when no new paths are found. Using the trip analogy (Figure 2-25), you would leave city A on any arbitrary road until you came to a dead-end. Then you would return to the previous village and try another road out of that village, hoping it will lead you to your destination. The danger in this approach is that you might go too far out of your way and eventually get lost before finding your destination. Notice from Figure 2-25 that if you left city A on road 3 you would never reach city B because a dead-end road is never encountered.

Depth-first Searches. The idea behind a depth-first search is illustrated by Figure 2-26. Here a given branch is followed until it ends. Alternate paths are only considered when the leaf node is not a goal node. Once a given path has been explored to its end without finding a goal node, the search backs up to the closest branch and continues in this manner until a goal node is found.

Depth-first searches are risky, especially if infinite paths exist within a tree. Furthermore, depth-first searches do not always find the shortest path to a goal node. This can be seen from Figure 2-26. However, the advantage of depth-first searching is that it minimizes backtracking within the search tree. This becomes extremely important when backtracking requires complex calculations that use large amounts of computer time and space.

In general, depth-first searches are only used if it is known that the search tree is small and there are many goal nodes within the tree. An alternative to both breadth-first and depth-first searching is a technique called progressive-deepening searching.

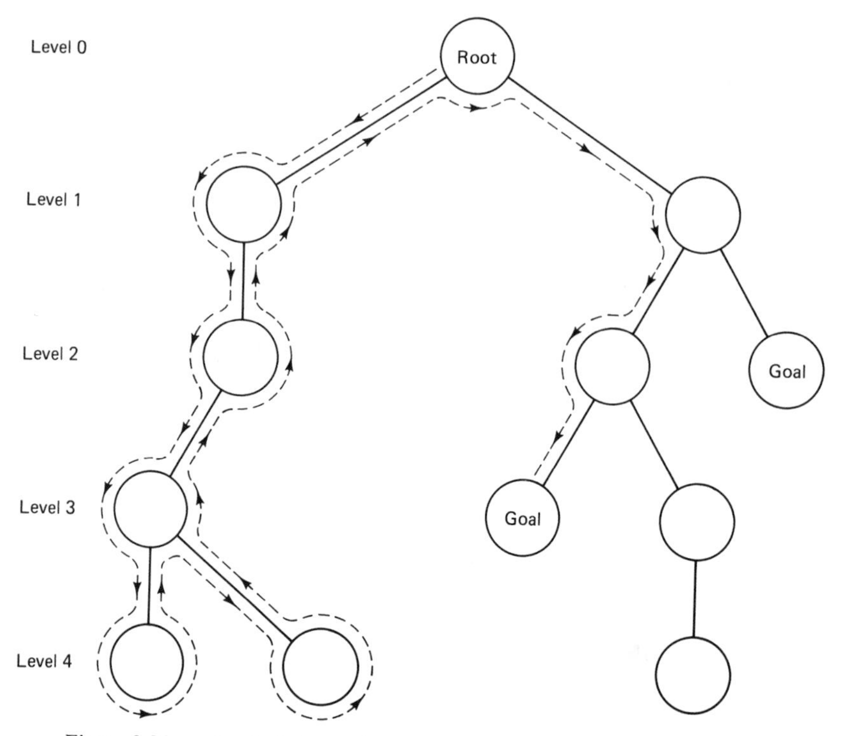

Figure 2-26 A depth-first search continues to explore a path until it ends. Alternate paths are only considered when a leaf node is not a goal node.

Progressive-deepening Searches. Progressive-deepening searching, sometimes called ***depth-bound*** searching, is a compromise between breadth-first and depth-first searching, and incorporates the better qualities of these two techniques. With progressive deepening, a depth bound is set that limits the level to which a search can proceed. The search is begun like a depth-first search, but once the depth-bound level is encountered the search must backtrack to the closest branch and continue forward until the depth bound is reached again. This forces the tree to "spread out" and minimizes the problem of encountering an infinite path before a goal node is found. This idea is illustrated by the search tree in Figure 2-27. Here the depth bound is at level 2. The search proceeds like a breadth-first search until all level 2 nodes have been examined. Notice that a goal node is found at this level. If no goal node existed at this level, the depth bound would be increased to, say, level 4. All nodes through levels 3 and 4 would then be examined. If no goal were found, the depth bound would be progressively increased until a goal node was found or it was determined that no solution path exists.

Actually, you might think of a breadth-first search as a depth-bound search that is progressively incremented by one level. In the same way, a depth-first search can be thought of as a depth-bound search whose depth bound is infinite.

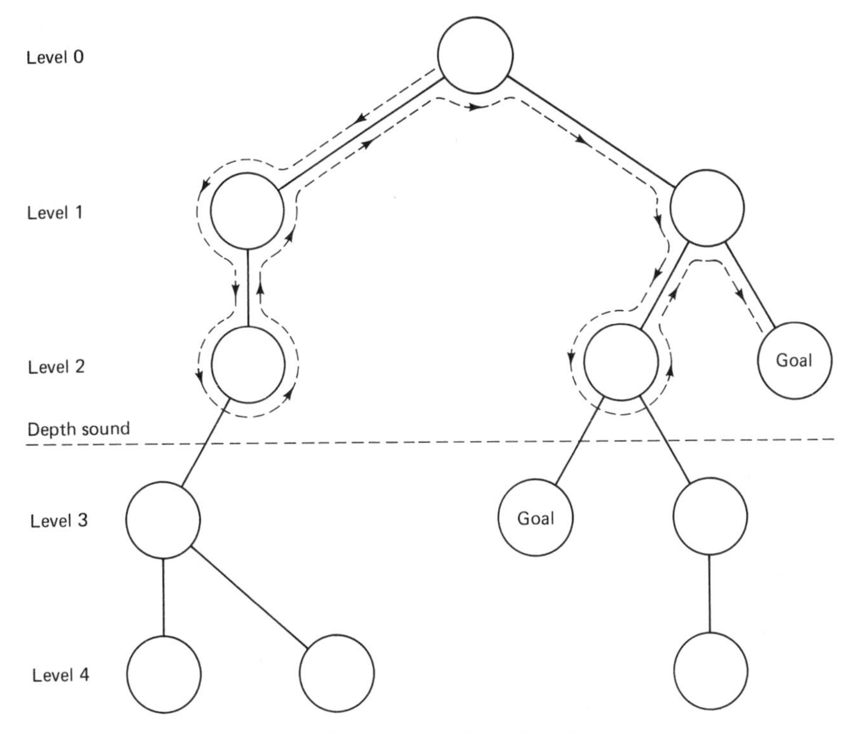

Figure 2-27 A progressive deepening, or depth-bound, search continues to explore a path until the depth bound is reached. The search then backtracks and explores an alternate path.

Like the breadth-first search, the depth-bound search is guaranteed to find a solution if one exists. In addition, it will also find the shortest-path solution. And like the depth-first search, it has the potential of finding the solution in the most efficient manner, but without as much risk. Consequently, most AI programs use the progressive-deepening strategy when blind searching for solutions to simple problems.

At this time, you might want to compare the search paths in Figures 2-24, 2-26, and 2-27. You should note that the depth-bound search produces the most efficient strategy for this particular problem.

Example 2-3

Given the state-space tree shown in Figure 2-28:

(a) Determine the minimum number of operators that must be applied to reach a goal node using each searching strategy. Include reverse operators in your count.

(b) Determine the maximum number of operators that must be applied to reach a goal node using each searching strategy.

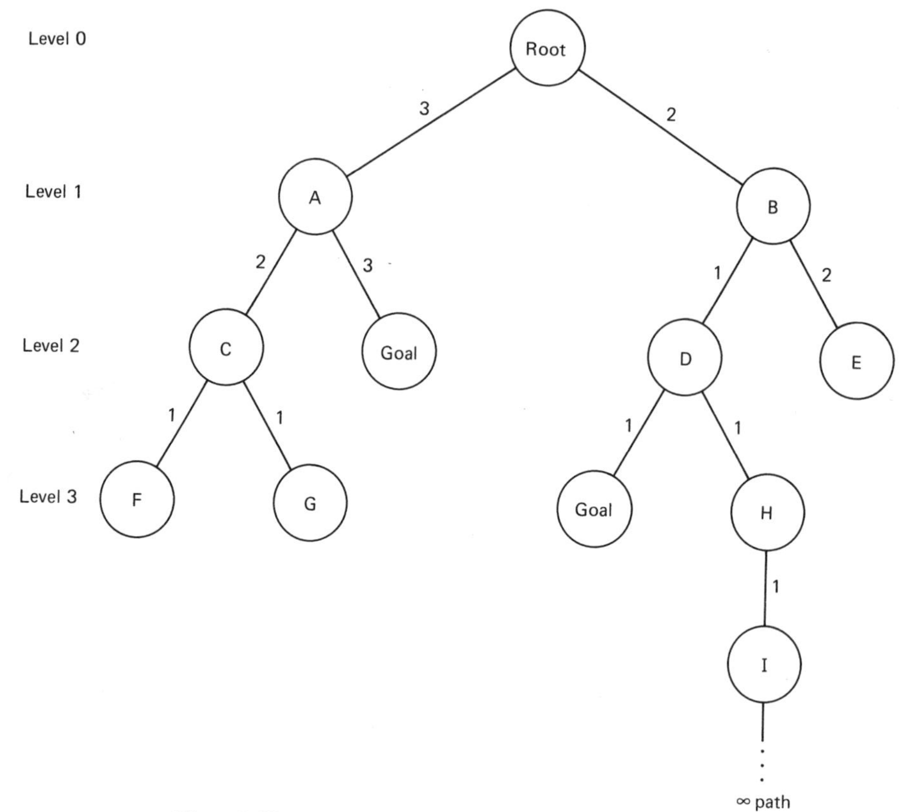

Figure 2-28 State-space search tree for Examples 2-3 and 2-4.

Solution

(a) Using breadth-first searching, the minimum number of operations is four. This minimum-search path would be *root* → *B* → *root* → *A* → *goal.*

Using depth-first searching, the minimum number of operations is two. This minimum search path would be *root* → *A* → *goal.*

With progressive deepening using a depth bound of two, the minimum path would be the same as that for depth-first searching, requiring two operations: *root* → *A* → *goal.*

(b) For breadth-first searching, the maximum number of operations is 12. This maximum search path would be *root* → *A* → *root* → *B* → *E* → *B* → *D* → *B* → *root* → *A* → *C* → *A* → *goal.*

Using depth-first searching, the maximum number of operations is infinite: *root* → *B* → *D* → *H* → *I* → · · · → ∞.

With progressive deepening using a depth bound of two, the maximum search path requires ten operations: *root* → *B* → *E* → *B* → *D* → *B* → *root* → *A* → *C* → *A* → *goal.*

Notice in Example 2-3 that both depth-first and progressive deepening require the minimum number of operations to get to a goal node. However, with depth-first searching you run the risk of an infinite search. Also notice that the possible range of required operations is best for the progressive-deepening search strategy.

Example 2-4

Suppose the numbers associated with the branches in Figure 2-28 represent the "cost" of performing a given branch operation. The same cost applies for both a forward and a backward operation.

(a) Determine the minimum-cost solution to the problem.

(b) What type of search should be used to find a solution most efficiently with the minimum amount of cost and risk?

Solution

(a) The minimum-cost solution path is *root* → *B* → *D* → *goal.* This solution path has a cost of 4.

(b) A progressive-deepening search with a depth bound of three should be used to find this minimum-cost solution path.

Notice in Example 2-4 that the shortest solution path (*root* → *A* → *goal*) is not the minimum-cost solution path. The minimum-cost solution path (*root* → *B* → *D* → *goal*) requires more operators to be applied, but each operator might require less computer time and space than does the shortest solution path. Furthermore, although a progressive-deepening search with a depth bound of three should be used to find the goal node, it does not guarantee a minimum-cost solution. It might be that the search will find a shortest-path solution that is not the minimum-cost solution.

Heuristic State-space Searches

The second general search category that must be discussed is the heuristic search. Clearly, the blind search strategies just presented can only be used efficiently for simple problem solving. As problems become more complex, the search tree grows enormously. For example, the size of the search space for a simple game of checkers is estimated to be around 10^{40} states. Even with a computer that could perform one operation every nanosecond, it would take 10^{31} seconds, or approximately 3.2^{23} years, to search the entire space. Obviously, a more powerful searching method must be used for such complex problems. Such a method must guide, or focus, the search based on knowledge about the problem. Both general knowledge about the problem and knowledge gained while the search is in progress must be included.

A search that uses additional information to guide the search, other than operator conditions and rules, is called a ***heuristic search.*** A heuristic search applies ***heuristic information*** when decisions must be made concerning the best path(s) to follow during the search process.

Webster defines the term heuristic as an adjective meaning "helping to discover or learn." In this text we will also use the term heuristic as a noun. When used as a noun, heuristic will mean a *rule of thumb* that is applied during a search to guide, or focus, the search, thereby drastically limiting the search space. However, the application of a given heuristic during a search does not guarantee the best solution or, for that matter, any solution at all. You might simply say that the application of a heuristic results in an educated guess as to the best path to follow based on the information currently available.

Ordered Searches. In AI programs, heuristic information can be applied at several different points to limit the search. The obvious place to apply a heuristic is when a decision must be made as to which node to expand next. A search that uses heuristic information to expand the *most promising* node next is called an ***ordered search.***

Aside from using heuristic information to expand the most promising node, it can also be used to make decisions about how the node is to be expanded. Recall that when a node is expanded its children are generated. Suppose the available heuristic information indicates that one or more of the children will not likely lead to a solution. Should all the children be generated arbitrarily, regardless of the likelihood of producing a solution? Of course not! Common sense would dictate that only those children most likely to produce a solution should be generated. The other children should be discarded, or ***pruned***, from the search tree. At this time, let's explore several techniques used to apply heuristic information for ordered searches in AI programs.

Minimum-cost Searches. In Example 2-4, you were acquainted with the idea of operational cost. You found that costs are incurred when applying an operator during node expansion or applying a reverse operator when backtracking during the search process. Such costs are normally associated with computer time and space. Since it is often desirable to find the minimum-cost solution, cost estimation can be a valuable heuristic during an ordered search.

Here's how it works. Suppose a node is about to be expanded. The program has been written to keep track of all operational costs up to the given node. That is, a cost factor has been assigned to each operation required for the search. The search program keeps a running total of all the operational costs incurred during the search process. Let's designate this running total with the letter t.

Suppose also that the program can estimate the cost required to get from any given node to a goal node. This estimate is calculated using an algorithm specifically designed for the problem at hand. We will designate the cost estimate with the letter e.

Let's define a heuristic, h, to be the sum of t and e. That is,

$$h = t + e$$

where t = total of all operational costs up to a given node
e = estimated operational cost from the given node to a goal node

When a node is to be expanded, a value of h will be calculated for each of its children. The child with the lowest overall cost estimate value, h, will be the child that is generated. All other children will be pruned from the search tree. If two children have the same cost estimate, the program must make an arbitrary decision as to which child to generate. Thus, the value of h becomes the heuristic that determines the most promising children.

As an example of this procedure, let's go back to the two-disc robot pick-and-place problem discussed earlier. The search tree for this problem is shown again in Figure 2-29. Since all operations are basically the same (i.e., remove a single disc from one peg and place it on another), we will assign a cost value of 1 for each operation. Consequently, the total, t, of all costs to a given node is equal to the depth, or level of the node.

The estimated cost value, e, from a given node to a goal node is a bit harder to define, but suppose we associate it with the number of discs that are on peg 3 at any given point in the search. If two discs are on peg 3, the problem is solved and no further operations are required. For this case, let the value of e be 0. If the large disc is on peg 3, at least one additional operation will be required to move the small disc to peg 3 and solve the problem. Let the value of e be 1 for this case. If no discs are on peg 3, the value of e will be 2, since at least two operations are required to reach the goal state. Finally, the worst cost results if only the small disc is on peg 3. For this case, at least three operations are required to solve the problem. So let the value of e be 3 for this case. The estimated cost assignment values are summarized in Table 2-2.

TABLE 2-2 Summary of Estimated Cost Values for the Two-disc Pick-and-Place Problem

Discs on Peg 3	Estimated Cost to Goal (e)
Both discs	0
Large disc only	1
None	2
Small disc only	3

With the preceding definitions for t and e, an overall cost estimate value, h, can be calculated for each node in the search tree. This value has been placed next to each node in Figure 2-30. Which path will the program take to the goal node? The solution path is shown in Figure 2-31. Notice that the heuristic value, h, has directed the search to find the minimum-cost solution path. In addition, the heuristic information has produced the most efficient search since no backtracking is performed. The heuristic function, $h = t + e$, that was used to direct this search is called an ***evaluation function.*** Evaluation functions are useful when trying to minimize the cost of a search since they produce an estimate of the cost to transform any given state into a goal state. However, such a function must be accurately defined for each particular problem.

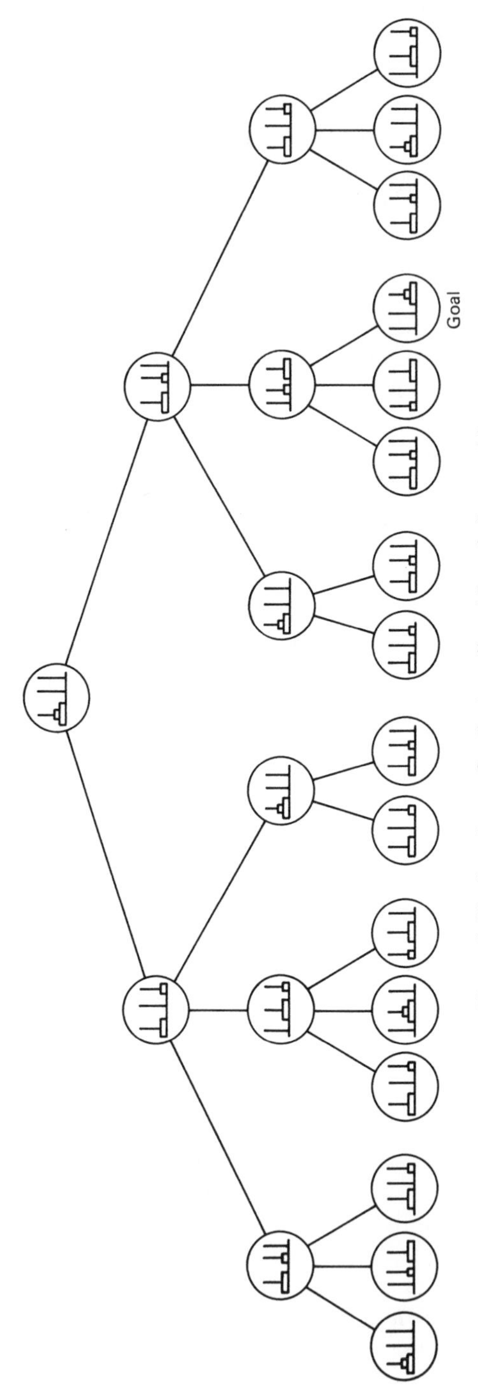

Figure 2-29 Search tree for the two-disc pick-and-place problem.

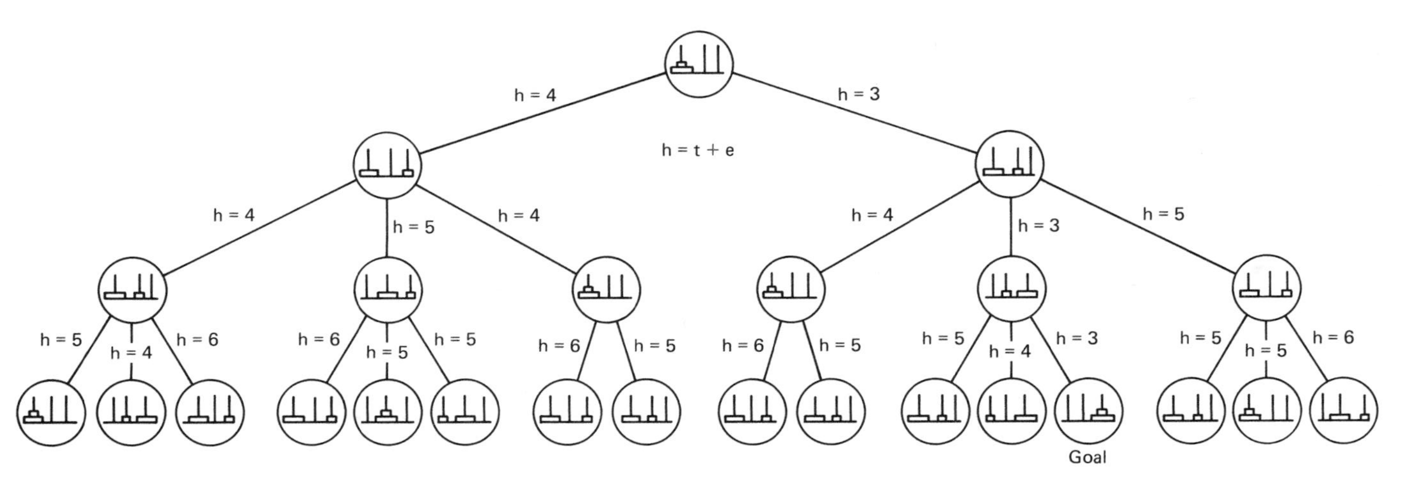

Figure 2-30 Two-disc search tree with heuristic cost information.

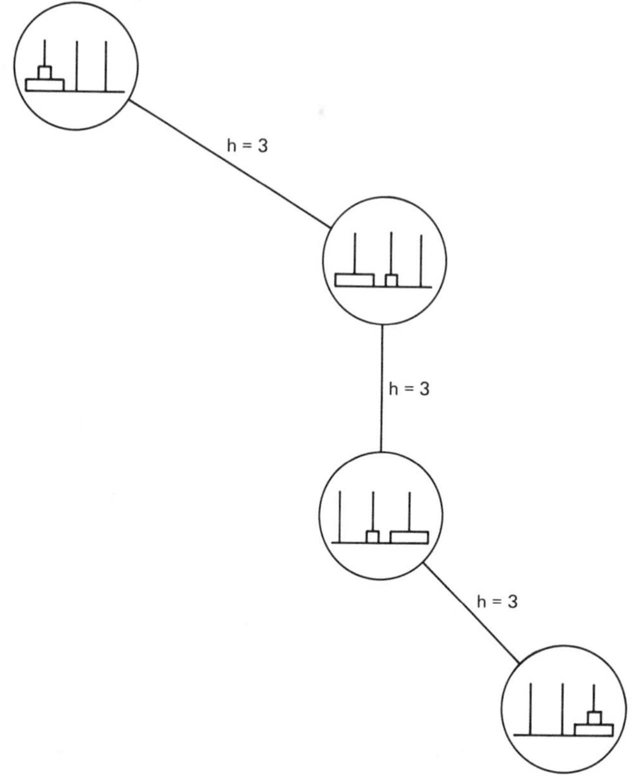

Figure 2-31 Solution path generated by application of the minimum-cost heuristic information.

Ordered Depth-first Searches. You might say that the minimum-cost technique is an ***ordered depth-first search***, since the search always progresses deeper into the tree, following the minimum-cost path. Ordered depth-first searches always select the most promising node as the next node to be expanded. Sometimes this technique is generally referred to as ***hill climbing.***

The term hill climbing is used since we can draw an analogy between climbing a hill and performing a search that always picks the most promising successor node next. Imagine climbing a hill and suddenly becoming fogbound. However, you must press on to reach your goal at the top of the hill. One way to progress to the top would be to take a single step in each direction and then advance in the direction that resulted in movement to a higher elevation. By repeating this process, you would eventually reach the hilltop. This is analogous to the ordered depth-first search, whereby a heuristic dictates the path to the most promising successor node. However, the hill-climbing analogy illustrates some of the pitfalls of ordered depth-first searching.

What about foothills, ridges, and plateaus? These are all problems encountered with a hill-climbing search procedure. The foothill problem results when a false, or intermediate, goal is reached. Such a goal, a ***local goal***, might be similar to the final problem goal, the ***global goal.*** However, the local goal does not completely solve the problem. Imagine yourself working your way to the top of the foggy hill only to find out that, when the fog lifts, it is a foothill to another larger hill across a valley, as illustrated in Figure 2-32(a).

The ridge problem, illustrated in Figure 2-32(b), also creates a false sense of accomplishment. If you ascend to a thin ridge, movement in any direction will either cause you to maintain the same elevation or result in a fast descent. In either case, you have not reached your goal. Finally, the plateau problem results in a wandering, or exhaustive, search for the goal, as shown in Figure 2-32(c). Such a search could be the result of an infinite path within the search tree.

How can the hill-climbing procedure be modified to produce a more efficient search? There are basically two ways: by using ***backtracking*** and by using a technique called ***difference reduction.*** Backtracking can be forced on the search by setting a depth bound. Recall that a depth bound causes the search tree to spread out. In the minimum-cost example discussed earlier, the subtotal cost value, t, from the initial node to any given node continues to increase as the search deepens. This results in a steadily increasing heuristic value, h. A limit can be set on h such that the search is forced to back up to the nearest branch and take the next minimum-cost path until a solution is found or the depth bound is reached again. In this way, the exhaustive search within a plateau region can be avoided.

Difference reduction requires that the search always progress in a direction that minimizes the difference between a given state and the goal state. An application for the difference-reduction technique is for pattern recognition. The shape of an object is digitally compared to several geometric shapes. The closest match is chosen; then the object is compared to several variations of the chosen geometric shape. Again the closest match is chosen, and so on, until a pattern is found that best matches the object. This application is illustrated in Figure 2-33. The difference-reduction technique minimizes the foothill and ridge problems encountered in a hill-climbing-type search.

Best-first Searches. Finally, I should point out that the ordered depth-first, or hill-climbing, search technique always advances the search from a parent node to the most promising child node. Another type of search, called a ***best-first*** search, provides a variation to this technique, resulting in a more efficient search. With best-first searching, the search always advances to the *best* node within the partially developed search tree, regardless of its position within the tree. Consequently, the search is not required to advance in the parent–child direction. Rather, if the expansion of a parent node does not produce any promising children, the search will move to the most promising, or best, node within the tree, no matter where it is located. Of course, the "best" node is determined by the heuristics being applied at the decision point.

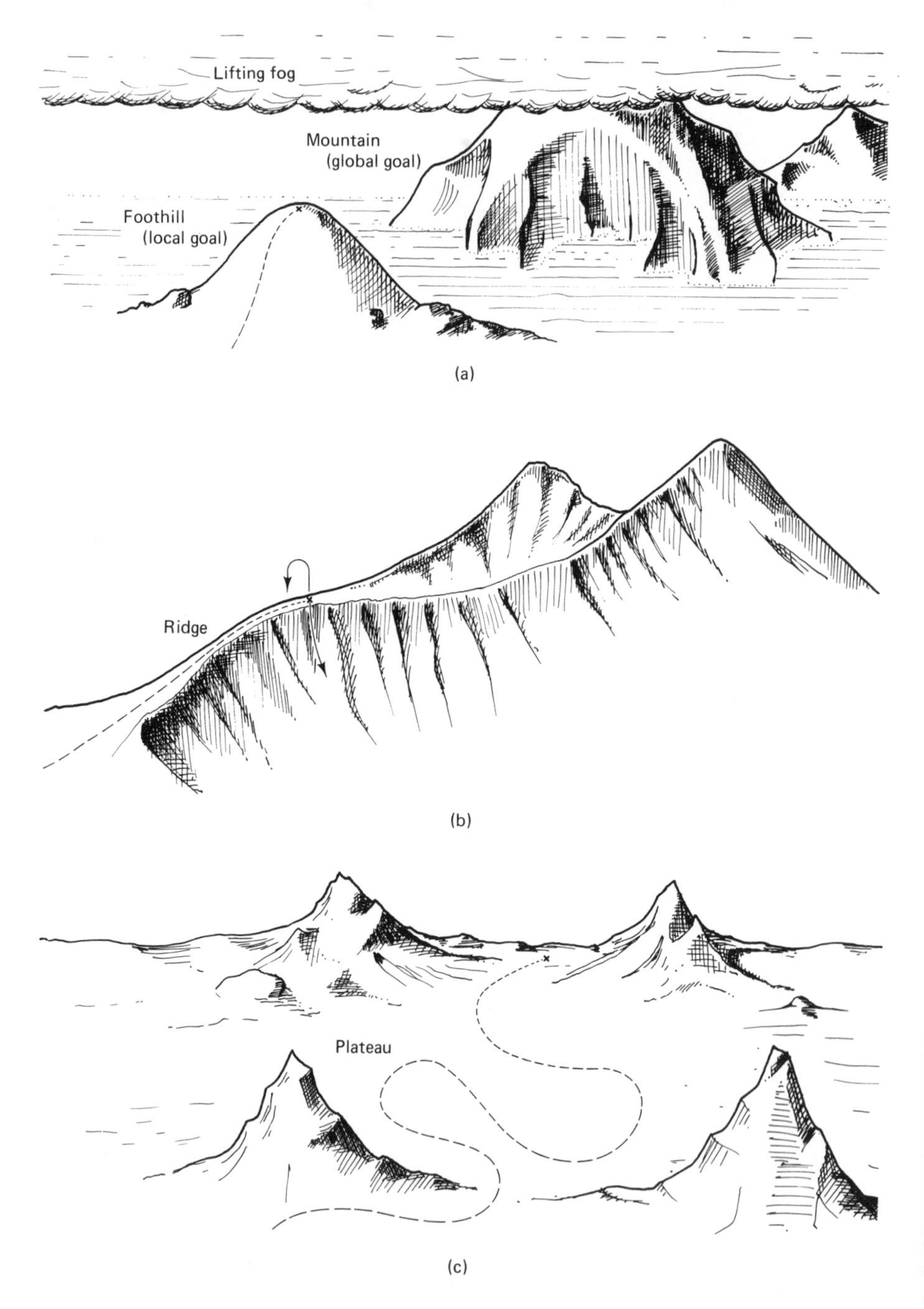

Figure 2-32 (a) Foothills, (b) ridges, and (c) plateaus all create problems when using the hill-climbing search technique.

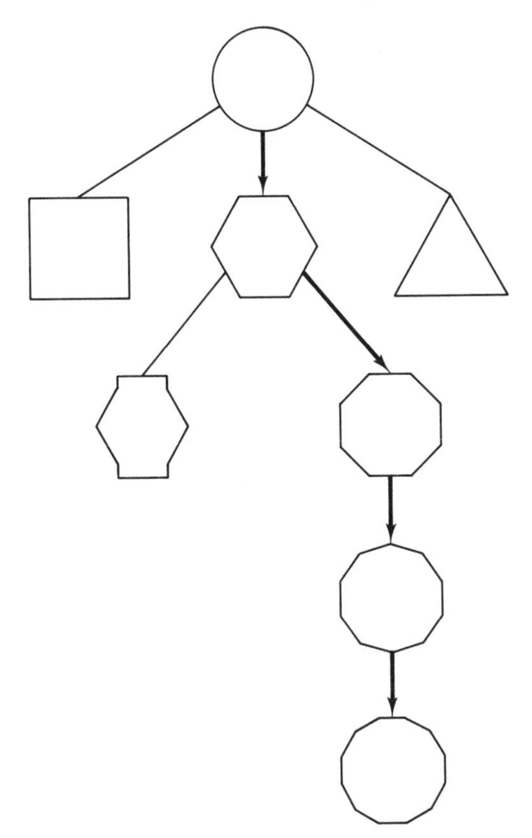

Figure 2-33 An application for difference reduction is for pattern recognition.

Problem-reduction Graph Searches

In Section 2-2, you found that complex problems could be broken down into a set of subproblems. Subproblems could be further broken down into a set of simpler primitive problems. Solving the primitive problems solved the subproblems, which in turn solved the overall problem. This technique was referred to as problem-reduction representation. Recall that the AND/OR tree is the graphical device used to implement problem-reduction representation. Basically, the searching techniques used for AND/OR trees are similar to those used for state-space trees. The familiar breadth-first, depth-first, progressive-deepening, and ordered search strategies can all be applied to AND/OR trees. However, a few considerations are unique to AND/OR searches.

First, AND/OR tree searches use backward reasoning. The search begins at the goal node and progresses through the tree until a set of primitive problems is generated whose solution solves the original problem. As the search progresses, AND/OR nodes are expanded to produce successor nodes. However, expanding an

AND/OR node differs slightly from the expansion of a node in a state-space search. In a state-space search, a given operator only produces one successor node, whereas an AND/OR tree operator may produce multiple successor nodes.

For example, consider the AND/OR tree in Figure 2-34. To expand node *A*, two operators must be applied: operator 1, which generates node *B*, and operator 2, which generates node *C*. Notice that operator 1 produces a single subproblem at node *B*. However, operator 2 not only produces subproblem *C*, but a whole set of subproblems consisting of nodes *F, G,* and *H*. In other words, operator 1 reduces problem *A* to a single subproblem (*B*), while operator 2 reduces the problem to a set of three subproblems (*F, G,* and *H*). Thus, when an AND node is generated, all its successors must also be generated. The reason for this is found in the solution criteria that must be applied to the AND/OR nodes.

A node in an AND/OR tree is solved if:

- It is an OR node and *any one* of its successors is solved.
- It is an AND node and *all* its successors are solved.
- It is a primitive problem (no successors) that is solved.

From these criteria, you can see that node *A* in Figure 2-34 is solved if subproblem *B* OR *C* is solved. However, to solve subproblem *C*, subproblems *F, G,* and *H* must also be solved. Thus, expanding node *A* requires that nodes *B, C, F, G,* and *H* be generated.

Another difference between AND/OR searches and state-space searches is determining when the search should be terminated. State-space searches apply forward reasoning until a goal node is found. However, AND/OR tree searches

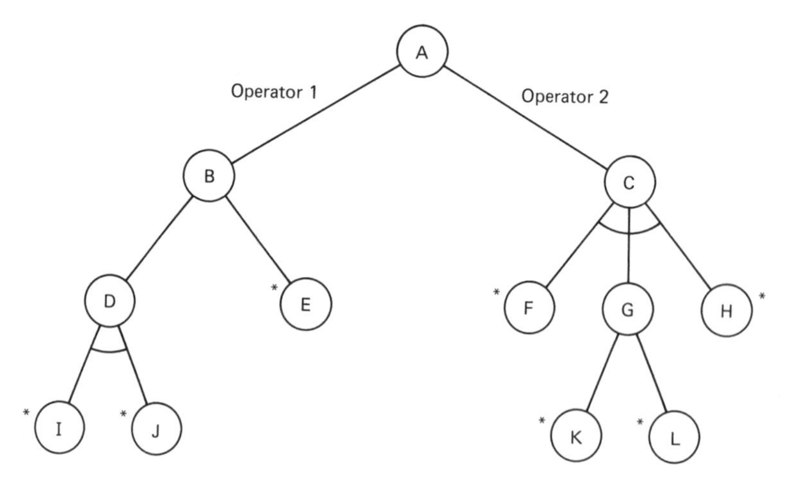

Figure 2-34 Operator 1 generates subproblem *B*, while operator 2 generates a set of subproblems: *F, G,* and *H.*

apply backward reasoning until a set of primitive subproblems is generated whose solution proves, through the AND/OR operations within the tree, that the goal node is solved. A solution to an AND/OR tree is a ***subgraph*** of the tree that demonstrates that the goal node is solved. For instance, Figure 2-35 shows all the possible solution subgraphs for the AND/OR tree in Figure 2-34. This assumes that nodes *I, J, E, F, K, L,* and *H* are all solvable primitive subproblems. Notice that, if node *I* were unsolvable, subgraph 1 could not be a solution path. Likewise, if node *E* were an unsolvable primitive problem, subgraph 2 could not be a solution path. In the

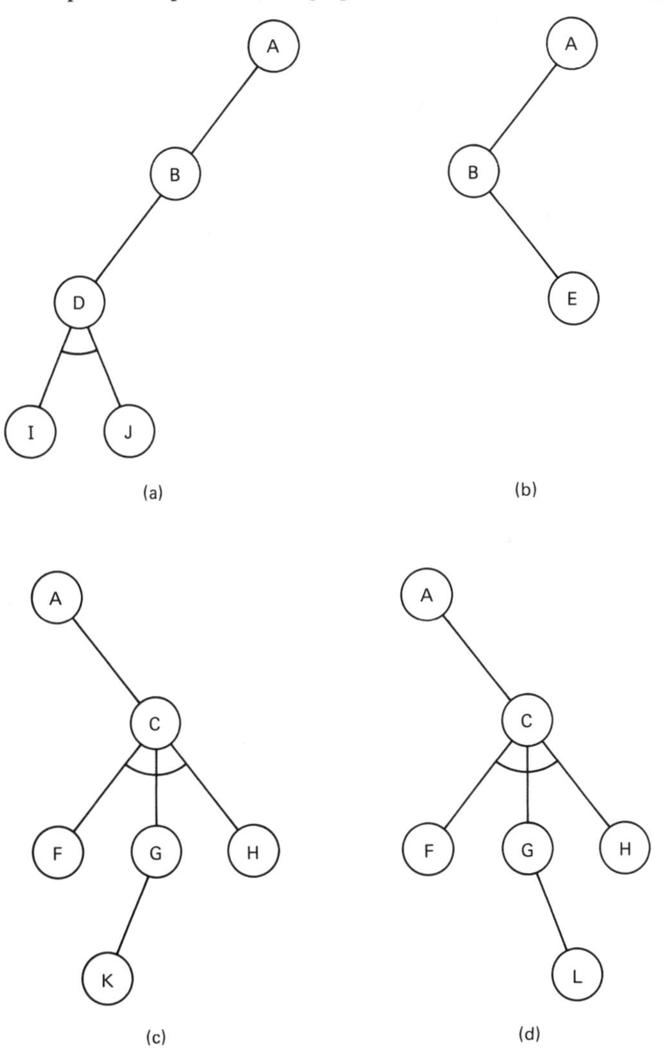

Figure 2-35 All possible solution subgraphs for the AND/OR tree in Figure 2-34: (a) solution subgraph 1; (b) solution subgraph 2; (c) solution subgraph 3; (d) solution subgraph 4.

same way, if node *F* were unsolvable, subgraphs 3 and 4 could not be solutions to the problem.

While the search progresses through the tree, checks must be made at the primitive nodes to determine if the goal is solvable or unsolvable. In the preceding example, suppose the search first progressed through node *C* to primitive subproblem *F*. If *F* were found to be unsolvable, the goal node (*A*) could not be solved by this solution path. To go on and examine nodes *G, H,* and *L* would be a waste of time. Rather, the search should back up and explore other possible solution paths through node *B*. If the search then determined that primitive nodes *I* and *E* were also unsolvable, the search must be terminated and report that no solution is possible.

Example 2-5

Given the AND/OR tree shown in Figure 2-36. The letter *s* beneath a primitive node means that the node is solved, while the letter *u* means that the node is unsolvable. Does a solution path(s) exist? If so, sketch the solution subgraphs(s).

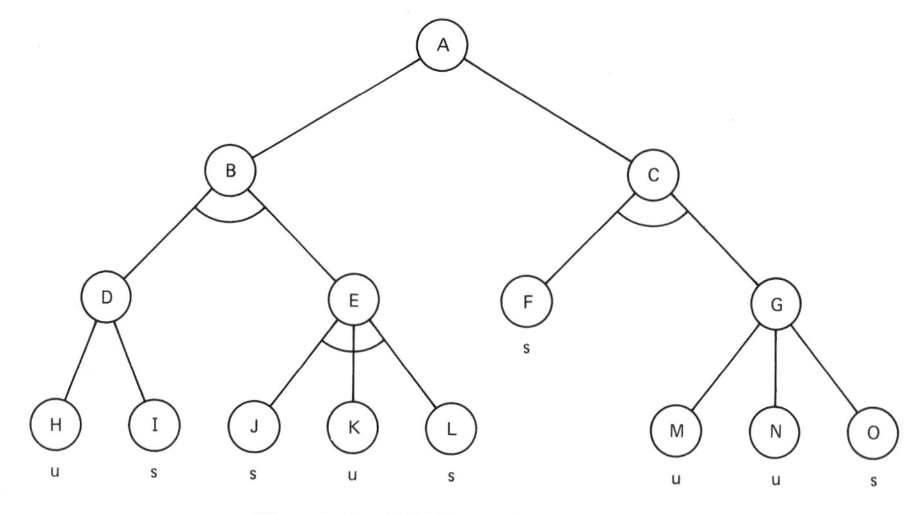

Figure 2-36 AND/OR tree for Example 2-5.

Solution

The unsolvable primitive nodes are *H, K, M,* and *N*. Tracing back through the left side of the diagram, you find that node *D* is solved since node *I* is solved. Notice that node *D* is an OR node. Thus, solving node *I* solves node *D*. However, node *E* is unsolvable since it is an AND node and one of its successor nodes (node *K*) is unsolvable. Consequently, a solution path cannot exist through node *B* since it is an AND node that has an unsolvable successor node. Now, tracing through the right side of the diagram, you find that nodes *F* and *G* are solved. Node *G* is solved since it is an OR node and one of its successors (node *O*) is solved. Since both nodes *F* and *G* are solved, node *C* is solved. Thus, a solution path exists through node *C*. The resulting solution subgraph is shown in Figure 2-37.

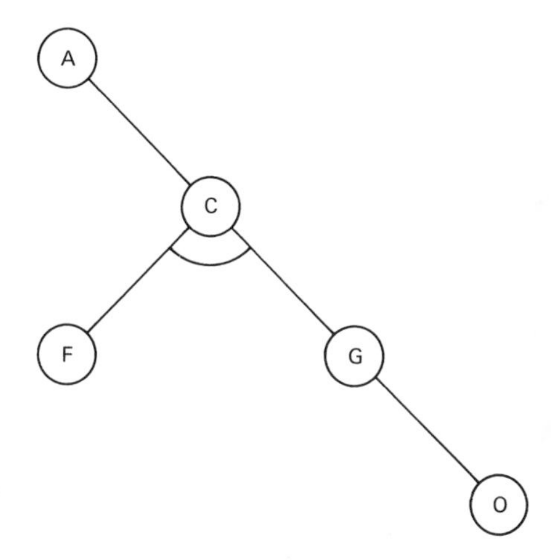

Figure 2-37 Solution subgraph for Example 2-5.

Like state-space searches, heuristics can be applied during AND/OR searches to streamline the search process. Such ordered searches usually apply an evaluation function at each node to estimate the minimum-cost solution path. Again, application of heuristic information simply directs the search using "educated guesses" and does not guarantee that a solution exists.

SUMMARY

The operation of the human brain is thought to exhibit a digital character at the neuron-synapse level. If this is true, the basic elements of the human thought process can be likened to integrated transistors within a digital computer. Consequently, once elements of the human thought process are understood, there is no reason why computers cannot be programmed to be truly intelligent. Although the fundamental nerve-fiber action of the human brain can be described and understood, the combined effect of many nerve fibers performing an intelligent act is not fully understood. For this reason, some AI researchers study how to make computers smarter by attempting to understand the human thought process.

Artificial intelligence is sometimes defined as "the science of making machines do things that would require intelligence if done by men." However, this definition leads to a more important question: What is intelligence? Intelligence has been likened to perception, creativeness, inductive reasoning, learning, and just plain commonsense. Thus, AI programming must eventually incorporate these attributes into computers if they are to become smarter, and not be viewed as simply stupid machines.

Problem solving in AI currently reduces to states, or models, of the problem within the computer, operators to manipulate the states, and a control strategy that applies the operators to produce a problem solution. Two general methods used to represent problems are state-space representation and problem-reduction representation. State-space representation divides a problem into states, operators, and goals.

Problem-reduction representation reduces a problem into subproblems whose solution solves the original problem.

Both state-space and problem-reduction networks are represented by graphs called search trees. State-space trees are basically OR trees that solve problems using forward reasoning, from the initial state to a goal state. Problem-reduction representation reduces to AND/OR trees that solve problems using backward reasoning, breaking the goal down to a set of primitive problems whose solution solves the goal state.

Search trees can be defined explicitly or implicitly. Implicit definitions are used by computers to expand nodes within the tree as the tree grows. A search tree grows as the search for a solution progresses through the tree. The tree-searching procedure should accomplish three tasks:

1. Always find a solution if one exists.
2. Always find the best solution.
3. Always find the most efficient solution.

Tree-searching techniques fit into two general categories: blind searches and heuristic searches. In a blind search, the expansion of the tree nodes is more or less arbitrary. Heuristic searches utilize additional knowledge about the properties of a problem to control the search process. Consequently, heuristic searching, sometimes called ordered searching, reduces and focuses the search. As a result, heuristic searches are much more efficient for solving complex problems than blind searches.

Searching techniques that fit into the two general search categories include breadth-first, depth-first, progressive-deepening, minimum-cost, hill-climbing, ordered depth-first, difference-reduction, and best-first searches.

PROBLEMS

2-1 Explain why the human brain exhibits a digital character.

2-2 What are synapses and how do they respond like digital gates to nerve impulses?

2-3 List several characteristics of the human brain.

2-4 List several characteristics of digital computers.

2-5 Define artificial intelligence, or AI.

2-6 State-space networks divide a problem into ______________, ______________, and ______________.

2-7 Problem-reduction representation graphs divide a problem into ______________ ______________.

2-8 Describe the difference between forward and backward reasoning and state where each type of reasoning is used.

2-9 Explain the overall problem-solving approach used in AI.

2-10 List the three general tasks that must be accomplished during a tree-searching procedure.

2-11 A robot must navigate within a room from point X to point Y using the track layout shown in Figure 2-38.

(a) List all possible paths that an unintelligent robot might take.

(b) Develop a search tree from the list of possible routes in part (a).

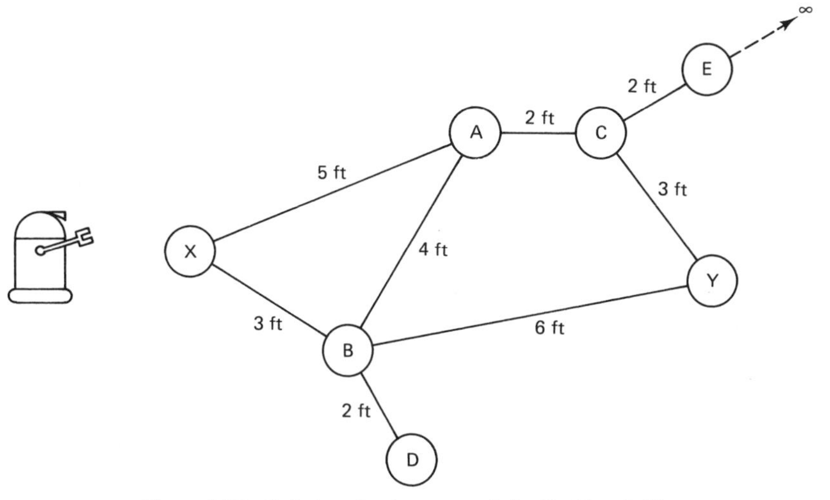

Figure 2-38 Robot navigation network for Problem 2-11.

2-12 From the search tree you developed in Problem 2-11, determine:

(a) The minimum number of paths that the robot must take to reach point Y using both breadth-first and depth-first searching. Include reverse paths in your count.

(b) Determine the maximum number of paths that could be taken to reach point Y using each of the search strategies in part (a).

2-13 The numbers associated with the paths in Figure 2-38 represent the distance, in feet, from one point to another.

(a) Determine the minimum distance path from point X to point Y.

(b) What type of search strategy would the robot use to find this minimum distance path most efficiently and with minimum risk?

2-14 Two implicit operations are defined as follows:

$$p_1 = (n+1)! \quad \text{and} \quad p_2 = \frac{(n+2)!}{2}$$

where $1 \leqslant n \leqslant 6$. If the initial node value is 1, generate an explicit state-space tree using these operations on each node.

2-15 Using the problem-reduction technique, divide a four-disc pick-and-place (tower of Hanoi) problem into three subproblems, one of which is a primitive problem.

2-16 One of the subproblems in Problem 2-15 was to move the three smaller discs (A, B, C) to peg 2. Construct a tree that will reduce this subproblem to a set of primitive problems.

2-17 A five-puzzle problem is shown in Figure 2-39. Given the initial state of the problem in Figure 2-39(a), the idea is to move the number tiles around until they appear in the order shown in the goal state [Figure 2-39(b)]. Construct a state-space tree of depth 3, beginning with the given initial state. Each operation within the tree must only move one number.

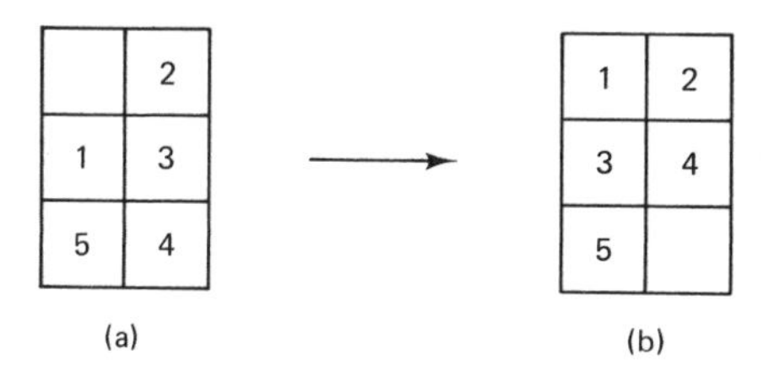

Figure 2-39 Five-puzzle problem for Problem 2-17: (a) initial state; (b) goal state.

2-18 Using the tree you developed in Problem 2-17, attempt to find the goal state by conducting a blind depth-first search. Flip a coin when an arbitrary decision must be made as to which path to follow. Did your search find the goal state? If so, did it produce the most efficient solution path? Again, using your coin to make decisions, conduct a blind breadth-first search. Then conduct a progressive-deepening search using a depth bound that increments by 2. Which of these searches produced the solution most efficiently?

2-19 Develop a heuristic evaluation function that could be used to solve the five-puzzle problem in Problem 2-17.

3 Elements of Knowledge Representation

In Chapter 2, you were introduced to artificial intelligence and how problems are defined and solved in AI programs. Once a problem is defined, an AI system must acquire knowledge about the problem. Two types of knowledge must be present in an intelligent system: (1) knowledge about the problem domain, and (2) knowledge about how to solve the problem. For example, a vision system must have knowledge about the kinds of objects it sees and know how to analyze various shapes in order to classify those objects. Thus, an AI system must have knowledge about its environment, or ***domain.*** Without such domain knowledge, a vision system could not understand what it sees. Clearly, the intelligence of the system is directly related to the amount of knowledge it possesses about its domain.

In addition to knowledge about its domain, an AI system must know how to process that knowledge in order to make intelligent decisions and conclusions. It does no good to supply a system with knowledge if it cannot process that knowledge in some meaningful way. This text is a basic source of knowledge, but without you, the reader, it is simply words printed on paper. Thus, the two major components of

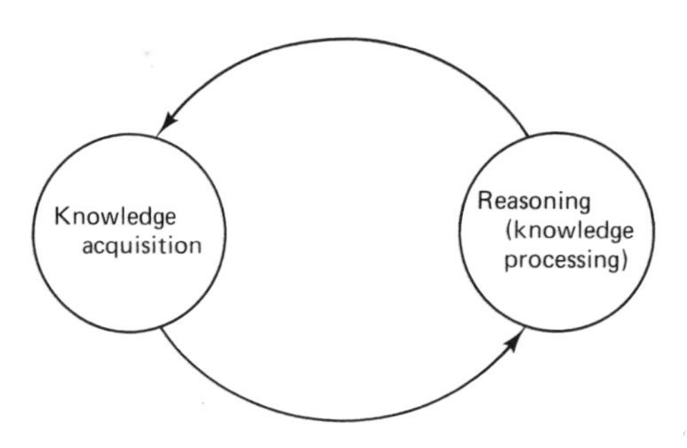

Figure 3-1 Knowledge acquisition and processing are separate elements of knowledge representation, but they are totally dependent on each other.

knowledge representation are ***knowledge acquisition*** and ***knowledge processing.*** In AI systems, the processing of knowledge is called ***reasoning.*** As you can see from Figure 3-1, knowledge acquisition and reasoning are separate elements of knowledge representation, but they are totally dependent on each other for intelligent system operation.

It has been said that the science of knowledge representation is "the glue that binds much of AI together." How do you judge the intelligence of a person? Usually by what the person knows and what he or she does with that knowledge. The same is true of computers. Experience has shown that good knowledge representation schemes are the key to making computers more intelligent. For this reason, knowledge representation has been one of the most active areas of AI research and development in recent years.

In this chapter, you will explore the four major knowledge-representation techniques used in AI: ***logic, production systems, semantic networks,*** and ***frames.*** Although other methods are used to represent knowledge, most can be reduced to one of these four major techniques. As you are about to discover, no unique knowledge representation technique fits all applications. Some techniques work well for certain tasks, while others do not. In many cases, a combination of the four major techniques is required to best satisfy the application. In any event, the application will always dictate the appropriate knowledge-representation scheme to use.

After a thorough examination of knowledge-representation techniques, you will then be introduced to the field of knowledge engineering and see how intelligent systems are engineered using the principles of AI presented in this text. Now, let's get on with the science of knowledge representation.

3-1 LOGIC

It is only fitting that we discuss logic first, since it was one of the first knowledge-representation schemes used in AI. Logic is a formal procedure whereby implications, called ***inferences***, are made from a set of known facts. For example, suppose a robot data base contains the fact that

All rectangles have four sides.

As you will soon discover, such statements can be translated into a mathematical formula using the rules of logic. Once translated, this knowledge can be easily included as a part of a computer's data base. Suppose your robot also "knows" that

A square is a rectangle.

Using logic rules of inference and the preceding two statements, the robot could determine that

A square has four sides.

Thus, the robot has logically deduced the following: If all rectangles have four sides, and if a square is a rectangle, then a square has four sides.

An important feature of logic is that *when logical rules of inference are applied to facts that are known to be true, any resulting new facts are guaranteed to be true.* For this reason, logic is often used in AI to verify that a given deduction or conclusion is correct. As a result, logical reasoning can be used during the search process discussed in Chapter 2 to determine when a given goal state is reached.

Two fundamental systems are used to represent and deduce knowledge using logic: ***propositional calculus*** and ***predicate calculus.*** Here the term calculus simply refers to a method of logic calculation and has no formal relationship to the differential and integral calculus that you might be familiar with. Let's explore each of these two logical systems in more detail.

Propositional Calculus

Propositional calculus gets its name from the idea that knowledge can be represented using a series of statements, called ***propositions.*** Propositions can be either true or false. Some examples of propositions are:

1. All rectangles have four sides.
2. All birds have wings.
3. All triangles have two sides.
4. Birds can fly.

Notice that the first two propositions are always true. Such propositions are called ***tautologies.*** Proposition 3 is a ***fallacy*** since it is always false and can never be true. The fourth proposition could be either true or false, depending on the circumstances. If the bird in proposition 4 is a mature and healthy robin, it can fly. However, if it is an ostrich, it cannot fly. Such propositions are neither tautologies nor fallacies.

Connectives. Propositions are used to represent simple facts, or knowledge, about a given situation. However, by themselves, propositions cannot be used for complex knowledge representation or deductive reasoning. What is needed is a set of rules that can be used to combine, or connect, simple propositions to form more powerful knowledge statements and deduce new statements. Such rules are called ***connectives.*** Five connectives are commonly used in propositional calculus:

Connective	Symbols
Not	$\overline{\ }$ or $\sim$
OR	$+$ or $\vee$
AND	$\cdot$ or $\wedge$
Implies	$\rightarrow$ or $\supset$
Equivalent	$\equiv$

You are probably familiar with some of these connectives from previous exposure to digital electronics or computer science. However, let's review their meaning as related to propositional logic. Suppose we use the symbols A and B to represent two arbitrary propositions. Then, the NOT connective has the following meaning:

$\sim A$, or $\bar{A}$ is *true* if A is *false*
is *false* if A is *true*

This connective can be summarized in terms of a *truth table*, as follows:

A	$\sim A$, or $\bar{A}$
T	F
F	T

Notice that the NOT connective simply reverses the true/false meaning of the proposition. In other words, it acts like a simple inverter in digital electronics.

The OR connective has the following meaning:

$A \vee B$, or $A + B$ is *true* if either A or B or both are *true*
is *false* if both A and B are *false*

In terms of a truth table,

A	B	$A \vee B$, or $A + B$
T	T	T
T	F	T
F	T	T
F	F	F

Notice that the OR connective is inclusive rather than exclusive. If it were exclusive, one proposition being true would exclude any others from being true. Since it is inclusive, more than one proposition can be true. Thus the OR connective acts like an OR gate in digital electronics and not an exclusive-OR (XOR) gate.

The AND connective has the meaning

$A \wedge B$, or $A \cdot B$ is *true* if both A and B are *true*
is *false* if either A or B is *false*

In terms of a truth table,

A	B	$A \wedge B$, or $A \cdot B$
T	T	T
T	F	F
F	T	F
F	F	F

Notice that the AND connective is analogous to a simple AND gate in digital electronics.

The implies (→) connective is a bit harder to nail down. It is much like the familiar IF/THEN statement found in high-level programming languages such as BASIC. Given two propositions A and B, the implies connective states that

If A is *true*, then B is *true.*

Clearly, this implication seems "logical." However, suppose A is false. Does this imply that B is also false? No! If A is false, nothing can be implied about B. In other words, if proposition A is false, proposition B could be either true or false. Thus, if A is false, the implication $A \rightarrow B$ must be true since it is permissible for B to be either true or false. Notice that I said the "implication" must be true. In terms of a truth table,

A	B	$A \rightarrow B$, or $A \supset B$
T	T	T
T	F	F
F	T	T
F	F	T

Notice from the truth table that the only time the implication can be false is when A is true and B is false. In this case, the implication is false, since a true proposition cannot imply a false proposition.

In summary, you can say that the implies connective ($A \rightarrow B$) only *guarantees* B to be true when A is true. If A is false, the state of B cannot be determined.

Finally, the equivalence connective simply says that two propositions have the same truth values. If A and B are both true, then $A \equiv B$ is true. Likewise, if A and B are both false, then $A \equiv B$ is true. However, if A and B have different truth values, then $A \equiv B$ is false. The resulting truth table is

A	B	$A \equiv B$
T	T	T
T	F	F
F	T	F
F	F	T

The application of the connective rules to simple propositions allows the formation of logical statements and produces the "calculus" in propositional logic. It might be helpful at this point to look at some examples.

Example 3-1

Prove that $\bar{A} \vee B \equiv A \rightarrow B$.

Solution

In logic, you can prove that two expressions are equivalent if their truth tables are the same. Thus, to prove equivalency, a truth table is generated separately for each side of the expression. If the two truth tables are the same, the equivalency holds. You already know the truth table for $A \rightarrow B$ from the preceding discussion. So the problem reduces to finding the truth table for $\bar{A} \vee B$. To do this, simply substitute all possible true/false combinations for A and B into the expression and apply the given connectives.

$$\bar{A} \vee B = \overline{\text{true}} \vee \text{true} = \text{false} \vee \text{true} = \text{true}$$

$$\bar{A} \vee B = \overline{\text{true}} \vee \text{false} = \text{false} \vee \text{false} = \text{false}$$

$$\bar{A} \vee B = \overline{\text{false}} \vee \text{true} = \text{true} \vee \text{true} = \text{true}$$

$$\bar{A} \vee B = \overline{\text{false}} \vee \text{false} = \text{true} \vee \text{false} = \text{true}$$

Thus, the resulting truth table is

A	B	$\bar{A} \vee B$
T	T	T
T	F	F
F	T	T
F	F	T

Notice that this truth table is identical to the implies truth table given earlier.

Example 3-2

Given the following propositions:

A = The sky is blue.

B = It is not raining.

Show that $A \rightarrow B$.

Solution

The implication A implies B ($A \rightarrow B$) is true if A is true and B is true. Consequently, if the sky is blue, then it is not raining. However, if A is false, the implication says nothing about proposition B. In other words, if the sky is not blue, there is no way of knowing if it is raining or not. In either case, the respective implication would be true. Finally, if A is true and B is false, the implication must be false since it would be raining with a blue sky. (Think about it!)

Rules of Inference. Recall I stated earlier that the important feature of logic is that new facts are guaranteed to be true if they have been deduced from true facts. So, given a set of known true facts, how do you deduce new facts that are guaranteed to be true? In other words, given a set of true statements, what rules must be applied to arrive at a new statement that is guaranteed to be true? The answer to this question is provided using logic ***rules of inference.*** A rule of inference is simply a rule that allows you to deduce new logical statements from given statements. Let's take a look at some of the more common rules of inference.

Conjunction. Conjunction is probably the simplest rule of inference since it relates directly to the AND connective. Let's define a ***premise*** to be a proposition or logic statement that is assumed to be true. The conjunction rule simply states that, if two propositional statements, or premises, are true, then the combined statement is also true. In other words, if A and B are premises, then the conclusion A AND B must be true. The conjunction rule can be represented graphically as follows:

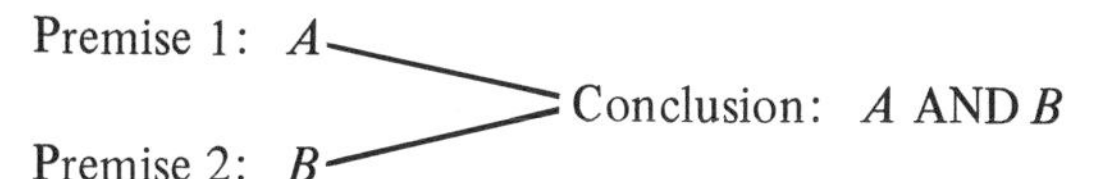

For example,

Premise 1: The sky is blue.
Conclusion: The sky is blue and it is not raining.
Premise 2: It is not raining.

The conjunction rule should be obvious and can be easily proved using the truth table for the AND connective. In fact, a rule of inference is valid if its truth table yields a true result, or conclusion, each time all the propositional statements, or premises, are true.

Modus Ponens. The modus ponens (MP) rule of inference is probably the best known inference rule. The modus ponens rule states that if proposition A is true and $A \rightarrow B$, then proposition B must also be true. That is,

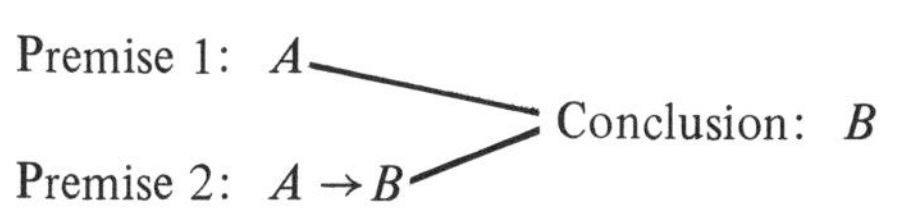

In terms of an example, given the premises that "The sky is blue" and "If the sky is blue, it is not raining," then MP allows you to conclude that "It is not raining." Thus,

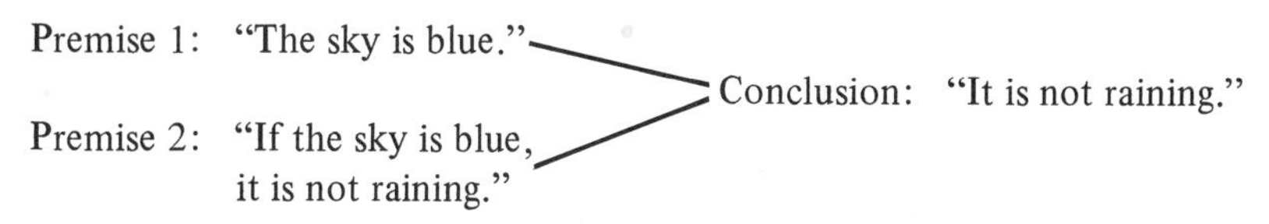

The modus ponens rule can be proved by observing the truth table for the implies connective. Notice from the truth table that if $A \rightarrow B$ is true and A is true, then B must also be true.

The application of the modus ponens rule allows you to replace two statements with a single statement. Notice that, when A and $A \rightarrow B$ are replaced with the single statement B, the implies connective is eliminated. For this reason, the modus ponens rule is sometimes called the ***implies elimination rule.***

Resolution. The resolution inference rule can be stated as follows: Given two propositional statements A and B, the premises $A \vee B$ and $A \vee \bar{B}$ can be combined to produce a single conclusion, A. Graphically, the resolution rule looks like this:

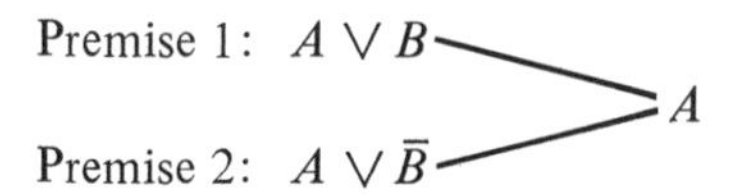

In other words, the combined premises are resolved, or *refined*, into a simpler propositional statement. Using the preceding notation, you could say that combining the two premises results in the B and $\bar{B}$ statements being canceled out.

Let's prove the resolution rule. Again, we will use a truth table for the proof. Given the two propositions, A and B, all possible truth values are entered in the left side of the table. Then the corresponding truth values for premises $A \vee B$ and $A \vee \bar{B}$ are entered on the right side of the table. Now notice that, when both premises ($A \vee B$ and $A \vee \bar{B}$) are true, the conclusion (A) is also true. These cases have been circled in the following truth table:

A	B	$A \vee B$	$A \vee \bar{B}$
T	T	T	T
T	F	T	T
F	T	T	F
F	F	F	T

Since the truth table yields a true conclusion each time all the premises are true, the resolution inference rule is valid.

Applying the Resolution Rule. As you have just seen, rules of inference are proved using truth tables. The idea is to determine if the inference rule produces a true conclusion each time all the premises are true. Computers can be programmed to verify logical conclusions using the truth table method. In fact, any inference rule in propositional calculus can be proved using this technique. However, given n different propositions, there are 2^n possible true/false combinations of the given propositions. In verifying a logical conclusion, a computer must generate and test all possible combinations. Such a procedure becomes extremely time consuming and thus inefficient, especially when the number of proposition variables is large. For this reason, most computers apply the resolution rule to verify logical conclusions. This is called the ***resolution principle of logical deduction.***

Simply stated, the resolution principle is proof by *contradiction.* Proof by contradiction begins by assuming that a given conclusion is false. Then, using this false assumption along with the premises that produced the conclusion, a statement is generated that is totally impossible. Such a statement is called a contradiction. Since the contradiction results from assuming that the original conclusion is false, the conclusion must be true.

The following six-step procedure must be used when applying the resolution principle to verify a conclusion.

1. Negate the conclusion. That is, assume the conclusion is false.
2. Put all premises and the negated conclusion into ***conjunctive form.*** A logical expression is in conjunctive form if:
   a. Only individual variables are negated, not connected expressions.
   b. OR connectives are only used between individual variables.
3. Divide each conjunctive expression into ***clauses.*** Clauses are terms within the expression that are ANDed together, but do not contain any ANDing operations. For instance, the expression $(A \vee B) \wedge C \wedge (A \vee \bar{B})$ has three clauses: $A \vee B$, C, and $A \vee \bar{B}$.
4. From all the clauses generated in step 3, select two clauses such that one clause contains a given variable and the other clause contains the negation of the given variable. Two such clauses might be $A \vee B$ and $A \vee \bar{B}$.
5. Using the resolution rule, reduce the two clauses into a new clause. Thus, $A \vee B$ and $A \vee \bar{B}$ would reduce to A using the resolution rule.
6. Add the new clause obtained from step 5 to the list of clauses and repeat steps 4 and 5 until a contradiction results, or new clauses can be generated. If a contradiction is generated, the original conclusion is true by the resolution principle. If no new clauses can be generated, the original conclusion is false.

A contradiction results when two clauses completely cancel out each other and you are left with nothing, sometimes called the ***empty set.*** For example, when A and $\bar{A}$ are combined using the resolution rule, they cancel and you are left with the empty set, which produces a contradiction. A box is used to designate this condition, as follows:

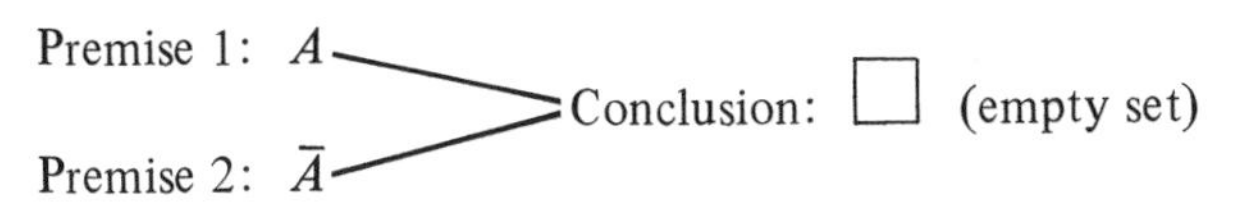

Example 3-3

The following is an excerpt from a hypothetical small town police report:

> A domestic robot sounded an alarm when it detected an intruder in the basement of a house. The robot detected a sound that was created by a window breaking in the basement. The sound could be heard throughout the house. When the police arrived, they found a person in the basement. The person told the police that he did not hear the window break and therefore could not be the suspected intruder.

A judge subsequently concluded that the person lied and found him guilty of breaking and entering. Prove that the judge's conclusion was correct.

Solution

Let's apply the resolution principle. From the police report, you could make the following propositions:

A. The robot's sensors were correct.
B. A person was in the basement.
C. The person was in the same house as the robot.
D. The person found in the basement heard the window break.
E. The person told the truth and did not break the window.

Note that these propositions might be true or false; it does not really matter. For example, if the person lied, then proposition E would be false, but $\bar{E}$ would be true. Now, given the preceding propositions, you can make the following premises:

Premise 1: $A \rightarrow B$
If the robot's sensors were correct, there was a person in the basement of the house.

Premise 2: $B \rightarrow C$
If there was a person in the basement, the person was in the same house as the robot.

Premise 3: $C \rightarrow D$
If the person was in the same house as the robot, the person heard the window break.

Premise 4: $E \rightarrow \bar{D}$
If the person told the truth and did not break the window, the person did not hear the window break.

Conclusion: $A \rightarrow \bar{E}$
If the robot's sensors were correct, the person did not tell the truth and broke the window.

After making these premises, the next step is to apply the resolution principle.

Step 1. Assume that the conclusion is false. That is, negate the conclusion: $\overline{A \rightarrow \bar{E}}$.

Step 2. Put all the premises in conjunctive form. For a premise to be in conjunctive form, only variables can be negated and OR connectives can only be used between individual variables. Recall the equivalency that was proved in Example 3-1:

$$A \rightarrow B = \bar{A} \vee B$$

Let's first apply this equivalency to all the preceding premises, as follows:

$$\text{Premise 1:} \quad A \to B \equiv \bar{A} \vee B$$
$$\text{Premise 2:} \quad B \to C \equiv \bar{B} \vee C$$
$$\text{Premise 3:} \quad C \to D \equiv \bar{C} \vee D$$
$$\text{Premise 4:} \quad E \to \bar{D} \equiv \bar{E} \vee \bar{D}$$
$$\text{Negated conclusion:} \quad \overline{A \to \bar{E}} \equiv \overline{\bar{A} \vee \bar{E}}$$

Notice that each of the equivalent expressions is now in conjunctive form.

Step 3. Divide each conjunctive expression into clauses. A ***clause*** is defined as a term within an expression that does not contain any ANDing operations. The preceding premises are all clauses since they do not contain any ANDing operations.

Step 4. From all clauses, select two such that one contains a given variable and the other contains its negation. Suppose you choose the first two premises: $\bar{A} \vee B$ and $\bar{B} \vee C$. Notice that premise 1 contains B and premise 2 contains its negation, $\bar{B}$.

Step 5. Apply the resolution rule to reduce the two clauses into a new clause.

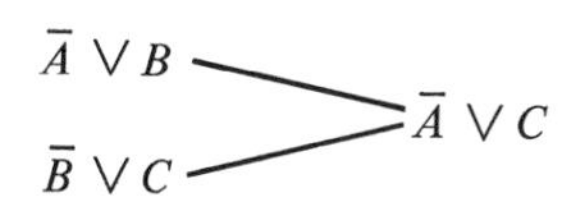

Step 6. Add the new clause to the list of clauses and repeat steps 4 and 5 until a contradiction results or no new clauses can be generated. The repeated application of steps 4 and 5 is shown next.

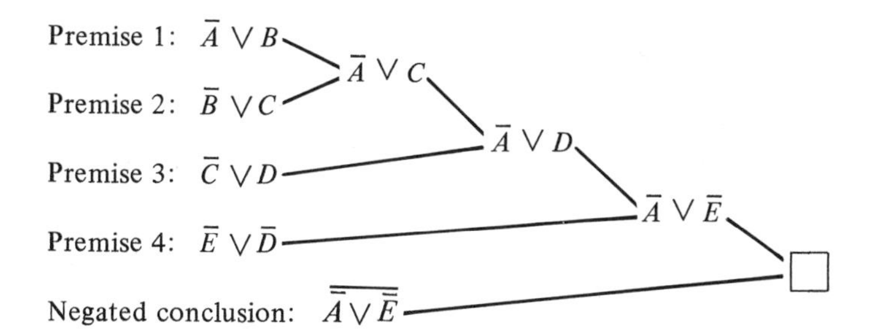

From this reduction, you can see that repeated application of the resolution rule generates the empty set, or contradiction. Therefore, the conclusion must be true: If the robot's sensors were correct, the person did not tell the truth and broke the window.

Predicate Calculus

It is now time to extend the idea of propositional calculus into a more powerful logic representation system called predicate calculus. In fact, predicate calculus is sometimes referred to as the BASIC language of logic. Like propositional calculus, predicate calculus deals with logic statements and inferences that can be labeled true or false. However, predicate calculus is much more flexible since it allows the use of ***variables, quantifiers,*** and ***functions*** within the logic statement. These characteristics allow logic statements to be broken down such that objects within the statement can be analyzed to determine why a statement is true or false and if additional relationships exist between propositions other than simple true/false relationships.

Predicate Statements. A logic statement in predicate calculus consists of two parts, a ***predicate*** and an ***argument***, as follows:

PREDICATE(ARGUMENT)

The argument is contained in parentheses and supplies the objects, or ***individuals***, to be acted on. Individuals within the argument are usually nouns or pronouns, such as the robot, the box, you, I, this excellent book, and so on. The predicate part of the statement is usually a verb or function that makes a declaration about the individuals within the argument. Examples include the following: is big, greater than zero, American, equals, is in. Thus, when representing English sentences in predicate calculus, the predicate is associated with the verb of the sentence and the argument is associated with the subject or object of the verb. For example, suppose you wish to represent the following English sentences using predicate calculus notation:

The robot is in the kitchen.
Janet is an American.
The value of x is greater than the value of y.
The value of x equals the value of y.

To represent these sentences in predicate calculus, the verb of the sentence becomes the predicate, while the subject and object of the verb become the individuals within the argument. Consequently, these sentences would be represented as follows:

IS IN(ROBOT, KITCHEN)
IS AMERICAN(JANET)
GREATER(x, y)
EQUALS(x, y)

These representations are called ***atomic formulas.*** Notice how the atomic formulas are interpreted. For instance, GREATER(x, y) is read "x is GREATER than y."

This method of reading a formula is illustrated by the arrows in the following statement:

GREATER(x, y)

Using this notation, the predicate is written first, followed by the individual objects to be acted on inside the parentheses. The statement

IS AMERICAN(JANET)

is called a ***one-place predicate*** since only one individual, JANET, is being acted on. The statement, GREATER(x, y) is called a ***two-place predicate*** since it involves two individuals. ***Three-place predicates*** involve three individuals, and so on.

Constants and Variables. The predicate part of the statement is always written using uppercase letters. However, the argument portion of the statement can be written using either upper- or lowercase letters, depending on how the individuals within the argument must be interpreted.

Uppercase letters are used to designate individual constants, such as ROBOT, KITCHEN, and JANET. Individuals are called ***constants*** if they are names for individuals and have only one interpretation. A predicate statement that contains all constants is always true or always false, depending on the statement.

On the other hand, lowercase letters such as x, y, and z are used to represent individual ***variables*** within the predicate statement. As with conventional algebra, variables allow you to substitute different values into the expression.

You might be wondering how predicate statements differ, except for notation, from propositional statements. Let's consider the four example statements given previously. The first two predicate statements,

IS IN(ROBOT, KITCHEN)
IS AMERICAN(JANET)

are also propositional statements since they can always be labeled either true or false. However, the last two statements,

GREATER(x, y)
EQUALS(x, y)

cannot be propositional statements. Notice that each statement contains the variables x and y. When a proposition contains a variable, it must always be true or always false for all possible values of the variable. A predicate statement differs in that it can be either true or false, depending on what is substituted for the variable. Thus, the last two statements qualify as predicate statements but not propositional statements. However, the first two statements can be classified as either predicate or propositional statements. For this reason, you could say that propositional calculus is a subset of predicate calculus.

By using variables, predicates can become more general and flexible than propositions. For example, the atomic formula

$$\text{MARRIED}(x, y)$$

could be used to represent all married couples in the world. It is true if the individuals, x and y, are married, but false if x and y are two individuals that were never married.

Connectives Used in Predicate Calculus. Predicate calculus makes use of the same connectives that are used in propositional calculus. The use of connectives allows more informative statements to be represented. Connectives are used in predicate calculus to combine atomic formulas and create statements that are called ***well-formed formulas***, or ***wffs.*** For example, compound sentences can be represented by the AND ($\wedge$) connective. The sentence "Andy lives in Randolph, New York" could be represented by the formula

$$\text{LIVES(ANDY,RANDOLPH)} \wedge \text{LOCATED(RANDOLPH,NEW YORK)}.$$

Likewise, the OR ($\vee$) connective can be used in a formula as follows:

$$\text{IN(DAVID,SCHOOL)} \vee \text{AT(DAVID,HOME)}.$$

This formula would translate to "David is in school or he is at home."

The implies connective is used when an IF/THEN statement is required. For example, the formula

$$\text{OWNS(RON,ROBOT)} \rightarrow \text{COLOR(ROBOT,BLUE)}.$$

could be used to represent the statement "If the robot belongs to Ron, then it is blue."

Finally, the Not operation could be used as follows:

$$\overline{\text{IN}}\text{(DAVID,SCHOOL)} \rightarrow \text{AT(DAVID,HOME)}.$$

This formula would translate to "If David is not at school, then he is at home."

Quantifiers. The use of variables in a predicate statement is governed by quantifiers. A ***quantifier*** determines the extent to which a predicate statement involving variables might be true or false. Two quantifiers are used in predicate calculus: ***there exists*** and ***for all.*** The "there exists" quantifier is represented by the symbol $\exists$. When the $\exists$ symbol is associated with a variable in a predicate statement, it means that "there exists" at least one value of the variable for which the statement is true. For example, the formula

$$(\exists x)[\text{IN}(x,\text{KITCHEN}) \wedge \text{ON}(x,\text{TABLE})]$$

would mean, "There exists at least one of something that is in the kitchen and on the table."

The "for all" quantifier is represented by the symbol $\forall$. When this symbol is associated with a variable it means that the predicate statement is true for all values of the variable. Thus, the formula

$$(\forall x)(\forall y)[\overline{\text{EQUAL}(x, y)} \rightarrow \text{GREATER}(x, y) \vee \text{LESS}(x, y)]$$

must be true for all values of x and y. This formula translates to "If x is not equal to y, then x is greater than y or x is less than y."

By using quantifiers, you are defining the range of the variables. For example, consider the following four formulas:

1. $(\forall x)[\text{PERSON}(x) \rightarrow \text{LIKES}(\text{JANET}, x)]$
2. $(\forall x)[\text{PERSON}(x) \rightarrow \text{LIKES}(x, x)]$
3. $(\forall x)(\forall y)[\text{PERSON}(x) \wedge \text{PERSON}(y) \rightarrow \text{LIKES}(x, y)]$
4. $(\forall x)(\exists y)[\text{PERSON}(x) \wedge \text{PERSON}(y) \rightarrow \text{LIKES}(x, y)]$

The first formula says that "Janet likes all people." The second statement says that "All people like themselves." The third statement says that "Everybody likes everybody." Finally, the fourth statement says that "Everybody likes somebody."

If a variable within a formula is quantified, it is said to be a ***bound variable.*** Variables that are not quantified are called ***free variables.*** If a formula has all of its variables bound, then it is called a ***sentence.***

Functions. A very powerful feature of predicate calculus is found in the use of functions. Recall from algebra that a function is simply a mathematical rule that returns a value when a value is substituted for a variable. For example, the function $f(x) = x + 2$ returns the value 3 when the value 1 is substituted for x. The idea is the same for functions in predicate calculus. Suppose that the variable x is used to represent any student and t is a function used to represent "the teacher of." Then the symbol $t(x)$ would represent the teacher of student x. Therefore, if substituting David for x in the preceding function returns Mr. Smith, or t(David) = Mr. Smith, then you know that Mr. Smith is David's teacher. Likewise, $t(t(x))$ would be used to represent the teacher of student x's teacher. For instance, if the function $t(t$(David) = Mrs. Murphy, then Mrs. Murphy taught David's teacher.

Rules of Inference. Recall that a rule of inference is used to produce new logic statements from given statements. The rules of inference that you studied in propositional calculus were conjunction, modus ponens, and resolution. Since propositional calculus is a subset of predicate calculus, these same rules can be used to create new predicate formulas from given formulas. For example, suppose that the following two formulas are given:

$$\text{FISH}(\text{TUNA})$$
$$(\forall x)[\text{FISH}(x) \rightarrow \text{SWIM}(x)]$$

From these two formulas, the modus ponens rule produces the formula SWIM(TUNA), or, in other words, since a tuna is a fish and all fish can swim, then a tuna can swim.

Earlier you saw that the modus ponens rule was derived directly from the implies truth table. Now, just for fun, suppose that you don't have the modus ponens rule but you do have the resolution principle at your disposal. Using this principle, prove that "If a tuna is a fish and all fish can swim, then a tuna can swim." Recall that to apply the resolution principle you must first assume that the conclusion is false. Thus, assume that "a tuna cannot swim." This translates to the formula

$$\overline{\text{SWIM}}(\text{TUNA}).$$

The next step is to put all the premises in conjunctive form and separate them into clauses. The first premise, FISH(TUNA), is already a clause since it does not contain any ANDed expressions. The next premise

$$(\forall x)[\text{FISH}(x) \rightarrow \text{SWIM}(x)]$$

reduces to the clause,

$$(\forall x)[\overline{\text{FISH}}(x) \vee \text{SWIM}(x)]. \quad \text{(See Example 3-1)}$$

Since this formula is true for all values of x, you can substitute TUNA for x and get

$$\overline{\text{FISH}}(\text{TUNA}) \vee \text{SWIM}(\text{TUNA}).$$

This substitution process is called ***unification*** in predicate calculus. At this point you have the following predicate formulas to work with:

Premise 1: FISH(TUNA)

Premise 2: $\overline{\text{FISH}}(\text{TUNA}) \vee \text{SWIM}(\text{TUNA})$

Negated conclusion: $\overline{\text{SWIM}}(\text{TUNA})$

Notice that the substitution of TUNA for x in premise 2 has created a situation that allows you to apply the resolution rule. Applying the resolution rule to premises 1 and 2 produces the formula

$$\text{SWIM}(\text{TUNA}).$$

But this formula contradicts the negated conclusion and by the resolution rule produces the empty set. Thus, the conclusion that "a tuna can swim" must be true.

Our example illustrates how the resolution principle can be applied in predicate calculus. It also illustrates the importance of unification. Substituting TUNA for x in premise 2 allowed application of the resolution principle. To solve more complex problems, AI programs attempt to find substitutions for variables that will allow clauses to be resolved using the resolution principle. In addition, if a given substitution results in two identical clauses, one of them can be dropped. This operation is called ***factoring*** in predicate calculus. In general, the process of searching for variable substitutions that will result in resolution or factoring is called unification.

Pros and Cons of Logic

You should now have an understanding of the basic principles used in AI programs that incorporate logic techniques for knowledge representation and reasoning. Of course, there is much more to logic than I have covered here. A complete discussion of logic would require a separate textbook. My intent here is to provide you with a "feel" for how logic can be used to produce intelligent programs. Before we go on, however, you should be aware of some general characteristics of logic that must be considered when it is applied to AI.

First, the biggest advantages of logic are its consistency and precision. All deductions made using the rules of logic are guaranteed to be correct. Because of its preciseness it is impossible to prove a false statement. Conversely, logic allows any true statement to be proved, given the required knowledge. Because of its consistency and formal structure, logic is also easily mechanized. As a result, AI programs that incorporate the rules of logic are relatively easy to write and understand. However, because of this same formal structure, the reasoning process is more constrained and not as flexible as it is when using other forms of knowledge representation.

Proponents of logic also argue that the expression of a problem in logic often resembles our intuitive understanding of the problem. This characteristic of problem representation is referred to as ***naturalness.***

Modularity is another feature of logic. This characteristic allows knowledge to be added to the data base independent of existing knowledge. Thus, the knowledge base can grow incrementally as new knowledge becomes available.

The biggest disadvantage of using logic is that it is not suitable for representing general knowledge, sometimes referred to as *commonsense* knowledge. Logic is not capable of commonsense reasoning since it makes no provision for dealing with uncertainty. Logic conclusions are either true or false. Different levels of truth are not possible. For instance, the statement "Most women are beautiful" could not be represented using predicate logic since the terms "most" and "beautiful" are not well defined or measurable. Such terms are sometimes referred to as ***fuzzy*** terms. A special type of logic, called ***fuzzy logic***, has been developed to handle this type of knowledge.

Another disadvantage of logic is that it does not lend itself to the application of heuristics in searching for a solution to a problem. As a result, too much knowledge leads to combinatorial explosion during the search process.

Critics of logic will also argue that if computers are to reason they must reason the same as we do, and obviously we do not reason using mathematical logic. Logic supporters counter this argument by contending that computer reasoning is "artificial" intelligence and, therefore, it is not important that computers reason the same way that we do.

Regardless of such arguments, logic must play an important role in AI programs, if only to confirm the results of thinking, rather than thinking itself. Now, let's progressively move on to less precise and more general knowledge representation techniques in an attempt to satisfy some of the critics of logic.

3-2 PRODUCTION SYSTEMS

The production-system technique for knowledge representation was originally developed as a model of human reasoning. Today, production systems find many applications in AI, especially in ***expert systems.*** These applications will be discussed shortly; however, let's first see how knowledge can be represented using a production system.

You can think of a production system as simply a collection of IF/THEN statements, or rules, called ***productions.*** The IF part of the production represents the conditions, or situation, that must be present to activate the THEN portion of the production. Simple productions can be illustrated using the popular "animals" example, as follows:

IF the animal has feathers, THEN it is a bird.
IF the animal has hair and gives milk, THEN it is a mammal.

Notice that the conditional part of the production provides a list of ***elements***, or facts, that must be present in order to make a conclusion. In the first production, there is a single condition element, "feathers." The second production has two condition elements, "has hair" and "gives milk."

In practice, a computer searches through a list of productions and tries to match the production elements against its data base. When all the elements of a given production are found, the production is ***fired.*** Firing a production carries out the action portion of the production. Once a production is fired, its conclusion is used to update the computer data base.

Using the animals example, a computer would search its data base looking for animal characteristics to classify a given animal. If it found "hair" and "gives milk," it would conclude that the animal is a mammal. The computer would then add "mammal" to its data base and continue to search for elements that would match more specific animal characteristics, such as "has teeth" and "has claws," in an attempt to fire other productions that would further classify the animal. This matching and firing process should eventually lead to the specific animal in question.

Productions are sometimes referred to as pattern → action, situation → action, or premise → conclusion pairs. The respective terminology should be self-explanatory. Regardless of what they are called, productions allow a computer to deduce new facts from known facts. Thus, facts flow through a series of productions until a desired conclusion is reached. This process is illustrated in Figure 3-2. From the figure, you could say that the reasoning process is from left to right. Consequently,

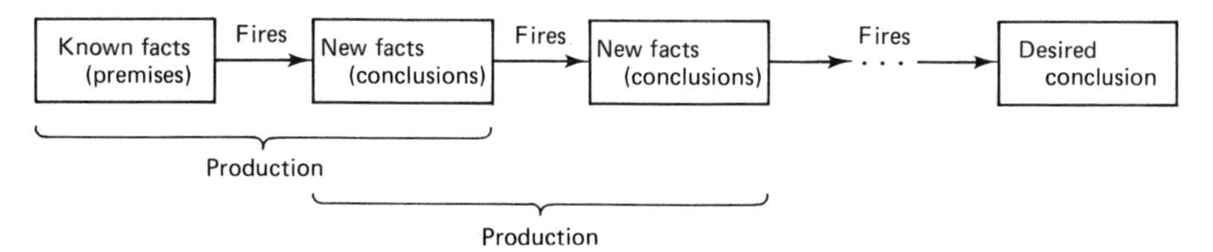

Figure 3-2 Facts flow through a series of productions until a desired conclusion is reached.

the conditional portion of a production is sometimes called the ***left-hand side***, or ***LHS***, and the action part of the production is called the ***right-hand side***, or ***RHS***.

Parts of a Production System

As shown in Figure 3-3, a production system is divided into three major parts: the ***rule base, context,*** and ***interpreter.*** The rule base consists of all the production rules required for the system. Hundreds, even thousands, of production rules make up the rule base of current AI programs used in expert systems. For this reason, production systems are sometimes called ***rule-based systems.*** In general, the larger the rule base, the more intelligent the system. However, larger rule bases increase the search time, resulting in decreased efficiency.

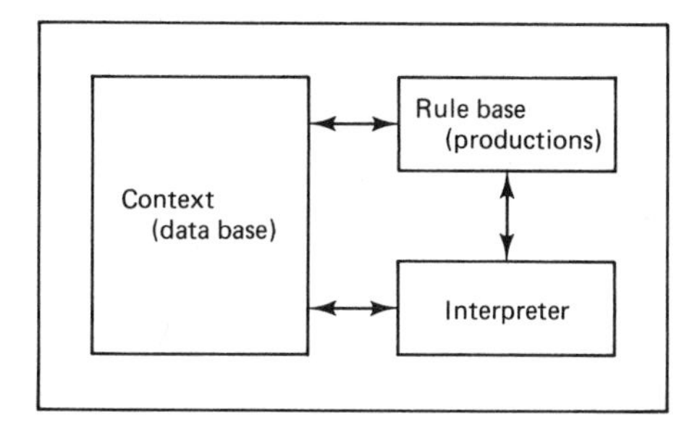

Figure 3-3 A production system is divided into three major parts: rule base, context, and interpreter.

The context is the computer data base that contains the system knowledge, or facts, available at any given time. These facts are the elements that are used to satisfy the conditional portions of the productions. Thus, before a production in the rule base can fire, its conditional elements must be present in the context. The firing of a production creates new facts that update the context so that other production conditions might be satisfied.

The interpreter controls the matching and production rule selection process. It determines which productions match a given set of elements and then decides the order in which the productions are to be fired. For example, suppose the system contains the context and rule base given in Figure 3-4. The interpreter must determine which production should be applied first. A forward-reasoning strategy

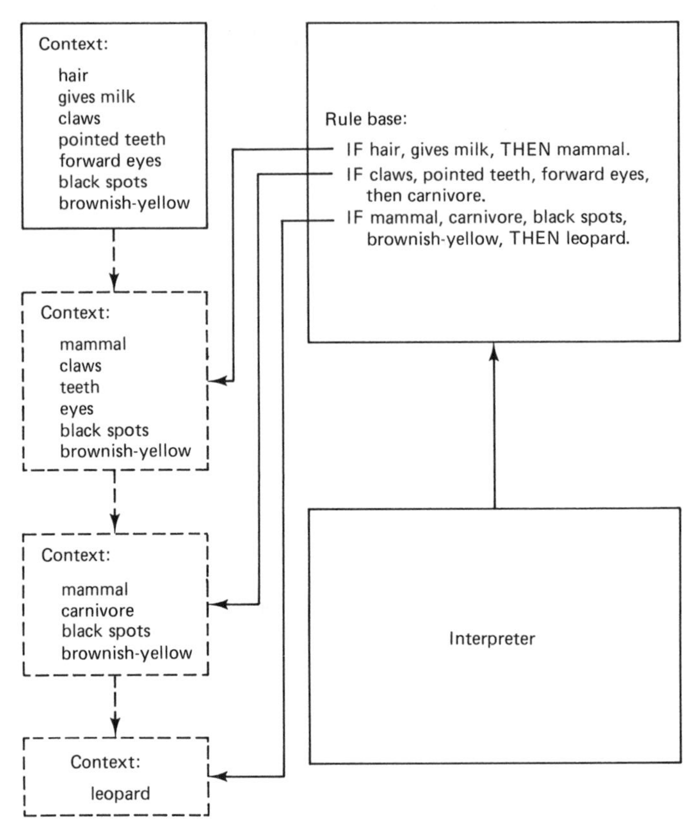

Figure 3-4 Operation of a production system.

would dictate that the interpreter direct the computer to search the context for "hair" and "gives milk." Upon finding these elements in the context, the first production is fired. At this point, if more than one production is applicable, the interpreter must decide which to fire first. This is called ***conflict resolution*** and will be discussed shortly. Note, however, that only one production is applicable in our simple example. After this first production is fired, the context is updated as shown in the figure. Notice that the elements "hair" and "gives milk" have been replaced by a new element, "mammal." Thus, the firing of a production has produced a new context. The new element, "mammal," is added and the old elements are dropped out to prevent repeated firing of the same production.

Next, the interpreter directs a search for "claws," "pointed teeth," and "forward eyes." Since these elements match the second production, it is fired and the context is updated again as shown in the figure. Finally, the interpreter directs a search for "mammal," "carnivore," "black spots," and "brownish-yellow." Upon finding these elements in the context, the third production is fired to produce the desired conclusion, "leopard."

Production Cycles

From the preceding discussion, it is evident that the production system operates in cycles. Each cycle consists of three phases: ***matching, conflict resolution,*** and ***action.*** The production system repeatedly executes this cycle, as illustrated in Figure 3-5. The matching and action phases do not need further explanation. However, let's take a closer look at the conflict-resolution phase.

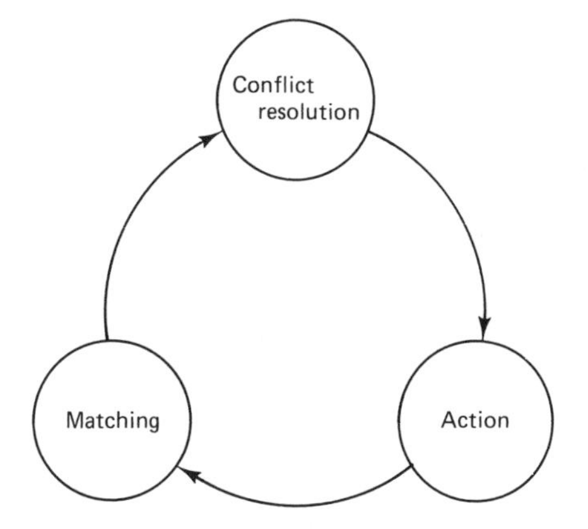

Figure 3-5 Production systems repeatedly execute the matching/conflict resolution/action cycle.

The interpreter must provide conflict resolution when more than one production is matched with a given set of conditional elements in the context. When this happens, the applicable productions are placed in a separate rule base called the ***conflict set of rules.*** From this set, the interpreter decides which production to fire first, based on a predetermined set of criteria. Sometimes, the interpreter simply fires the first matching production encountered. However, many times the production that matches the longest list of context elements is selected. This is sometimes called the ***toughest match.*** In addition, some elements might have higher priority than others. If this is the case, the production that matches the longest list of high-priority elements is selected. Such productions are called ***privileged productions.*** Another way to resolve conflicts is to select the production that matches the most recent elements placed in the context. This assures forward movement toward a desired conclusion. As you can see, conflict resolution is an important phase of the cycle and must incorporate additional knowledge about the problem. Therefore, heuristics are often applied during the conflict-resolution phase to guide the reasoning process.

AND/OR Trees in Production Systems

Production systems are easily visualized using AND/OR trees. The best way to see this is by using an example. The AND/OR tree in Figure 3-6 illustrates the animals example discussed earlier. Notice that the production rules are now represented graphically using the tree. Each fact, or element, in the original context

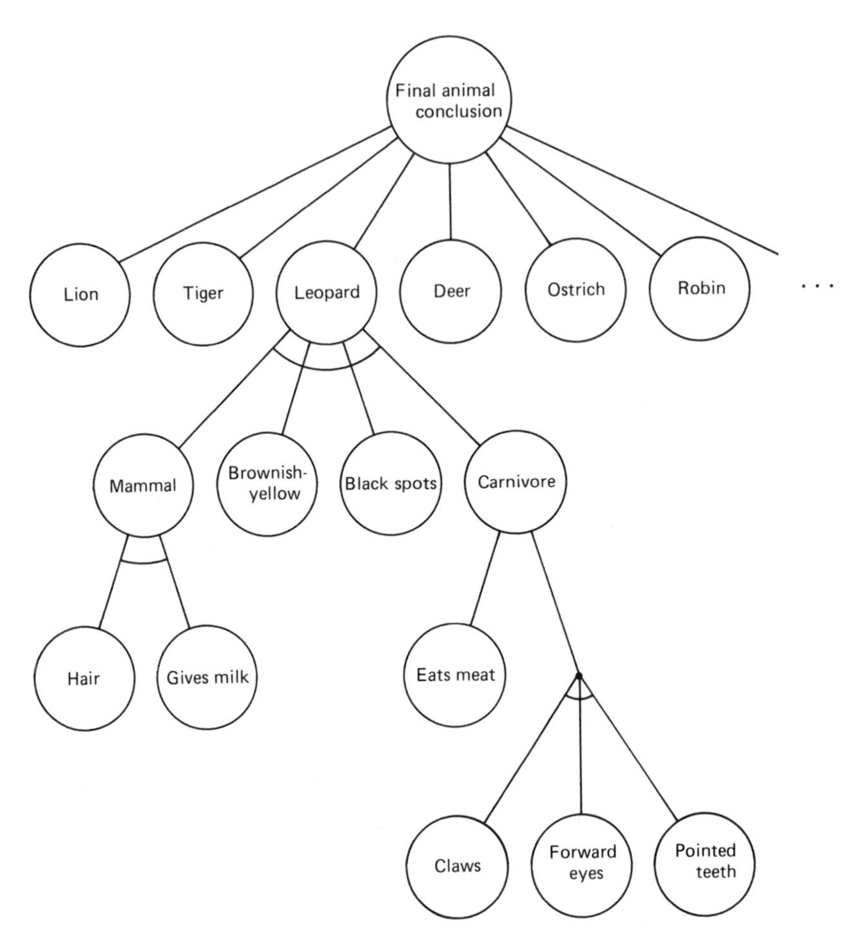

Figure 3-6 The animals production system can be illustrated using an AND/OR tree.

knowledge base is a terminal, or leaf, node. Each intermediate node represents a subgoal to be achieved during the reasoning processes. Nodes are formed using the production system rules as follows:

- An AND node is formed by a production that has more than one conditional element. The conditional elements form the branches of the node as shown in Figure 3-7(a).
- An OR node is formed when more than one production has the same action, or conclusion. The conditional elements of the individual productions form the branches of the OR node, respectively, as shown in Figure 3-7(b).

For example, consider the subgoal "carnivore" shown in Figure 3-6. This subgoal can be reached using either of the following two productions:

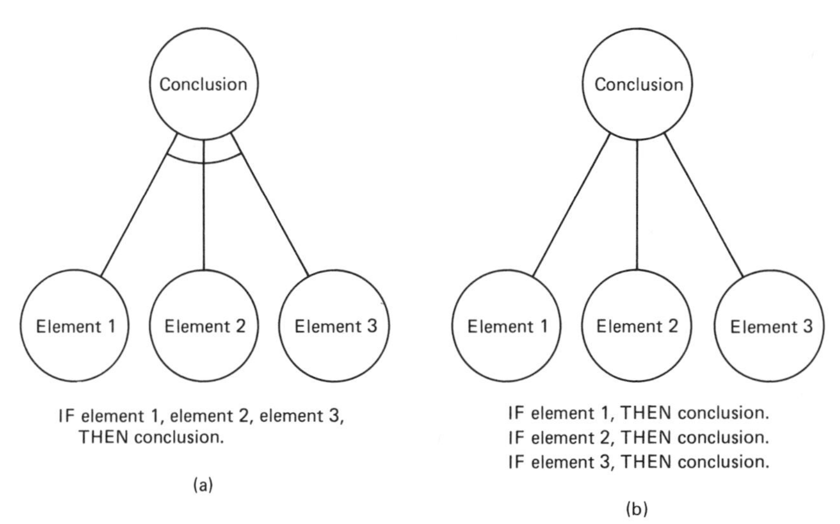

Figure 3-7 Production rules create natural AND/OR trees.

IF the animal eats meat, THEN it is a carnivore.

OR

IF the animal has claws, AND pointed teeth, AND forward eyes, THEN it is a carnivore.

Both of these productions generate the same conclusion. Thus, their elements form an OR node. On the other hand, the second production forms a separate AND node, since several facts must be present before it can be used to generate the conclusion "carnivore."

In summary, you could say that an OR node represents a production with only one conditional element, while an AND node represents a production with several conditional elements.

By defining the production system using an AND/OR tree, the reasoning strategy is clearer and can be better understood. In addition, you can immediately see how changes in the reasoning strategy will affect the process of finding a solution. Many times, sections of the AND/OR tree can be displayed on a graphics terminal so that you can visualize the production system. But most important, the AND/OR tree allows for *verification* and *explanation* of a given solution.

Now let's look at how a computer might use a production system AND/OR tree to reason. Suppose the production system for the animals example is programmed into a computer using the AND/OR tree shown in Figure 3-6. You see an animal and the system must figure out what kind of animal it is. So the system begins to ask the following questions:

Question: Does the animal have hair?
You Respond: Yes.

Question: Does the animal give milk?
You Respond: Yes.

Question: Does the animal eat meat?

You Respond: Yes.

Question: Is the animal brownish-yellow in color?

You Respond: Yes.

Question: Does the animal have black stripes?

You Respond: No.

Question: Does the animal have black spots?

You Respond: Yes.

System Responds: The animal is a leopard!

From this dialogue you can see that the system first assumes that the animal is a mammal, and then verifies the assumption by asking you if the animal has hair and gives milk. Once this subgoal is verified, the system attempts to verify that the animal is a carnivore. Once the system knows that the animal is a carnivorous mammal, it assumes that it is a tiger and tries to verify this assumption. Since the animal does not have black stripes, the tiger assumption is wrong. Finally, the system assumes that the animal is a leopard and verifies this by asking if the animal has black spots.

Notice that the reasoning process is backward. That is, the system assumes a production conclusion and then attempts to verify the assumption by asking you if the production elements exist. The system was simply following the AND/OR tree during its questioning procedure. Each response you gave allowed the system to verify a given subgoal in the tree until the final goal was verified. Of course, "you" could be replaced by sensor inputs, thereby creating a truly intelligent machine.

As you can see, once the solution path has been determined, the AND/OR tree structure allows for verification and explanation of a given conclusion. This is one of the most important features of a production system. The AND/OR tree allows the following questions to be answered:

- *How?* How was a given conclusion reached?
- *How not?* How was a given conclusion not reached?
- *Why?* Why was a given fact used in the reasoning process?
- *Why not?* Why was a given fact not used in the reasoning process?

For example, suppose you ask the system, "How was it concluded that the animal is a leopard?" Using its AND/OR solution path, the system could respond with, "It was concluded that the animal is a leopard since it has a brownish-yellow color and is a carnivorous mammal with black spots." If you ask the question, "How was it concluded that the animal is not a tiger?," the system could respond with, "It was concluded that the animal is not a tiger since it does not have black stripes."

Similar questions could be asked of the system to verify and explain why a given fact was used, or not used, in the reasoning process. All such questions could be satisfactorily answered using the AND/OR tree structure.

This verification and explanation characteristic of a production system is extremely important for expert-system applications. In a disease-diagnosis system, a doctor must be able to verify how the system reached a given diagnosis and how other diagnoses were not possible. Without the traceability provided by the AND/OR tree, the doctor would have to assume that the system always produced the correct diagnosis. Would you be willing to stake your life on the system diagnosis alone? Or would you rather have your doctor interact with the system to be sure of a given diagnosis?

Production System Reasoning Strategies

In Chapter 2, you found that basically two reasoning strategies can be applied to most problem-solving tasks: forward and backward reasoning. When applied to production systems, the reasoning strategy controls the direction in which facts flow through production rules.

Forward reasoning within a production system is called ***forward chaining***, or ***bottom-up***, reasoning. This type of reasoning is data driven and works from the known facts to a conclusion. With forward reasoning, the system deduces new facts from a set of known facts. The data base, or context, is updated as new deductions are made. This allows additional production rules to fire, and so on, until a desired goal is reached. I began this section with examples of forward reasoning, whereby the reasoning process flows from left to right within the set of production rules. Such a reasoning process seems very logical; however, it has two distinct disadvantages. First, it encourages combinatorial explosion, since it can result in the needless accumulation of irrelevant facts in searching for a desired goal. Of course, if the goal is to discover all possible conclusions from a given set of facts, then a forward-chaining strategy must be employed. Second, forward chaining does not lend itself to verification and explanation of the conclusion. In other words, forward chaining does not generate AND/OR trees.

Backward chaining, on the other hand, reduces the possibility of combinatorial explosion and generates AND/OR trees that provide verification and explanation abilities. You discovered this when we applied backward chaining to the animals example via an AND/OR tree. Backward chaining begins with a hypothesized conclusion, then "unwinds" the production rules to determine if the context elements support the conclusion. Given a set of productions with a common goal, the elements of these productions become subgoals that are concluded from other productions whose elements become subgoals, and so on, until elements are found that either support or reject the hypothesized conclusion. In other words, the productions are executed in reverse, from conclusion to premise. The chaining ends successfully when enough facts are found to support the conclusion. The chaining ends unsuccessfully when not enough facts exist to support the conclusion, or if some required fact conflicts with a known fact.

The backward-chaining process is illustrated in Figure 3-8. Notice how backward chaining generates a natural AND/OR tree. It reduces combinatorial explosion since only those facts and conclusions required to support the goal are explored.

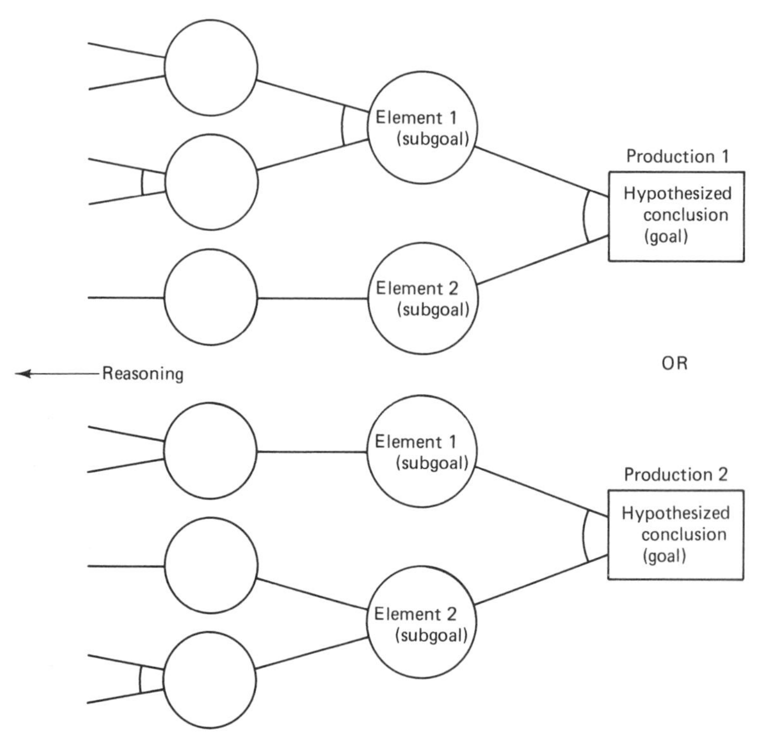

Figure 3-8 The backward-chaining process unwinds the production rules, from conclusion to premises.

Thus, the needless accumulation of irrelevant data is prevented. Consequently, backward chaining should be used when a particular conclusion must be verified or rejected.

Pros and Cons of Production Systems. The major reason that production systems are popular, especially in expert systems, is that human knowledge is easily represented using the IF/THEN production rules. In fact, some researchers believe that production systems provide the best model of the human reasoning process. These researchers believe that humans think by recognizing elements, or patterns, in a ***short-term memory***, or ***STM.*** Such elements trigger production systems in a ***long-term memory*** that carry out a particular thought process. The short-term memory is updated by sensory information and the results of long-term memory productions. As the short-term memory is updated, old information is passed to the long-term memory or forgotten. This thought process is illustrated in Figure 3-9.

Of course, another big advantage to production systems is that they lend themselves easily to verification and explanation of results and conclusions. Recall that backward reasoning using production rules generates AND/OR trees that can be traced by the system to answer questions about how a certain conclusion was reached or not reached, and why a given fact was used or not used.

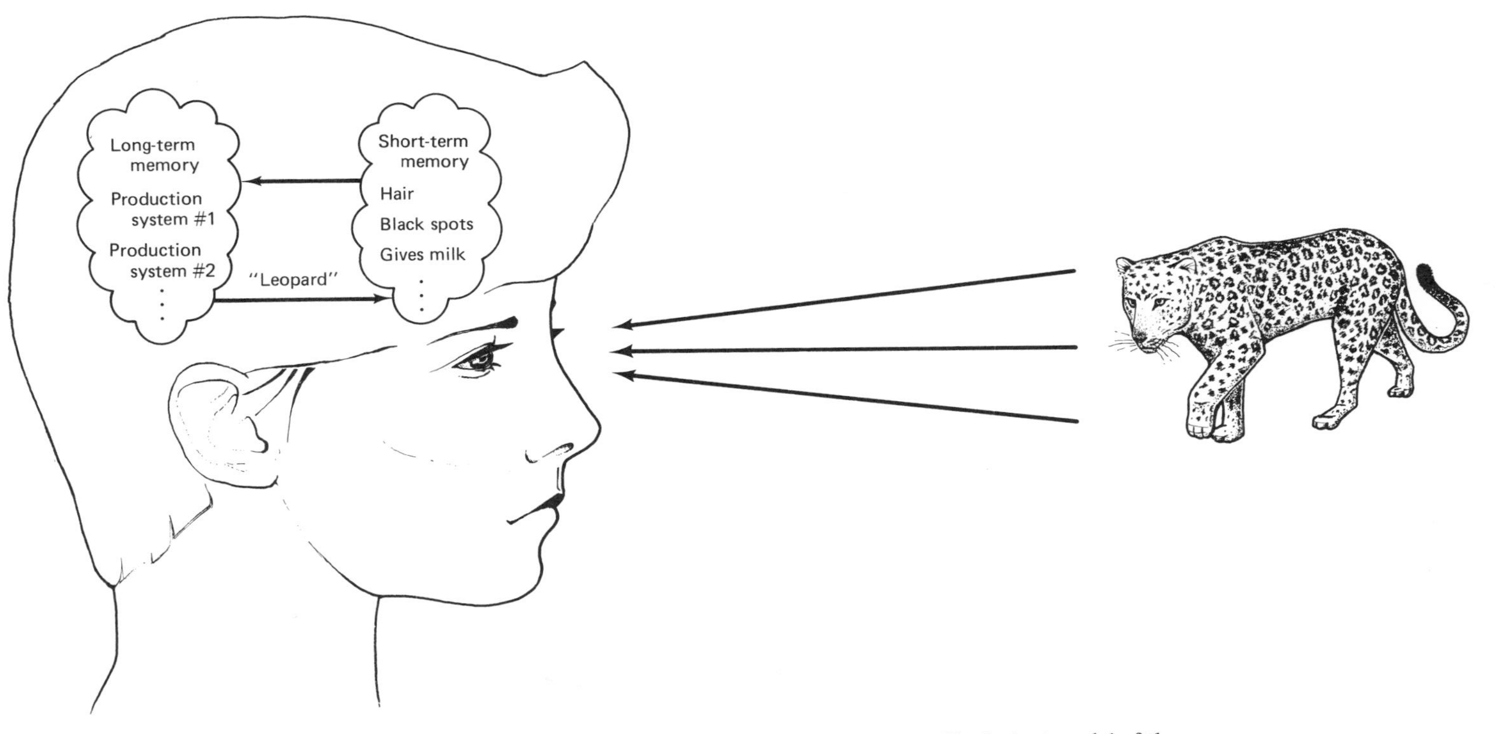

Figure 3-9 Some researchers believe that production systems provide the best model of the human thought process.

Another advantage of production systems is that they are manageable. As a knowledge-representation technique, production systems allow knowledge to be added, deleted, or changed without consequence to the rest of the system. In other words, production rules can be added, deleted, or changed without affecting other rules within the system. Thus, production rules act like independent pieces of knowledge, which do not depend on each other. This feature of a system is called ***modularity.*** The only thing common to the production rules is the data base, or context. This makes creating and changing the knowledge base a much easier task.

Production systems also force knowledge to be represented in a uniform manner within a system. All reasoning is carried out in a well-defined, organized manner. This uniformity permits the verification and explanation of system deductions.

The biggest disadvantage of production systems is that large amounts of knowledge require too many production rules. As a result, problem solving becomes inefficient. A solution to this problem is to partition the productions into manageable subsystems. Some expert systems use this approach. A particular method of system partitioning is called the ***blackboard model.*** This particular model will be described shortly, when we discuss expert systems.

3-3 SEMANTIC NETWORKS

The term semantic refers to the relationships that exist between symbols and what those relationships mean. A ***semantic network***, or ***semantic net***, is a network of symbols that describes relationships between elements of knowledge. In other words, a semantic net is a graphical representation of knowledge.

Semantic nets were originally developed as a descriptive model of the way in which the human brain makes associations between objects and concepts. For example, you "know" that all mammals have hair and nurse their young with milk. You also know that a leopard has hair and gives milk. These two pieces of knowledge allow you to conclude that

A leopard is a mammal.

How might this association be described graphically? Looking at this statement, it is obvious that the objects being related are *leopard* and *mammal.* The relationship between these two objects is described by the verb phrase "is a." Graphically, the objects of a relationship can be represented by nodes. Relationships between objects are then represented by arcs connecting the nodes. Thus, the relationship between leopards and mammals can be represented graphically as follows:

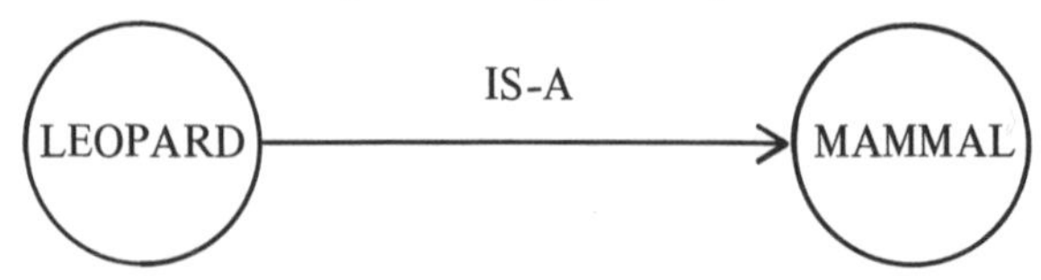

More knowledge can be added to the graph by linking other objects, as shown in Figure 3-10. What can you deduce from this network? Clearly, you can say that a

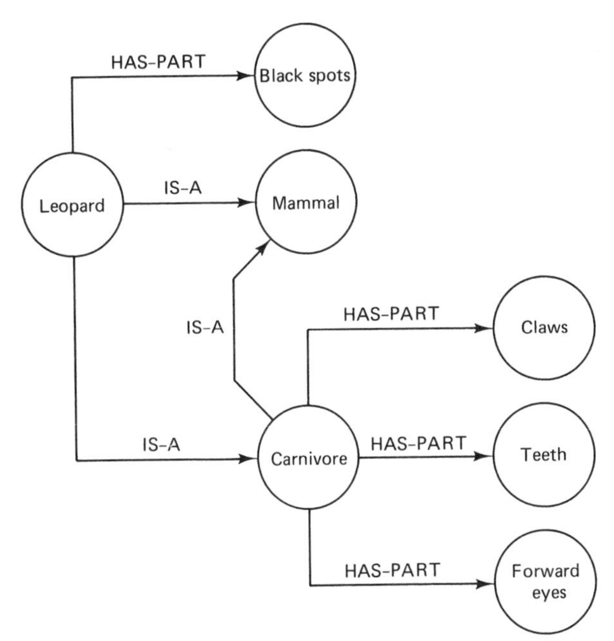

Figure 3-10 A typical semantic net consists of objects represented by nodes and relationships between the objects represented by arcs.

leopard is a mammal and a leopard is a carnivore. In addition, the network shows that all carnivores have claws, teeth, and forward eyes. Furthermore, the network makes the inference that a leopard has claws, teeth, and forward eyes.

The graph in Figure 3-10 is a typical semantic network. In general, semantic networks consist of nodes and arcs that link the nodes. The nodes contain objects, concepts, or situations in the problem domain. The arcs, sometimes called ***links***, represent relationships between the nodes. Thus, a semantic network is a node-and-link system.

Nodes and Links

It is obvious that objects can be represented by nodes in a semantic net system. Less obvious is the idea that semantic nets allow you to represent variables, concepts, and situations as nodes. This idea is illustrated by the semantic net shown in Figure 3-11. This network relates objects such as Bob, robot, gripper, box, and table. In addition, several concepts are represented. For example, the ideas that robots have grippers and tables have four legs are general concepts that can be deduced from the network. The situation that "The robot is in the kitchen" is also represented by the network. Aside from representing simple facts, such as "Bob is a robot" or "The kitchen is part of a house," the network provides a means of deducing overall situations, such as "Bob, the robot, is in the kitchen, which contains a box, which is located on a table." Notice also that variables can be represented by nodes. In Figure 3-11, the box has a color x. This link implies that the color of the box on the table in the kitchen is variable to allow for different boxes to be on the

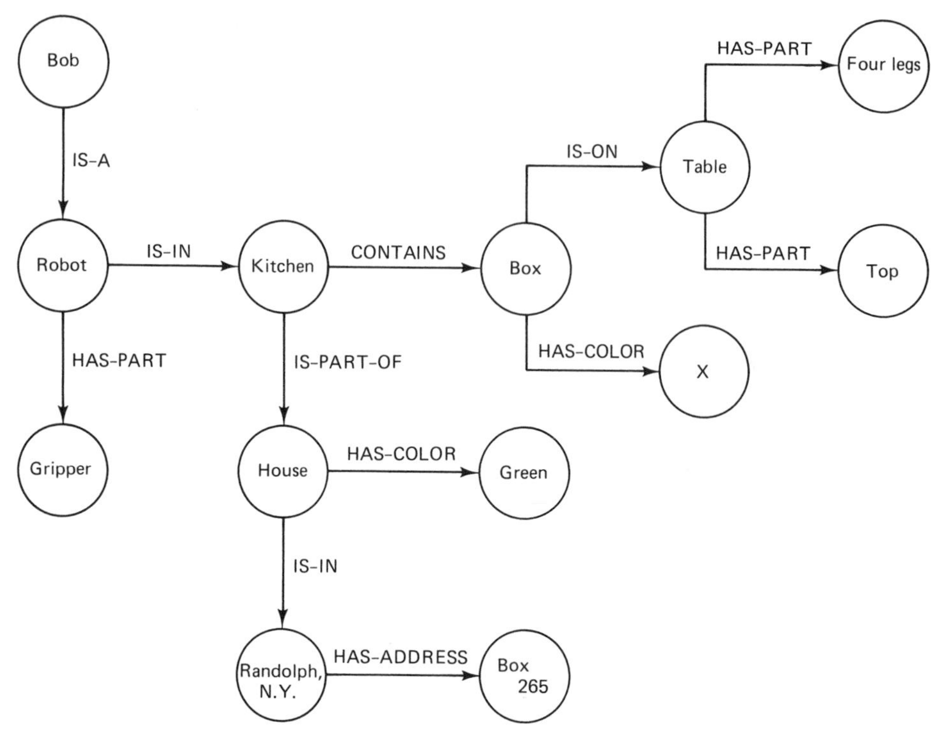

Figure 3-11 Semantic nets are used to represent concepts and situations, as well as relationships between objects.

table. So you can see that semantic nets provide a very flexible means of representing different types of knowledge.

The nodes of a semantic network can be divided into two categories: ***generic*** and ***individual.*** As the name implies, a generic node is very general and can apply to many individuals. In the animal identification problem, generic nodes would include mammal, carnivore, bird, fish, and the like. Individual nodes, on the other hand, are very specific. Usually, individual nodes represent descriptions that are applicable to a single individual. For example, "Bob" in Figure 3-12 is an individual node, while "robot" is a generic node. Individual nodes are very specific, whereas generic nodes can be more or less specific depending on where they are located in the network. Consequently, generic nodes occupy intermediate levels within the network, and individual nodes are the leaves of the network.

Clearly, the links between the nodes provide the real source of knowledge

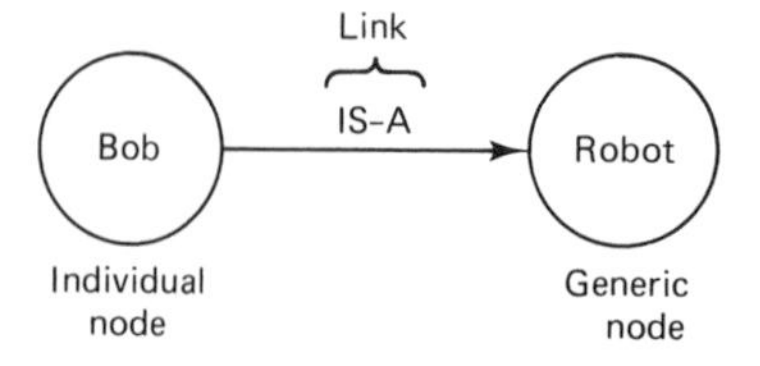

Figure 3-12 Individual nodes are very specific, whereas generic nodes are more or less specific, depending on their location within the network.

within the network. In the early days of knowledge engineering, it was realized that a great deal of knowledge could be represented using the IS-A link. Consequently, IS-A is the most common link used in semantic networks. Other common links include HAS-PART, IS-IN, and IS-ON. Additional links such as ADDRESS-IS and HAS-COLOR are usually specific to the problem at hand. Let's investigate the IS-A link a bit more, since it is common to most semantic networks.

The IS-A Link. The IS-A link is important since it can be used to develop relationships between general ideas, as well as between very specific objects. As a result, the IS-A link can be divided into two sublinks: one which relates two generic nodes and the other which relates an individual node to a generic node.

An IS-A link that relates two generic nodes is usually a subset/set relationship. For example, the IS-A link shown in Figure 3-13(a) relates carnivore to mammals. Both nodes are generic, since there are many different types of carnivores and mammals. However, all carnivores are mammals. Thus, the set of all carnivores is a subset of the set of mammals and the IS-A link is used to describe this subset/set relationship.

An IS-A link that relates an individual node to a generic node is an element/set relationship. In Figure 3-13(b), "Bob is a robot" means that Bob is a member, or element, of the set of robots.

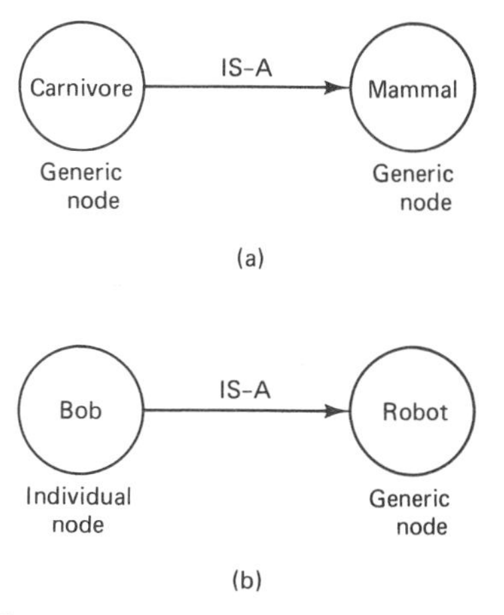

Figure 3-13 IS-A links can be used to relate (a) two generic nodes or (b) an individual node to a generic node.

Properties of Semantic Networks

In the preceding discussion, you found that some nodes are more generic than others and that IS-A links produce subset/set relationships. These natural subset/set relationships produce a hierarchical structure within the network. Links are made from very generic nodes to less generic nodes, and so on, down to individual nodes. Thus, a pyramid effect is produced by IS-A links, as illustrated by the animals network shown in Figure 3-14. Notice the natural hierarchical property of the IS-A link. In a computer program, these pyramids can be labeled and called like subroutines to get at particular facts.

The subset/set relationship of the IS-A link also produces another very im-

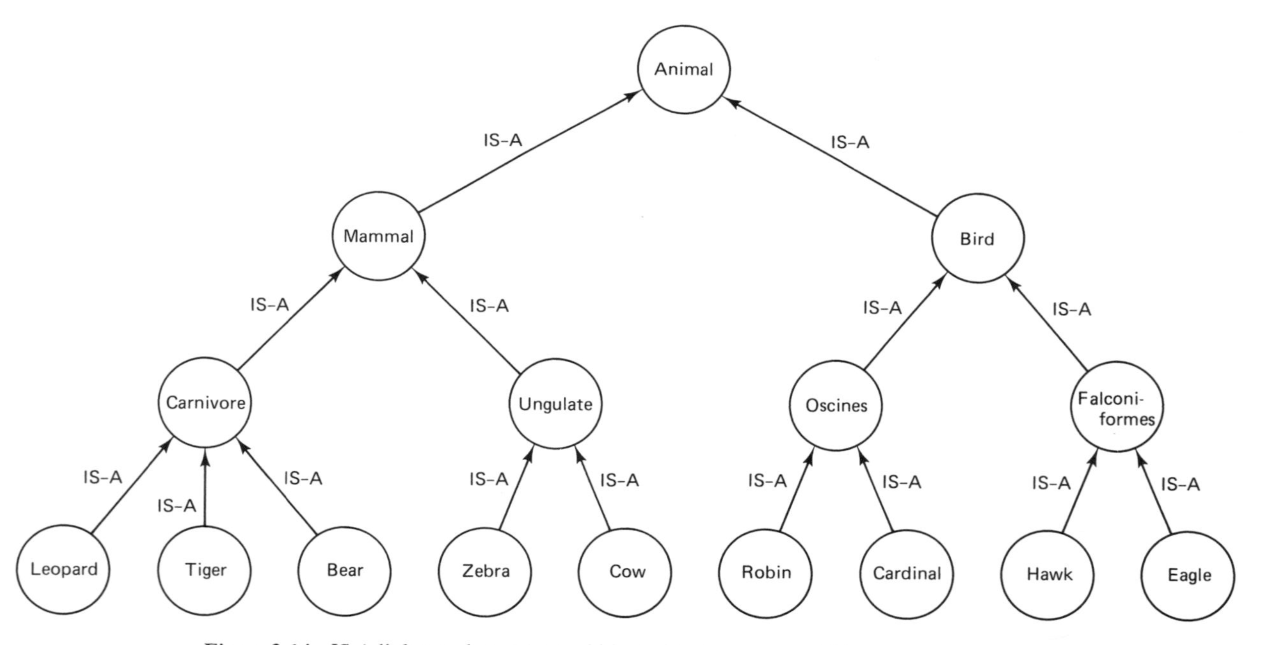

Figure 3-14 IS-A links produce a natural hierarchical structure within a semantic network.

portant property of semantic networks: ***inheritance.*** The term inheritance comes from the idea that any knowledge associated with higher, or more generic, nodes is passed down, or inherited, by lower, or less generic, nodes. For example, knowing that a carnivore is a mammal and a leopard is a carnivore allows a system to deduce that a leopard is a mammal. Thus, the mammal property is inherited by the leopard even though there is no direct link between leopard and mammal. For this reason, the IS-A link is referred to as a ***property-inheritance link.***

The inheritance property of semantic nets permits deductive reasoning and conserves memory space within a computer system. A computer system reasons by simply following the links. The links act as pointers within the program execution. The system starts at a given node, then follows the links to related nodes, then to more distantly related nodes, and so on, until a desired goal is reached. For example, using the network shown in Figure 3-11, suppose the system needs to answer the question "What is contained in the kitchen?" Looking at the question, it is clear that, to determine what is contained in the kitchen, the system must search for the object "kitchen" and the link "contains." A ***network fragment*** is constructed as shown in Figure 3-15. Notice that the fragment represents the object and link that must be found. The system then attempts to match the fragment against the network data base. In our example, it is looking for the "kitchen" node and the "contains" link. When a match is made, the "contains" link points to the data that will answer the question. Thus, the system can respond with "The kitchen contains an x-colored box and a table with a top and four legs." If no match were possible, the system might respond with "Nothing is in the kitchen." This matching process, coupled with the inheritance links, allows a system to reason using semantic nets.

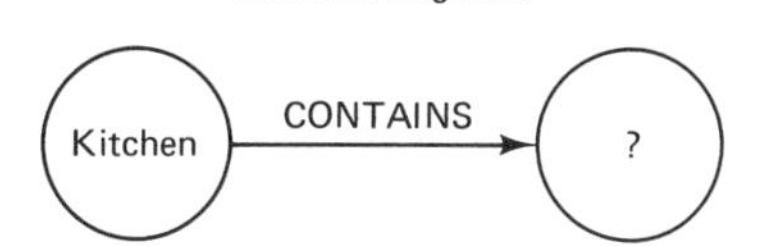

Figure 3-15 During the reasoning process, network fragments are constructed and attempts are made to match the fragment to the network data base.

The inheritance property also allows the system to jump from one related idea to another, much the same as we humans reason by association. Furthermore, memory space is conserved since separate links do not have to exist for each relationship. The IS-A link allows facts to be shared among many nodes within the network. In other words, a very large number of complex relationships can be represented with a small set of facts. This produces a very efficient means of knowledge representation.

Relationships to Other Knowledge-representation Techniques

Before leaving semantic nets, we should point out the ways in which they relate to other knowledge-representation techniques with which you are familiar. In particular, the IS-A link can be used to represent predicate logic statements. For example, consider the following logic formula:

$$(\forall x)[\text{LEOPARD}(x) \rightarrow \text{MAMMAL}(x)]$$

This logic statement can be represented in a semantic net as follows:

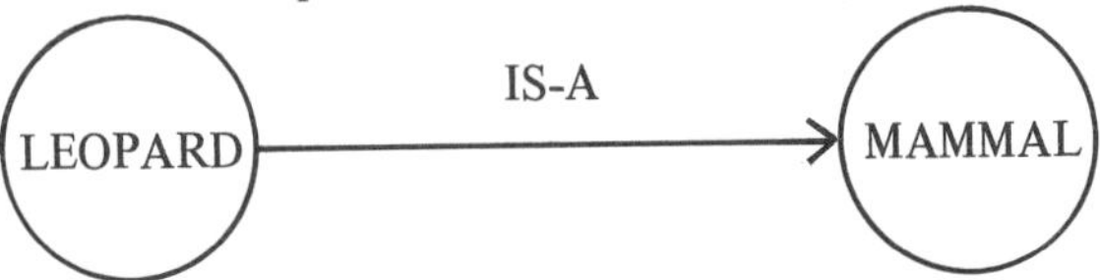

Likewise, the production rule:

IF x is a leopard, THEN x is a mammal

can be represented using the same leopard/mammal link. In fact, many semantic net systems make use of production rules to reason within the network structure. Moreover, logic can be used to verify that a given conclusion is true or false.

The difficulty with semantic nets is that no common principles can be applied to all networks, since different networks contain different link structures. In other words, the semantic net notation and reasoning process are not formalized as in logic. Of course, this can be an advantage or a disadvantage, depending on the reasoning task. Links such as IS-A and HAS-PART are common to many networks, but other links depend on the specific application of the network. The only thing that is common to all semantic nets is the node-and-link structure and notation.

Now let's expand the idea of semantic networks into an even less formal and more generalized knowledge-representation technique–frames.

3-4 FRAMES

The theory of frames is a relative newcomer to the science of knowledge representation. Frame theory was proposed by Marvin Minsky at the Massachusetts Institute of Technology in 1974 as a technique for representing large amounts of general-purpose knowledge, called ***commonsense knowledge.*** For a computer, or robot, to be truly intelligent, it must be knowledgeable about situations that might be experienced during the course of its everday activities.

If you think about it, your knowledge about everyday life consists of a series of stereotyped situations that are stored in your memory. For example, what items would you "expect" to find in a kitchen? You probably would expect to find a sink, refrigerator, stove, dishwasher, and so on, before even experiencing a specific kitchen situation. Now suppose you visit a friend's house and walk into a kitchen that you have never experienced before. You expect to see something that is similar to the kitchen that is stored in your memory. Of course, all kitchens are not the same, so you look around and your mind begins to adapt to the real situation by changing the details of the memorized kitchen scene. Your friend's kitchen might contain a microwave oven in addition to a stove. A side-by-side refrigerator/freezer might be present rather than the overhead refrigerator/freezer that you expected, and so on. Thus, the general kitchen expectation that you had in your mind is altered to fit the present situation.

This example illustrates the theory of frames. A ***frame*** is simply a data structure that consists of expectations for a given situation. Notice how the kitchen frame provides commonsense knowledge about a kitchen before a given kitchen is even experienced or analyzed. Likewise, additional frames can contain commonsense knowledge about other everyday situations. For instance, a living room frame would contain expectations such as a sofa, television set, fireplace, and so on. A winter frame might contain such commonsense knowledge as "snow is white," "icy roads are slippery," and "put your coat on before going outside."

Individual frames like these can be combined into a frame network, whereby all knowledge in a given environment can be represented. Thus, a ***frame network***, or system, provides a general knowledge-representation scheme through which new information is interpreted based on previous experience. If you think about it, this is much the same way as you and I go about performing everyday commonsense tasks. Let's explore the details of frame theory as a knowledge-representation scheme for computers.

Structure and Use of Frames

Frames generalize the semantic network idea. In frame theory, the important relationships and properties of semantic net nodes or concepts are bundled into a common data structure called a frame. A frame can consist of objects and facts about a situation, or procedures on what to do when a given situation is encountered. Thus, frames are used to represent two types of knowledge: ***factual*** and ***procedural knowledge.***

The knowledge associated with a frame is contained in structures called ***slots***, as shown in Figure 3-16. The slots of a frame are used to store factual and procedural knowledge associated with the frame. In other words, the frame slots contain expectations about a given situation as well as what to do in a given situation.

For example, the kitchen frame might contain slots labeled sink, refrigerator, stove, table, and so on, as shown in Figure 3-17. These slots might be terminal or provide links to subframes. As an example, the refrigerator slot could provide a link to a refrigerator subframe, as shown in Figure 3-18. Notice that the subframe contains slots that further describe the refrigerator. These slots could again form additional subframe links that are needed to completely describe the refrigerator, or its contents.

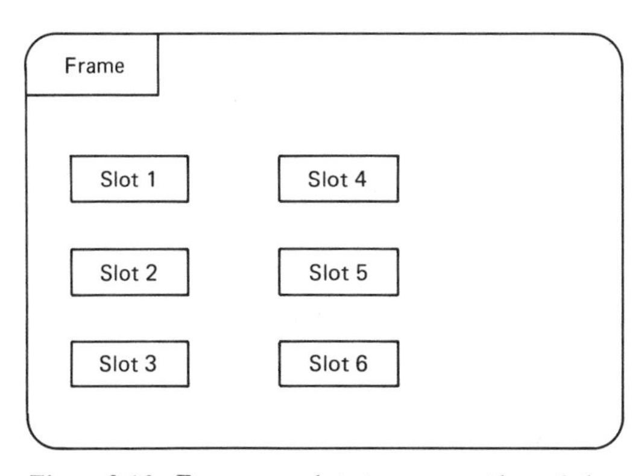

Figure 3-16 Frames use slots to represent knowledge.

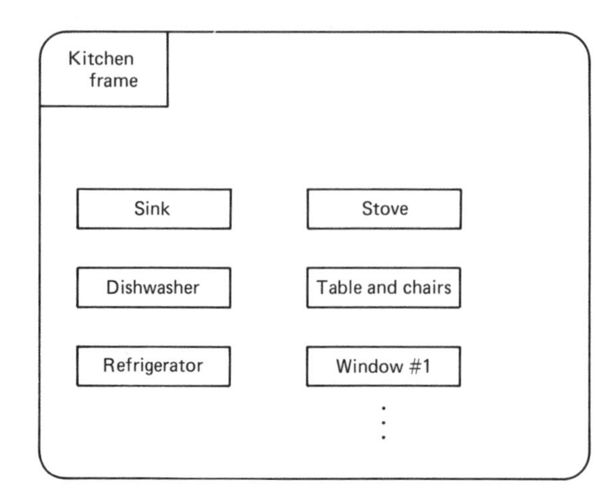

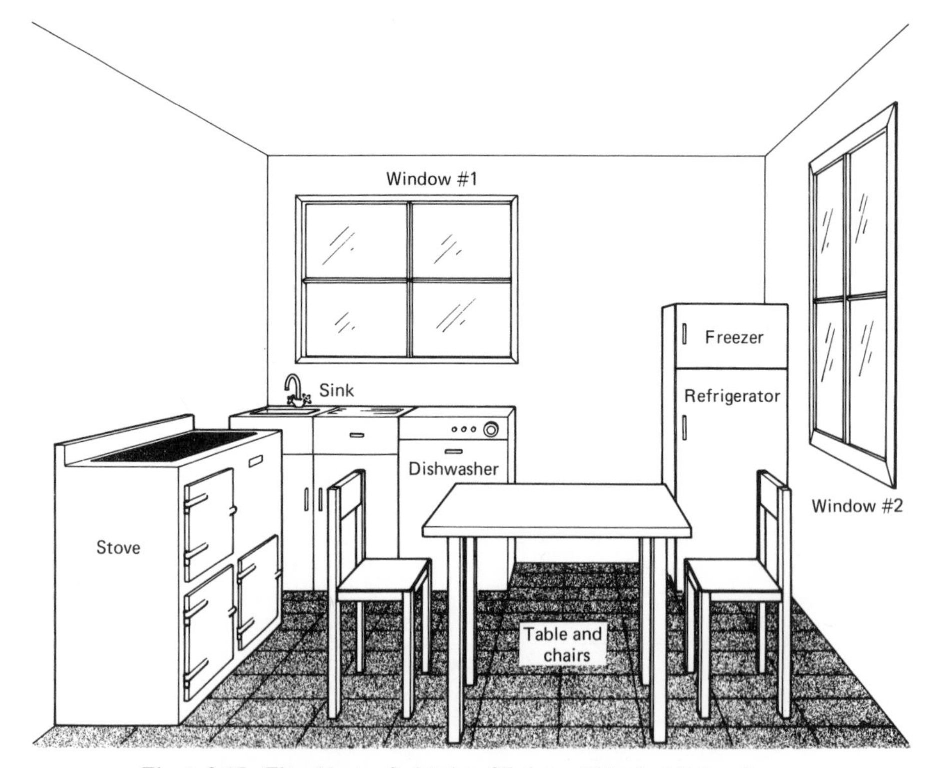

Figure 3-17 The objects of a kitchen fill slots within the kitchen frame.

In fact, our kitchen frame could be a subframe of a more general room frame, as shown in Figure 3-19, which illustrates the natural hierarchical structure of frames. At the top of the structure you find general, commonsense knowledge, while at the bottom of the structure you find very specific knowledge. As with

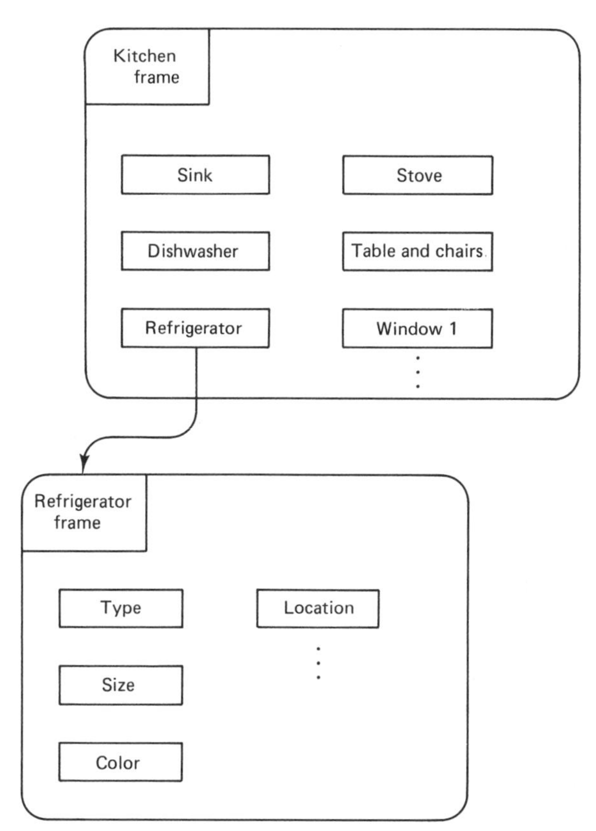

Figure 3-18 In many cases, slots produce subframes that further describe the situation.

semantic networks, the hierarchical structure of frames produces an inheritance property among the frames in a given family. Thus, information contained in a parent frame is inherited by its children. Frames that contain knowledge about a given situation are sometimes called ***situational frames.***

Example 3-4

Construct a family of situational frames that could be used to represent the dining room pictured in Figure 3-20.

Solution

Notice from Figure 3-20 that the dining room contains a table with two chairs, a china cabinet, a window, a chandelier, and a picture of Aunt Em. Each item in the room, except the table and chairs, can be associated with one of the walls or the ceiling. Consequently, a very general dining room frame contains slots labeled west wall, north wall, east wall, and ceiling, as shown in Figure 3-21. In addition, the general dining room frame contains a slot labeled table and chairs, since these items are located in the middle of the room.

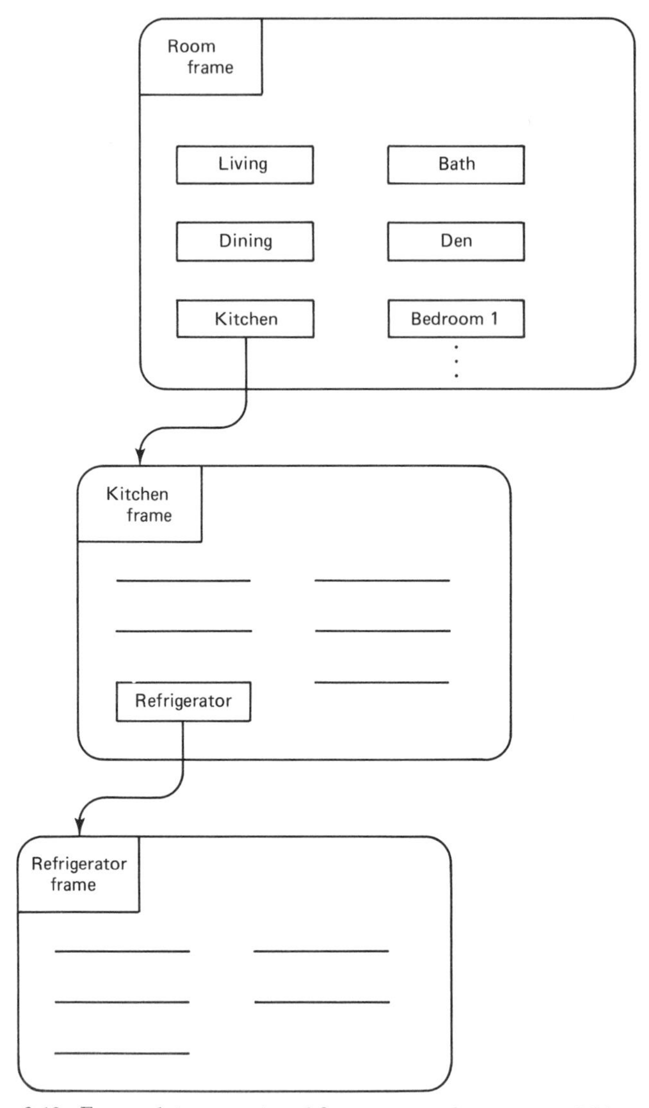

Figure 3-19 Frame slots generate subframes to produce a natural hierarchical structure.

To completely describe this particular dining room, you must construct subframes that describe the other items within the room. The required subframes are shown in Figure 3-22. Observe that the subframes not only describe the other items within the room, but they also show their location through links to the dining room frame. In addition, the subframes contain slots that provide information about the construction and appearance of the walls and ceiling. The wall and ceiling subframes will, in turn, link to additional subframes to further describe the items in the room. For instance, the left wall subframe will link to a china cabinet subframe that

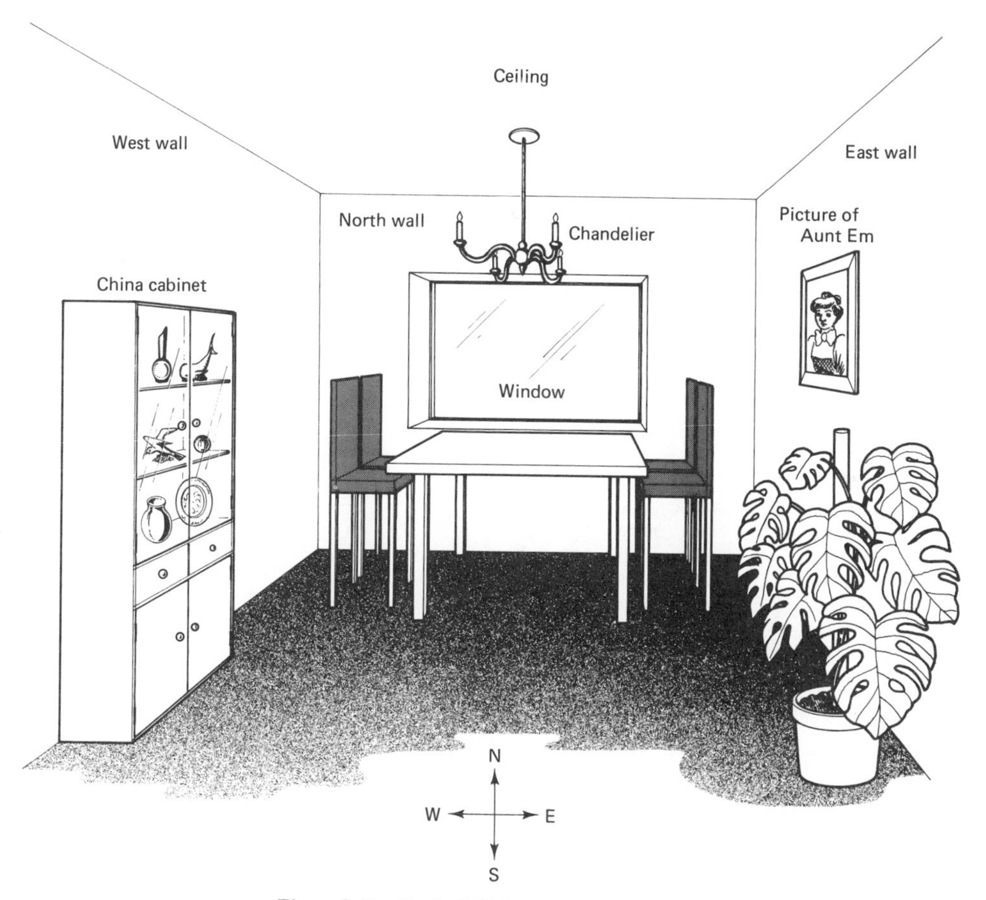

Figure 3-20 Typical dining room scene.

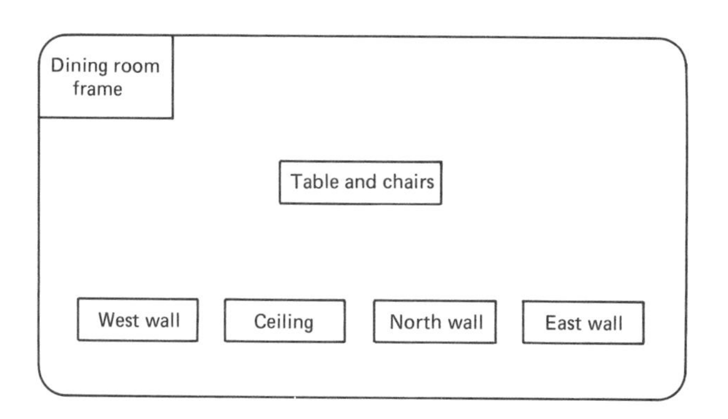

Figure 3-21 A general room frame might consist of slots for each wall and ceiling, as well as slots for items within the room.

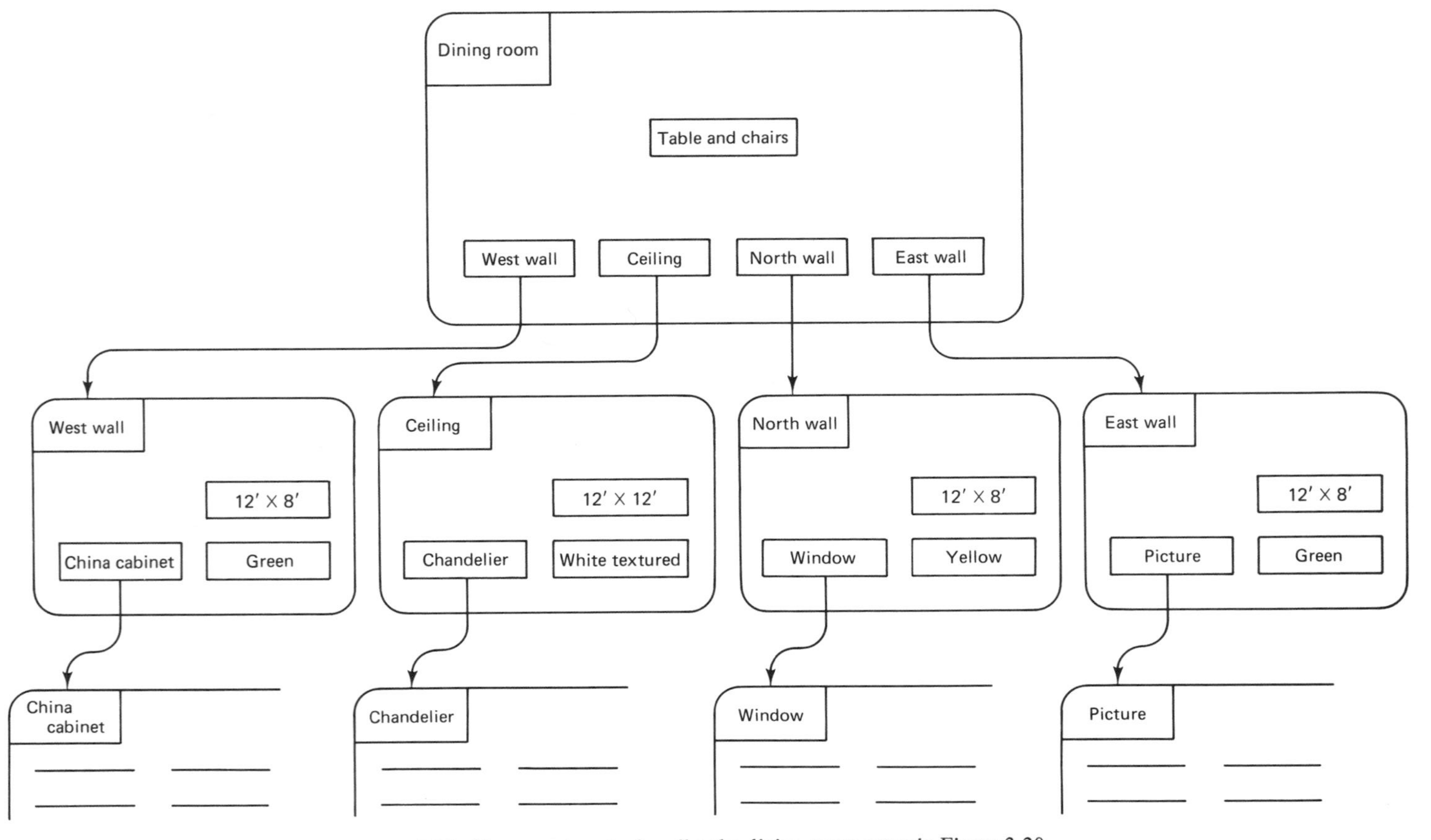

Figure 3-22 Frame system to describe the dining room scene in Figure 3-20.

describes the china cabinet. This subframe might link to additional subframes that describe the contents of the china cabinet, and so on, until each item in the room is completely described within the frame structure.

Aside from factual knowledge, frames are used to represent procedural knowledge. These frames are sometimes referred to as ***action frames.*** An action frame contains slots that describe individual tasks to be performed as part of an operation. For example, suppose a robot is to move a single disc from one peg to another. The action frame shown in Figure 3-23 describes the overall operation. Notice that the frame is made up of slots that describe the action to be performed. The ***actor slot*** defines who or what is to perform the operation. The ***object slot*** defines the item to be operated on. Knowledge about where the operation is to be performed is contained in the ***source*** and ***destination*** slots. Finally, the ***task slots*** generate the subframes required to perform the operation.

The action subframes for the pick-and-place operation are illustrated in Figure 3-24. Each subframe describes a particular task to be performed during the operation. Of course, these action subframes might generate other subframes that contain the operational details required for each individual task.

Action frames can be linked with situational frames to describe cause-and-effect relationships. Consider the three-disc robot pick-and-place problem discussed in Chapter 2. The initial problem state is represented by the situational frame shown in Figure 3-25. Now, recall that to solve the problem the robot must first move disc A to peg 3. This operation is described by the task slot in the initial state frame. The task slot produces the action subframe shown in Figure 3-26 to perform the required operation. Of course, the various pick-and-place tasks required for this operation are represented by slots that might generate additional subframes. Nevertheless, the result of the move disc operation is represented by the new situation

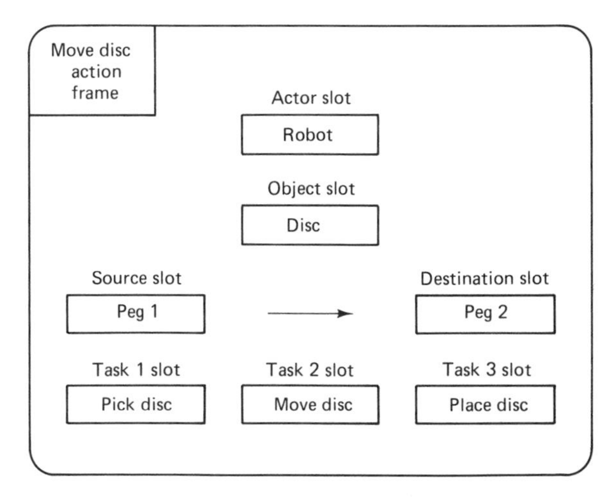

Figure 3-23 Action frames are used to represent procedural knowledge.

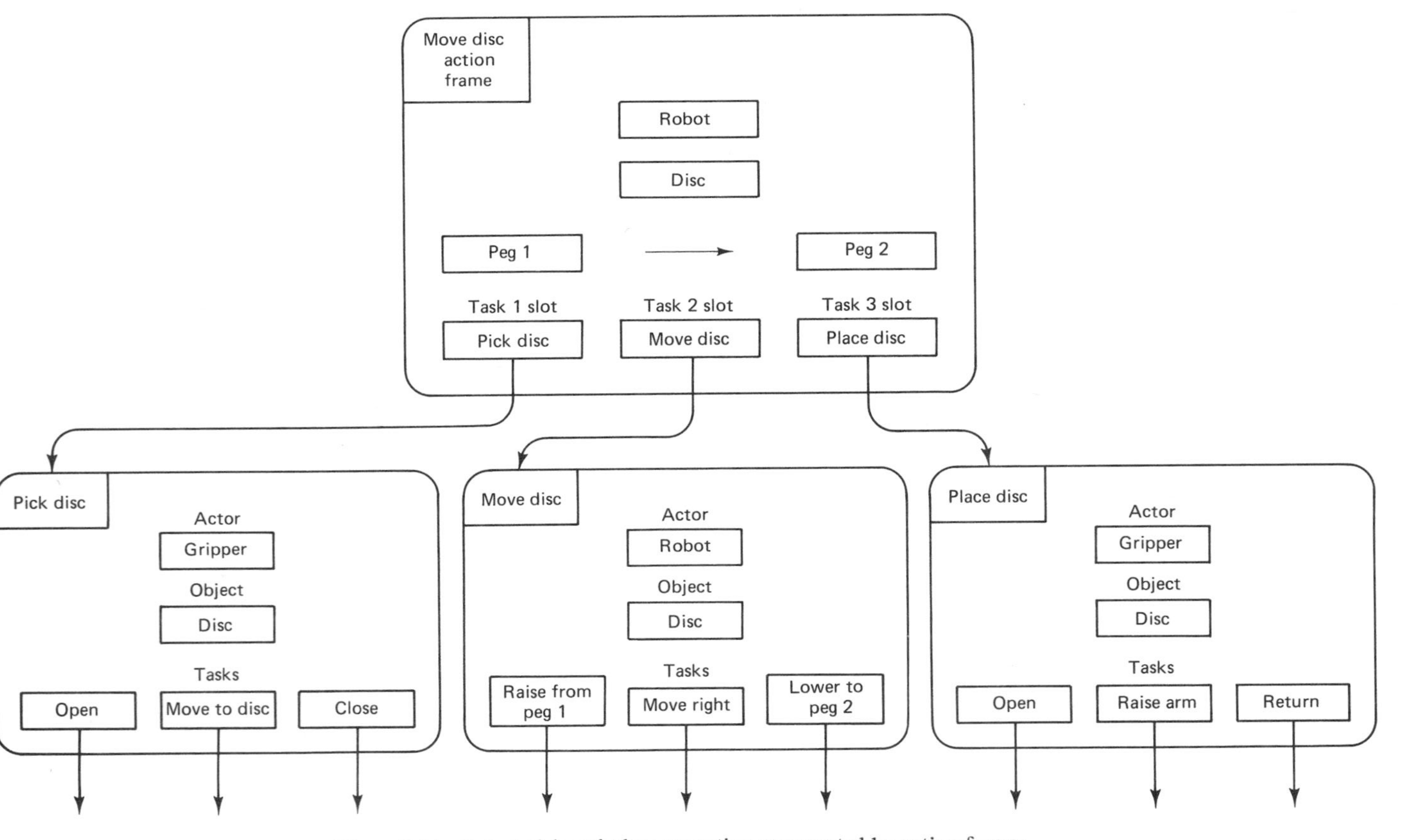

Figure 3-24 Robot pick-and-place operation represented by action frames.

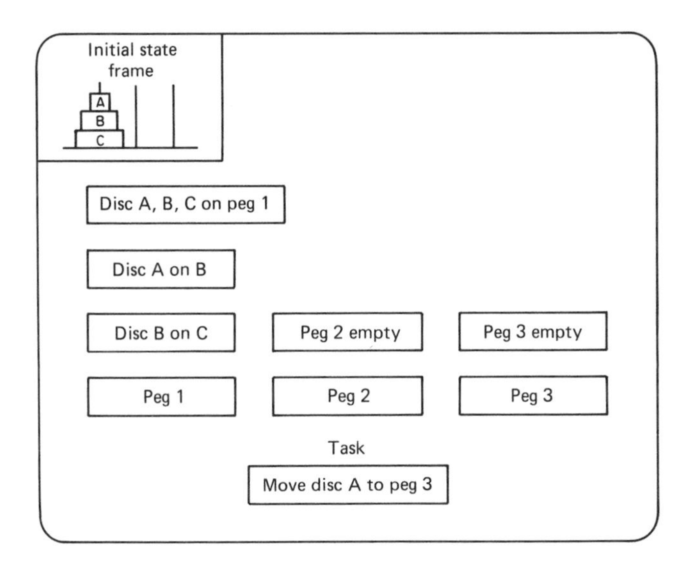

Figure 3-25 Initial state frame required to describe the three-disc problem.

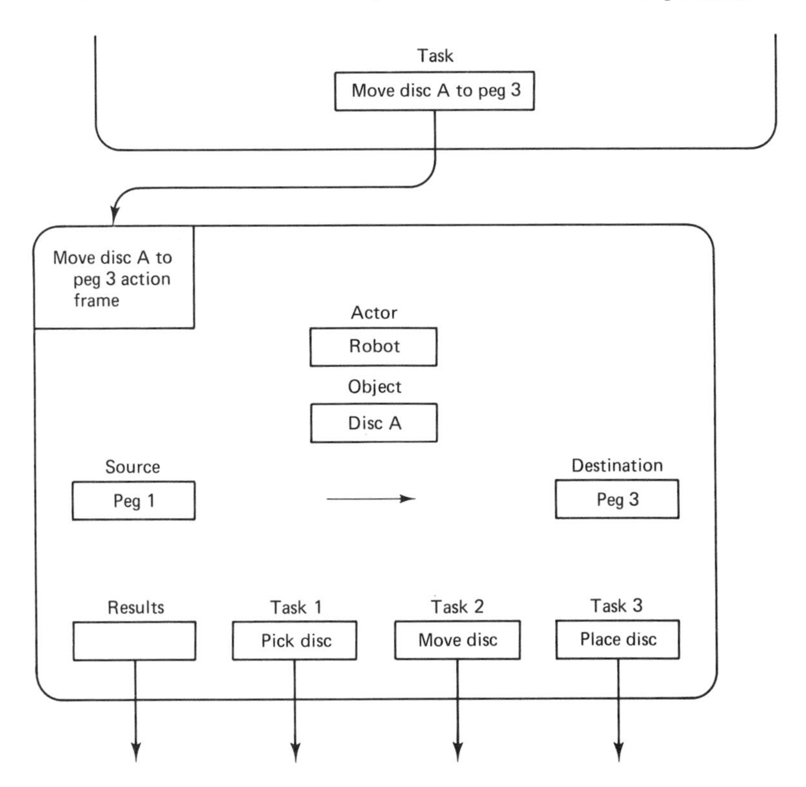

Figure 3-26 The initial state frame task slot produces an action subframe to perform the required operation.

frame illustrated in Figure 3-27. This frame would then be followed by another action frame to move disc B to peg 2, resulting in a new situation frame. The process would continue until the goal situation frame is produced. Notice how the action frames transform the problem from the original situation to the goal situation. As a result, the pick-and-place problem is represented and solved using a series of situation/action frames. In fact, any task can be represented with a sequence of situation/action frames.

Example 3-5

News stories are particularly suited for representation using frames. Consider the following weather-related news story.

Angry Weather Roars out of The Rockies

A violent storm stretching from the Rockies to the Gulf Coast wreaked a "miniature disaster" today with tornadoes in Kansas, heavy rains and hail in Louisiana, and 6-foot snowdrifts that closed major highways in Colorado.

Up to 15 inches of snow was on the ground early today in Burlington, Colo., near the Kansas border, where winds gusted to 67 mph, the National Weather Service said.

Already blamed for two deaths on Wyoming highways, the angry weather system was the fifth major wintry storm to roar out of the Rockies since spring began March 20.

In parts of Louisiana, as much as 4 inches of rain was reported in a 24-hour period ending Monday night, and . . . .

Cold air collided with warmer air over eastern Kansas on Monday night, spawning five tornadoes . . . .

By THE ASSOCIATED PRESS
THE POST JOURNAL
Jamestown, NY
Tuesday, April 3, 1984

Construct a family of situation/action frames that represent the above news story.

Solution

News stories lend themselves to frame representation since they rely on a straight-to-the-point method of conveying knowledge to the reader. Major headlines form frames, with the story details filling in the slots. The title of this story suggests a general weather frame, such as the one shown in Figure 3-28. Each slot within the frame represents a particular type of weather system. The weather system slots link to subframes, as shown in Figure 3-29, that provide the details of the weather story. In addition, an action subframe describes the weather action and links to a situational subframe that describes the results of the weather system. Notice how the frame network represents the important facts of the story. From this representation, a journalist could easily reconstruct the news story.

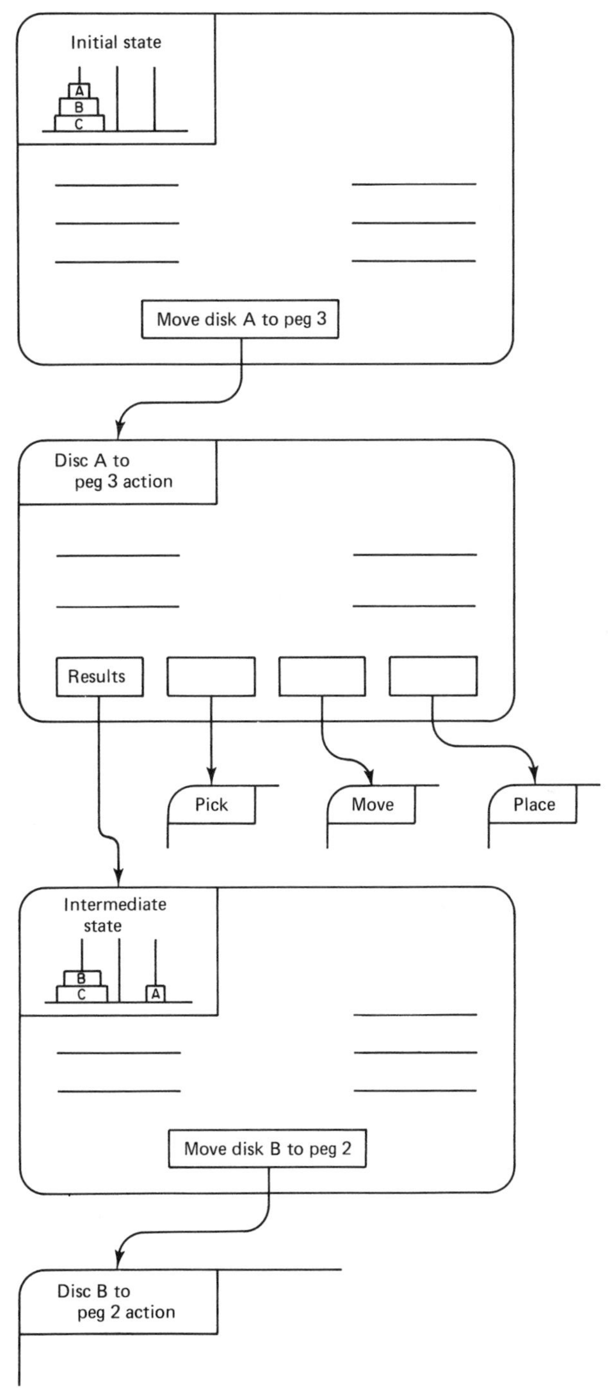

Figure 3-27 Problems are solved by a series of situation/action frames.

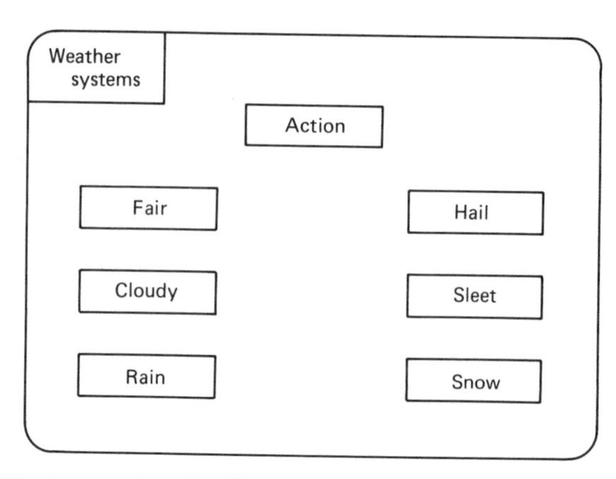

Figure 3-28 A news story headline suggests a general frame such as this for a weather story.

Reasoning with Frames

You have already seen how frame theory can be used to represent a given situation and perform operations. Now, how can the power of frames be utilized to perform commonsense reasoning? For example, if a robot enters a kitchen, how can it determine that it is actually in the kitchen and not some other room? In addition, how can the robot be programmed to learn from any new experiences gained while in the kitchen?

Let's suppose your robot is equipped with a vision system and programmed to navigate the house using knowledge stored in frames. In other words, your robot's memory contains a general room frame linked to a set of subframes that describes each room in the house. These frames are stored in mass memory and called when needed via subroutine or mnemonic labels.

The robot enters a room and loads the room frame into its working memory. Recall that the room frame consists of slots for each room in the house, as shown in Figure 3-30. To understand which room it is in, the robot must *match*, or *fill*, the appropriate slot.

Each room slot is linked to a subframe describing a given room. So the problem now is for the robot to decide which individual room slot to activate. Notice in Figure 3-30 that a heuristics slot is included in the frame along with the room slots. Heuristics slots are used to guide the reasoning process. Such a slot might contain a series of IF/THEN production rules. In our example, the robot would activate the heuristics slot first to decide which individual room slot to activate.

Suppose, upon activating the heuristics slot, the robot encounters the series of production rules shown in Figure 3-31. Notice that the production rules are fired according to the color of the room. Of course, we must assume the robot vision system is equipped with color filters in order to determine the room color. A given

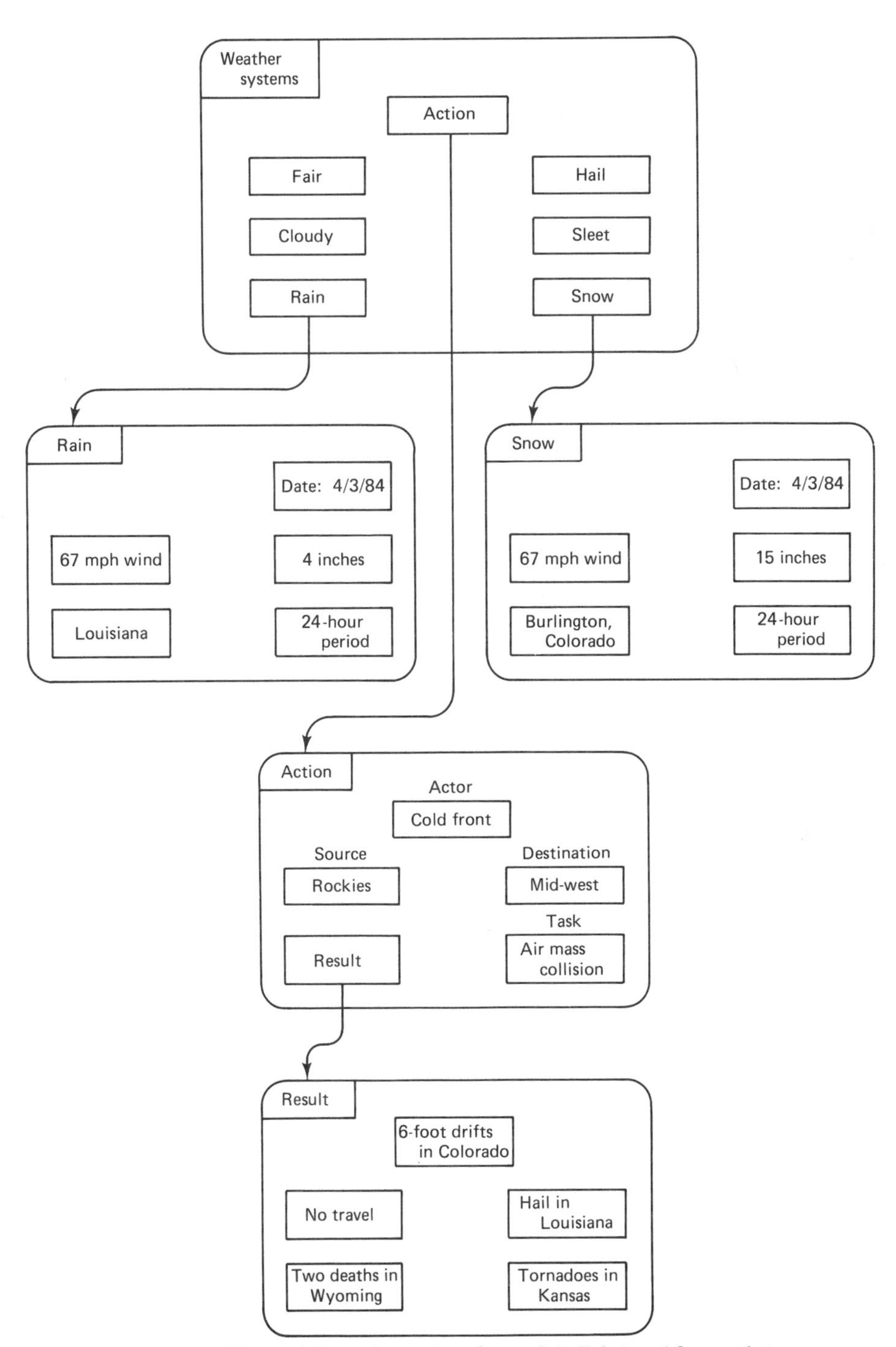

Figure 3-29 The general weather system frame slots link to subframes that provide the story details.

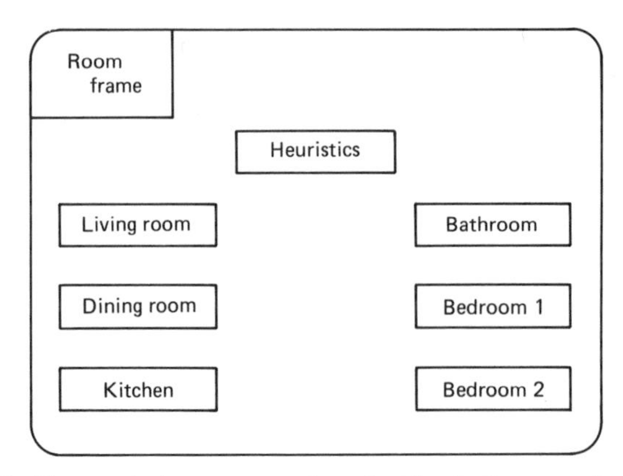

Figure 3-30 Upon entering a room, the robot loads the room frame into its working memory.

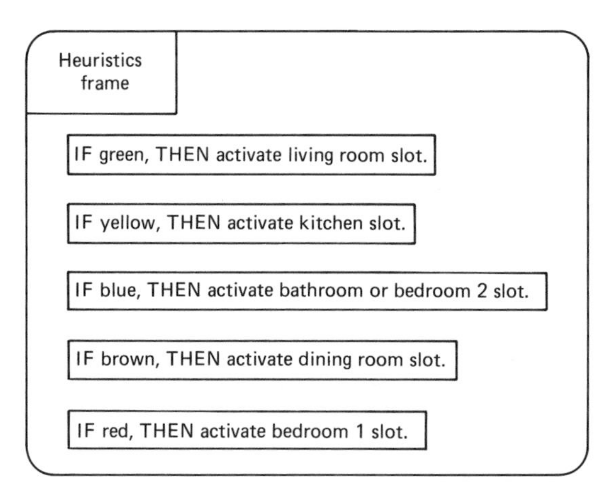

Figure 3-31 A heuristics frame contains slots filled with production rules that guide the reasoning process.

color causes a production rule to fire, which in turn activates a given room slot within the room frame. If two rooms have the same color, the robot must make an arbitrary decision as to which room slot to activate first.

Assuming the color of the room that the robot has entered is yellow, the heuristics slot directs the robot to activate the kitchen frame. Consequently, the kitchen frame shown in Figure 3-32 is loaded into the robot's working memory. The reasoning process now requires the robot to *fill* the slots within the frame.

Filling frame slots is the key to reasoning with frames. The slots are filled using a matching process. For example, to fill the refrigerator slot shown in Figure 3-33, the robot must activate the refrigerator subframe and attempt to match the items in the subframe slots. To do this, the robot must scan the room until it finds

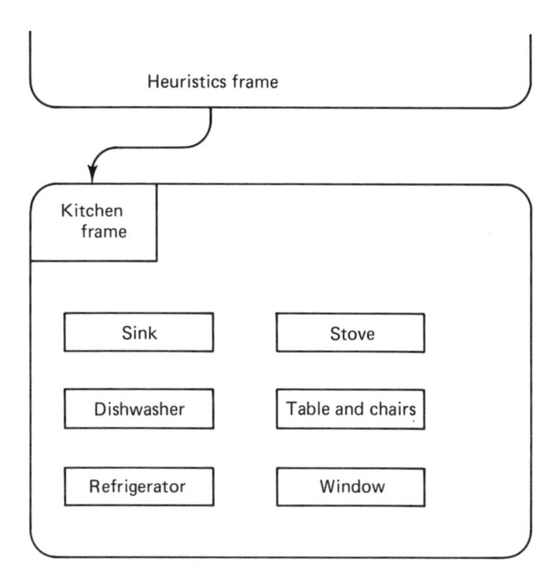

Figure 3-32 Upon detecting that the room is yellow, the heuristics frame directs the robot to activate the kitchen frame.

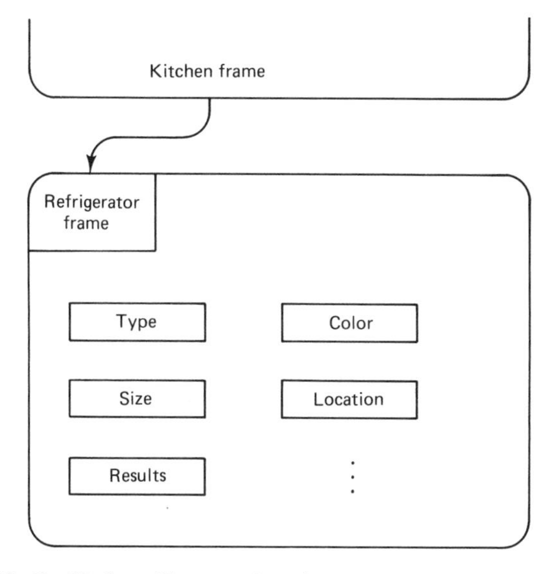

Figure 3-33 To fill the refrigerator slot, the robot must activate the refrigerator subframe and attempt to match the subframe slots.

an object that matches the size and color data stored in the refrigerator frame. Upon finding such an object, attempts are made to fill all the slots within the frame by matching the stored data to visual data being received. If a given slot characteristic is not present, the slot provides a *default value* for that characteristic. If a given slot characteristic is different from the observed, or actual, characteristic, the slot data is updated to conform to the current situation. For example, if the robot observes a side-by-side refrigerator/freezer rather than the overhead refrigerator/freezer it was looking for, the refrigerator "type" slot data would be changed to reflect the actual observation. In this way, the robot learns by experiencing a new situation in the same way that you and I learn through experience. This is a very important feature of frames.

Once the refrigerator subframe slots are filled, the robot attempts to fill the remaining kitchen frame slots in the same way. When all the slots are filled, the robot not only "understands" that it is in the kitchen, but has "learned" through experience any changes that were made to the kitchen. Of course, if none of the kitchen slots could be filled, the robot would understand that it is not in the kitchen and activate another room frame.

Our example illustrates the reasoning power of frame theory. You might say that the frame reasoning process is driven by expectations. Notice how filling frame slots confirms expectations, directs the reasoning process, and gathers information about the current situation. Furthermore, the reasoning process allows the system to determine which frames are applicable, since the frames try to match themselves to the current situation. If a frame cannot be matched, control is transferred to a more applicable frame.

To summarize, frame theory provides a means of representing general, or commonsense, knowledge. In addition, frames provide a natural means whereby computers can understand their environment and learn from experience, particularly in computer vision applications. For these reasons, frame theory promises to be a very powerful knowledge-representation technique. However, frame theory is an abstract idea, and thus requires much more research before it becomes a practical means of knowledge representation. It is often criticized for its lack of precision. As I stated earlier, the ultimate knowledge-representation scheme will most likely consist of a combination of logic, production systems, semantic nets, and frames.

3-5 EXPERT SYSTEMS

One of the largest applications of artificial intelligence is in the area of knowledge-based systems, popularly referred to as expert systems. A ***knowledge-based***, or ***expert***, system is a computer system that exploits the specialized knowledge of human experts to achieve high performance in a specific problem area, or domain.

From this definition, you could say that an expert system is actually a computerized consulting system. Such systems are used to diagnose diseases, evaluate mineral deposits, analyze chemical compounds, and troubleshoot circuits, just to mention a few present applications. Other application areas include the military, academic advisement, tax advisement, computer-aided design and computer-aided manufacturing (CAD/CAM), VLSI design, office automation, and engineering. In fact, the possible applications for expert systems are almost endless. Any field that requires specialized knowledge in a specific domain is a candidate for an expert system.

Some of the first efforts to apply artificial intelligence involved the application of general problem-solving techniques to a spectrum of problem areas, or domains. Such efforts proved to be impractical, since general problem-solving systems could not handle the complexity of everyday real-world problems. As a result, expert systems were born to limit the problem domain. As you will soon discover, these systems capture the knowledge of human experts and focus it with the speed and deductive reasoning power of a computer to solve problems in a very specific domain.

In this section, you will learn about expert systems. You will first be introduced to the basic structure of an expert system and then learn how they work. You will find your knowledge of production systems to be very helpful in understanding expert systems, since most of these systems use production rules to reason. Finally, you will explore several real expert programs in an effort to gain a "feel" for their capabilities and potential applications.

Structure of an Expert System

A block diagram of a typical expert system is shown in Figure 3-34. As you can see, the internal structure of the system consists of three major regions: a ***knowledge base***, a ***data base***, and a ***rule interpreter***. Notice the resemblance to the production-system structure discussed in Section 3-2 (see Figure 3-3).

The knowledge base contains the rules of inference that are used during the reasoning process. Most existing systems are based on production-system technology and therefore use the IF/THEN type of production rules to represent knowledge in the knowledge base. A typical expert system might contain from a few hundred to several thousand production rules, depending on the problem domain. The system might also incorporate elements of logic, semantic nets, and frame theory in the knowledge base. When this is the case, production rules are often used as heuristics to guide the reasoning process through the appropriate knowledge structure.

The data base is the system context. Recall from production systems that

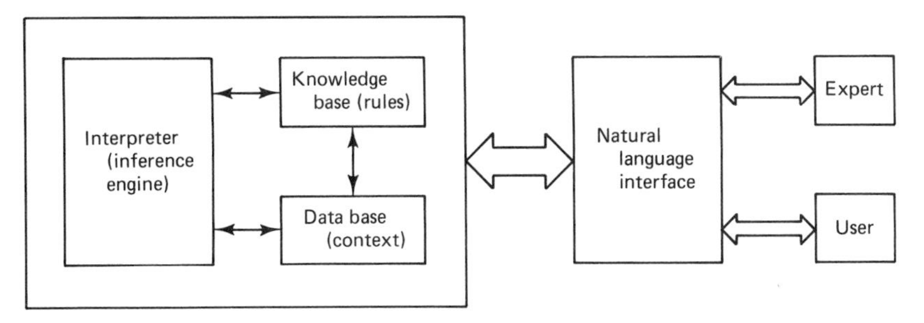

Figure 3-34 The structure of a typical expert system resembles that of a production system.

the context contains the facts available to the system at any given time. These facts are the elements that are used to satisfy the conditional part of the system rules, or productions. Before a rule is applied, its conditional elements must be present in the system data base, or context. The application, or firing, of a rule creates new facts that update the data base so that other rule conditions might be satisfied.

The rule interpreter, sometimes called an ***inference engine***, guides the reasoning process through the knowledge base by attempting to match the facts in the data base to the rule conditions. In addition to controlling the matching process, the inference engine provides conflict resolution when more than one production rule is matched with a given set of conditional elements in the data base.

Communication with the system is provided by a natural language interface. The natural language interface allows the expert or system user to communicate with the system using his or her natural language (English, French, German, etc.), rather than a specialized computer language. This enhances the friendliness of the system and allows the user to communicate with the system without the assistance of a programmer.

How Expert Systems Work

There are three major modes of operation for an expert system: the ***knowledge-acquisition mode***, the ***consultation mode***, and the ***explanation mode.*** Let's take a brief look at each of these operating modes.

Knowledge-acquisition Mode. To gain its knowledge, the expert system must interact with human experts. For example, a system that diagnoses infectious diseases must interact with an infectious disease specialist in order to build its knowledge base. Likewise, a minerals expert system must interact with a geologist, a chemical analysis system with a chemist, and so on.

The process of acquiring knowledge is the biggest bottleneck in developing any expert system. Because the human expert is not familiar with the system or its operation, knowledge acquisition requires the assistance of a system specialist called a ***knowledge engineer.*** A knowledge engineer acts as an intermediary between the human expert and the expert system during the knowledge-acquisition mode, as illustrated in Figure 3-35. He or she is responsible for formalizing the human ex-

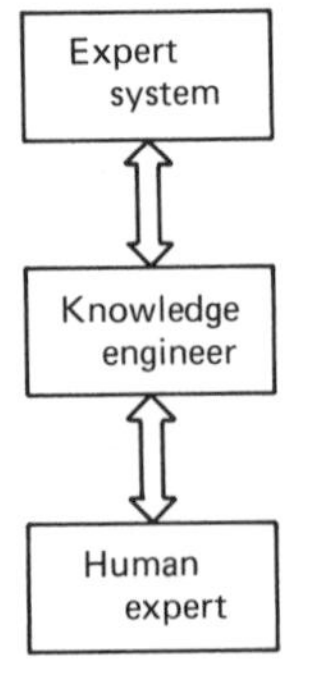

Figure 3-35 The knowledge engineer acts as an intermediary between the human expert and the system during the knowledge-acquisition mode.

pert's knowledge and integrating that knowledge into the system knowledge base. Knowledge that must be transferred to the system includes the following:

- Vocabulary, or jargon, used in the domain
- General concepts and facts used in the domain
- Problems that occur in the domain
- Solutions to problems based on experience in the domain
- Solution skills for solving problems in the domain

In other words, the knowledge engineer must "pick the brain" of the human expert in order to build the system knowledge base. Moreover, the knowledge engineer is responsible for testing the accuracy of conclusions made by the system through problem simulation.

Consultation Mode. The system is in the consultation mode when it is performing its function as a computerized consultant. In this mode, a nonexpert user interacts with the system by entering data and responding to system questions. The data entered by the user are placed in the system data base and accessed by the inference engine during the reasoning process. Recall that the inference engine attempts to match these data with conditional elements in the production rules. Answers to system questions during the consultation mode provide heuristic information that helps guide the reasoning process. For example, a typical user/system consultation session might go something like this:

System: What is the patient's name?

User: George Washington

System: What is the patient's age?

User: 252 years

System: Is the patient male or female?

User: Male

System: What are the symptoms?

User: Headache, dizziness, upset stomach, high fever.

System: When did these symptoms first appear?

User: December 13, 1799

.

.

.

System: I conclude that the patient has a viral infection. My recommendation is to take two aspirins, drink plenty of fluids, get plenty of rest, and call me in the morning.

P.S. Tell George that I don't make house calls.

(George Washington died on December 14, 1799, of a viral infection.)

Of course, this dialogue is fictitious, but it clearly illustrates the consultation mode of a real expert system. As you can see, the system carries on a consultation dialogue very much the same as you would with a doctor. Such a dialogue is possible because of the natural language interface and the production-rule knowledge base. The production-rule knowledge base lends itself easily to a consultation dialogue since the IF/THEN rules are structured similarly to the way people talk to solve problems.

Explanation Mode. The explanation mode provides one of the most important features of the expert system. It allows the system to explain its conclusions and reasoning process. This ability comes directly from the AND/OR trees created during the production-system reasoning process. As a result, most expert systems can answer the following why and how questions:

- Why was a given fact used?
- Why was a given fact not used?
- How was a given conclusion reached?
- How was it that another conclusion was not reached?

As an example, suppose that a doctor wishes to verify the conclusion and reasoning process of a disease diagnosis made by an expert system. The following is an explanation dialogue provided by a disease diagnosis program called MYCIN. Suppose that MYCIN has concluded that a given patient has contracted meningitis while in the hospital. The doctor who is using MYCIN now wishes to ask the system several why and how questions in order to verify MYCIN's conclusion. The dialogue goes something like this. While in the consultation mode, MYCIN has asked the doctor the following question:

MYCIN: Is meningitis a hospital-acquired infection?

At this point, the doctor wishes to know *why* the system is asking the question. So, the doctor responds:

Doctor: Why?

The system then goes into its explanation mode and responds with:

MYCIN: [2.0] This will aid in determining the organisms (other than those seen on cultures or smears) that might be causing the infection.
It has already been established that:
[2.1] the infection that requires therapy is meningitis, and
[2.2] organisms were not seen on the stain of this pending csf culture, and
[2.3] the type of infection is bacterial, and

[2.4] neurosurgery is not one of the relevant items from the history of pt100

[2.5] the age of pt100 is greater than 25 years.

Therefore, if

[2.6] the infection was acquired while the patient was hospitalized, then there is evidence that the organisms that might be causing the infection are E. coli (.75), staphylococcus-coag-pos (.3), pseudomonas-aeruginosa (.3), klebsiella-pneumoniae (.5).

Notice how MYCIN provides a detailed explanation of why it asked whether or not meningitis is a hospital-acquired infection. Now, suppose the doctor wished to know *how* item 2.3 was concluded. That is, *how* was it established that the type of infection is bacterial? The doctor must ask the question as follows:

Doctor: How 2.3.

The system then responds with:

MYCIN: The following rules concluded about the likelihood that the type of the infection is bacterial:

[2.7] RULE148 (.19)
[2.8] RULE500 (.51)
[2.9] RULE501 (.95)
[3.0] RULE502 (.97)
[3.1] RULE526 (.98)
[3.2] RULE504 (.97)
[3.3] RULE524 (.98)

Thus, it has been established that the type of infection is bacterial.

Notice that MYCIN responds with the actual production rules that were used in the reasoning process to determine that the infection is indeed bacterial. The system actually walks the doctor through the production system AND/OR tree to explain its conclusions. At this point, the doctor could add new rules or modify existing rules if he or she did not agree with the reasoning process.

You are probably wondering what the numbers in brackets and parentheses mean. The numbers in the brackets are simply line numbers for reference purposes. The numbers in parentheses are ***certainty factors.*** Since conclusions are seldom certain, MYCIN generates a certainty factor for each conclusion it makes. The certainty factors range from −1 to +1 and express the confidence that a given conclusion is true or false. A +1 means there is complete confidence that a conclusion is true. On the other hand, a −1 means there is complete confidence that a conclusion is false. Thus, from line [2.6] the doctor can determine that the system is 75 percent confident that the organism causing the infection is E. coli.

The preceding example should give you a good feel for the explanation abilities of expert systems. Now let's survey a few real expert systems in an effort to stimulate your imagination for the applications potential of these systems.

Some Real Expert Systems

Now that you know what an expert system is and how it works, let's take a closer look at two representative real-world expert systems. The two systems that we will discuss, MYCIN and HEARSAY, have been developed using the principles of artificial intelligence and knowledge representation discussed in this text.

MYCIN. As you are now aware, MYCIN is an expert system that was developed to help doctors diagnose infectious diseases. Many times, a doctor does not know the exact cause of an infectious disease owing to a lack of experience and/or time to investigate the illness. As a result, many doctors will prescribe broad-spectrum drugs to treat the disease. These drugs, like penicillin, will cover most of the disease possibilities. However, if the exact type of infection could be determined, the doctor could prescribe more disease-specific drugs, resulting in earlier recovery and less cost to the patient. Consequently, MYCIN was developed to provide quick analysis of bacterial infections, thereby allowing disease-specific treatment of the patient.

MYCIN begins in its consultation mode by asking for general information about the patient and his or her condition. Information such as the patient's history, symptoms, and lab results are entered into the system by the user in response to MYCIN's questions. The specific questions are formulated from an ongoing analysis of the answers to previous questions. The questions get more specific as evidence builds to a given diagnosis. When MYCIN is satisfied that it has enough information, it generates a qualified diagnosis and actually recommends a treatment for the disease.

This all might sound very complicated or mysterious to you, but you now have all the fundamental concepts for understanding how such a system works. As I stated earlier, MYCIN is a production-rule-based system. It uses a backward chaining reasoning process to verify a given disease, based on the input data. MYCIN starts from a list of approximately 100 possible disease diagnoses and works backward from each of these, trying to match the conditional elements of its production rules with the data being input by the system user.

MYCIN uses approximately 450 production rules during the reasoning process. The IF/THEN rules relate the symptoms and characteristics of a disease to the infecting organisms that cause the disease. The following production rule is typical of those found in MYCIN:

IF: the infection is primary bacteremia, and the suspected portal entry
of the organism is the gastrointestinal tract, and
the site of the culture is one of the sterile sites,

THEN: there is evidence (.8) that the organism is bacteroides.

An AND/OR tree is constructed during the reasoning process to allow the system to explain its results. As you are aware, backward chaining permits verification and explanation via the AND/OR tree structure. The various *how* and *why* questions are answered using the tree structure, as I discussed earlier.

In addition, MYCIN assigns certainty factors to each node within the AND/OR tree. The certainty factors are based on the reliability of the production rules. As the tree grows, individual certainty factors are combined to produce an overall certainty factor for a given conclusion. As you observed earlier, the certainty factor scale ranges from -1 to $+1$.

HEARSAY. A problem with production-rule-based expert systems is that, to make the system more intelligent, you must add more rules to the knowledge base. This results in combinatorial explosion during the reasoning process, thereby making the system extremely inefficient. A partial solution to this problem is to partition the production rules into manageable chunks of related rules. These rule chunks actually form subsystems that interact during the reasoning process. A given subsystem is only activated when the inference engine determines (via heuristics) that enough evidence exists to justify its activation. Such systems are called ***partitioned production systems.***

Another name for a partitioned production system is a ***blackboard system.*** A block diagram of a typical blackboard system is shown in Figure 3-36. The sys-

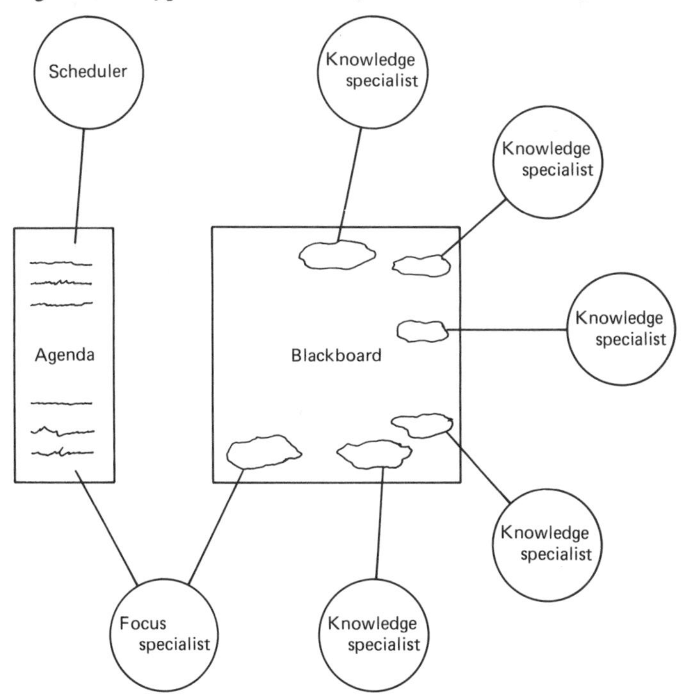

Figure 3-36 Typical partitioned production, or blackboard, system.

tem partitions are called ***specialists.*** Each specialist is a complete set, or system, of production rules in its own right. These specialists are scheduled for activation based on how promising they look at any given step within the reasoning process.

The term blackboard comes from the idea that the specialists communicate through a common data structure called a ***blackboard.*** The blackboard contains the initial data or goals, partial solutions produced by the specialists, and data that are passed between the specialists.

Notice from Figure 3-36 that one specialist is labeled the ***focus specialist.*** The focus specialist monitors the blackboard and identifies the best solution at any given time using heuristic-type production rules. From heuristic information, the focus specialist develops and assigns priorities to an agenda for activating the other specialists. The system ***scheduler*** then selects and activates the specialists from the agenda.

The HEARSAY expert system uses the blackboard idea. HEARSAY is a speech-recognition system that is about 90 percent accurate. The specialists used in HEARSAY are related to natural speech units such as phrases, words, syllables, phonemes, and signal characteristics, as shown in Figure 3-37. The specialist rules use the blackboard data to form the respective phonemes, words, and phrases of the spoken language.

Here is how it works. The phoneme specialist monitors the speech signal

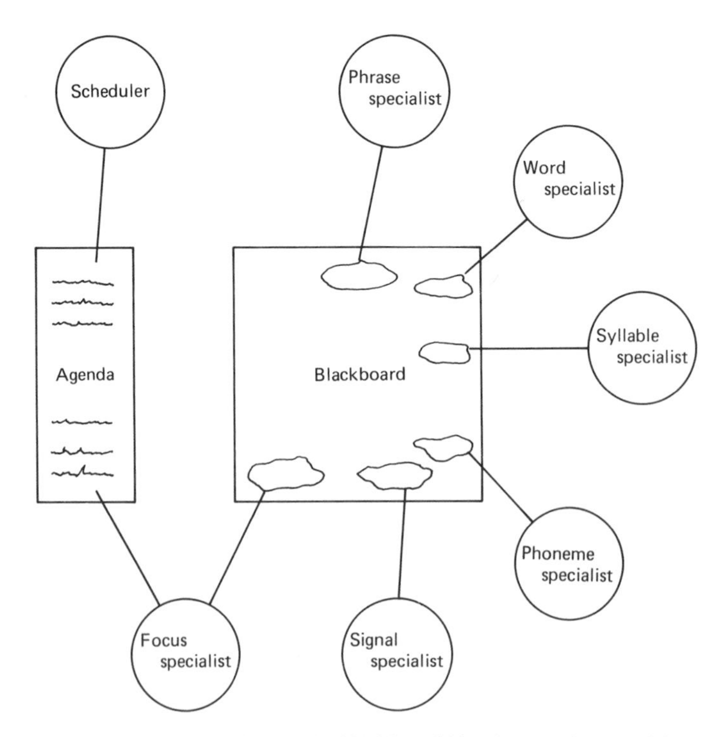

Figure 3-37 HEARSAY uses the blackboard idea for speech recognition.

data to form various phonemes. The resulting phonemes are placed on the blackboard, where the syllable specialist uses them to form syllables. The resulting syllables are placed on the blackboard and used to form words by the word specialist, which are finally put into phrases by the phrase specialist.

The HEARSAY system was one of the first expert systems to use the blackboard idea. Speech recognition lends itself easily to such a system since speech can be easily segmented into production partitions. However, any application that can be broken down into distinct chunks of knowledge is a candidate for a blackboard-type expert system. For this reason, the blackboard idea used in HEARSAY has since been applied to other AI applications, such as vision, electronic signal interpretation, mineral classification, and psychological modeling.

Other Expert Systems. I chose to discuss the preceding two expert systems since they are typical of most existing systems. In addition, MYCIN and HEARSAY were two of the first expert systems to be developed. The fundamental concepts used with these two systems have since been applied to the development of many subsequent expert systems.

Since MYCIN and HEARSAY, increased commercial interest in the application of expert systems to many diverse areas has resulted in several new systems developed to meet a variety of application needs. Table 3-1 summarizes some of the more popular expert systems that are on the job today.

TABLE 3-1 Summary of Expert Systems

System	Application
ACE (Automated Cable Expertise)	Analysis and automated decision making for telephone cable maintenance (Bell Telephone Laboratories)
PROSPECTOR	Mineral exploration (SRI International)
R1, XCON, XSEL, and XSITE	Computer configuration and order processing. (Digital Equipment Corp.)
Epistle	Mail reading and analysis (IBM)
Intellect	Business: financial, marketing, sales, and personnel (Artificial Intelligence Corp.)
Explorer	Oil and gas exploration (Cognitive Systems, Inc.)
Automated Will Writer and Estate Planner	Legal (Cognitive Systems, Inc.)
Automated Tax Assistant	Accounting (Cognitive Systems, Inc.)
CASNET	Diagnosis and treatment of glaucoma
DENDRAL	Chemical analysis
MYCIN	Infectious disease diagnosis (Stanford University)
HEARSAY	Speech recognition (Carnegie-Mellon Institution)

Table 3-1 is by no means complete. This is only the beginning. Future applications for expert systems are only limited by your imagination. It has been said that expert systems will offer the same productivity increase to white-collar workers as automation has to blue-collar workers.

3-6 PROGRAMMING LANGUAGES

To implement the problem-solving techniques and knowledge models discussed in the last two chapters, you must use a programming language. Although some high-level programming languages (like BASIC, Pascal, and FORTRAN) can be used to implement some AI concepts, they are not ideal since they are not ***symbolic*** languages. A symbolic language uses arbitrary symbols to represent and manipulate almost anything, not just numeric data.

Several symbolic languages have evolved from AI research, some of which are listed in Table 3-2. The "grandaddy" of them all is LISP, developed by John McCarthy in 1958. Second to FORTRAN, LISP is the oldest programming language presently in use. In addition, it has been the primary language used for AI research. Other AI languages, such as MACLISP, INTERLISP, and QLISP have obviously evolved from LISP. Still others, such as PLANNER, CONNIVER, SAIL, PROLOG, and FUZZY, were developed to improve on the features of LISP as applied to various specific areas, such as knowledge representation, logical reasoning, and commonsense reasoning.

TABLE 3-2 Summary of Current AI Programming Languages

Language	Date Developed
LISP	1958
QLISP	mid-1960s
POP-2	1967
SAIL	1969
PLANNER	1971
CONNIVER	1972
MACLISP	1966–1974
FUZZY	1977
INTERLISP	1978
PROLOG	1977–1981
FOL. PROLOG	1979

A thorough study of AI programming languages is beyond the scope of this text. The reason is threefold. First, the purpose of this text is to teach you those fundamental AI concepts that must be incorporated into any AI language, and how those concepts are applied to intelligent tasks such as voice communication, vision, navigation, and tactile sensing. The concepts learned here will provide a basis for a further study of AI programming languages. Second, as you can see from the list in Table 3-2, new languages are evolving continuously as a result of AI research.

Although LISP is presently the standard, it is destined to be replaced by a better standard as the field of AI evolves. Finally, I could not do justice to any AI programming language in the limited space that could be devoted to it here. If you want to learn LISP, for example, I suggest that you study a text dedicated to LISP.

SUMMARY

Good knowledge representation is the key ingredient of artificial intelligence. The two elements of knowledge representation are knowledge acquisition and reasoning. Although these are separate elements of knowledge representation, they are totally dependent on each other for intelligent system operation.

Most knowledge-representation schemes can be broken down into some combination of four major knowledge-representation techniques: logic, production systems, semantic networks, and frames. No unique knowledge-representation technique fits all applications. Some techniques work well for certain tasks, while others do not. The application will usually dictate the appropriate knowledge-representation scheme to use.

Logic is a formal knowledge-representation technique. Knowledge is represented in logic using propositional and predicate calculus. Propositional calculus represents knowledge using statements, called propositions, that are either true or false. Predicate calculus extends the idea of propositional calculus by using variables, quantifiers, and functions within the logic statement. In logic, inferences are made from known facts to produce new facts that are always guaranteed to be true. Two very powerful laws of logic that allow reasoning are resolution and unification.

The term semantic refers to the relationships that exist between symbols and what those relationships mean. A semantic network is a network of symbols that describes relationships between elements of knowledge. Semantic networks use a series of nodes and links to represent knowledge. The nodes contain objects, concepts, or situations in the problem domain. The links represent relationships between the nodes. The IS-A link is common since it can be used to develop relationships between general ideas as well as very specific objects. In addition, IS-A links produce a hierarchical structure within the network that creates a natural inheritance property. This inheritance property of semantic networks permits deductive reasoning and conserves memory space within a computer system. A disadvantage of using semantic networks for representing knowledge is that the notation and reasoning process is not formalized. Thus, no common principles can be applied to all networks.

A frame is a data structure that consists of expectations for a given situation. A frame contains objects and facts about a situation, as well as procedures on what to do when a given situation is encountered. The knowledge associated with a frame is contained in slots. The slots contain both situational and procedural information that sometimes forms links to subframes that are needed to completely describe a situation or task. Subframes are combined into hierarchical families of frames that contain inheritance properties. At the top of the family structure, you find general, commonsense knowledge, while at the bottom of the structure you find very specific knowledge. Within the family, action frames are linked to situational frames to describe cause-and-effect relationships. Reasoning is accomplished by filling the frame slots. Thus, the frame reasoning process is driven by expectations. Frame theory provides a means of representing very general, or commonsense, knowledge. In addition, frames provide a means whereby computers can understand their environment and learn from experience. However, frame theory is a very abstract idea, requiring more research before it becomes a practical means of knowledge representation.

Expert systems represent a direct application of artificial intelligence. An expert system is actually a computerized consulting system. Most existing expert systems use production systems for knowledge representation. As a result, the internal structure of an expert system consists of a knowledge, or rule, base, a data base, and a rule interpreter. Expert systems have three modes of operation: the knowledge-acquisition mode, consultation mode, and explanation mode. The specialized talents of knowledge engineers are required during the knowledge-acquisition mode to act as an intermediary between the human expert and the expert system. During the consultation mode, the user can ask questions of the system to verify its conclusions and reasoning process. Commercial expert systems are now being used for disease diagnosis, mineral exploration, telephone cable maintenance, chemical analysis, and speech recognition, just to mention a few application areas. Future applications of expert systems are only limited by your imagination.

PROBLEMS

3-1 Explain the difference between propositional and predicate logic.

3-2 Prove that $\overline{A \vee B} \equiv \bar{A} \wedge \bar{B}$.

3-3 The exclusive OR connective, $\forall$, is defined by the following truth table:

A	B	$A \forall B$
T	T	F
T	F	T
F	T	T
F	F	F

Show that $A \forall B \equiv A\bar{B} \vee \bar{A}B$.

3-4 Given the two propositions:

(1) A chicken is a bird.

(2) If a creature is a bird, then it has wings.

What logic rule of inference allows you to say that "A chicken has wings?"

3-5 Prove or disprove that:

IF B is true and A implies B, THEN A is true.

3-6 What inference rule allows premises in a logic statement to be canceled out?

3-7 Represent the following sentences using predicate calculus notation:

(a) David is a boy.
(b) David is a sick boy.
(c) The value of x is less than the value of y.
(d) All men are mortal.
(e) There is a red part in the bin that has a circular shape.

3-8 Use predicate calculus to prove the following statement:

IF Bob is a robot and all robots have manipulators, THEN Bob has a manipulator.

3-9 Given the animal tree in Figure 3-38, develop a series of production rules that can be used to reason that a given animal is a tiger.

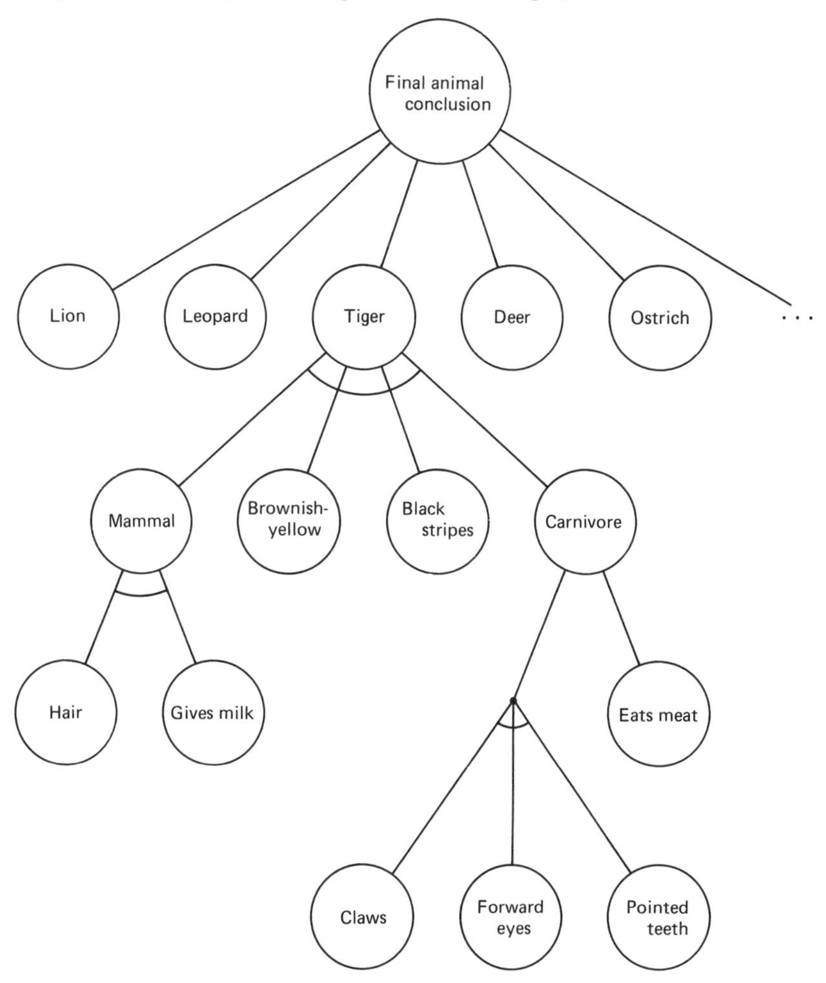

Figure 3-38 Animal tree for a tiger.

3-10 In what order would the production rules in Problem 9 be activated using a backward chaining reasoning strategy?

3-11 Explain how a production system performs backward reasoning.

3-12 What are the three major parts of a production system?

3-13 What are the three phases of a production-system operating cycle?

3-14 What is the major reason that production systems are popular, especially in expert systems?

3-15 What four questions can be answered using an AND/OR tree generated from a backward-reasoning production system?

3-16 Construct a semantic network for a tiger using the animals tree structure in Figure 3-38.

3-17 Represent the following logic statements using semantic network nodes and links.

(a) $(\forall x)[\text{EAGLE}(x) \rightarrow \text{BIRD}(x)]$

(b) $(\exists x)[\text{IN}(x,\text{KITCHEN}) \wedge \text{ON}(x,\text{TABLE})]$

(c) $(\forall x)(\exists y)[\text{PERSON}(x) \wedge \text{PERSON}(y) \rightarrow \text{LIKES}(x,y)]$

3-18 Construct a semantic network that could be used to represent the situation described in Example 3-3.

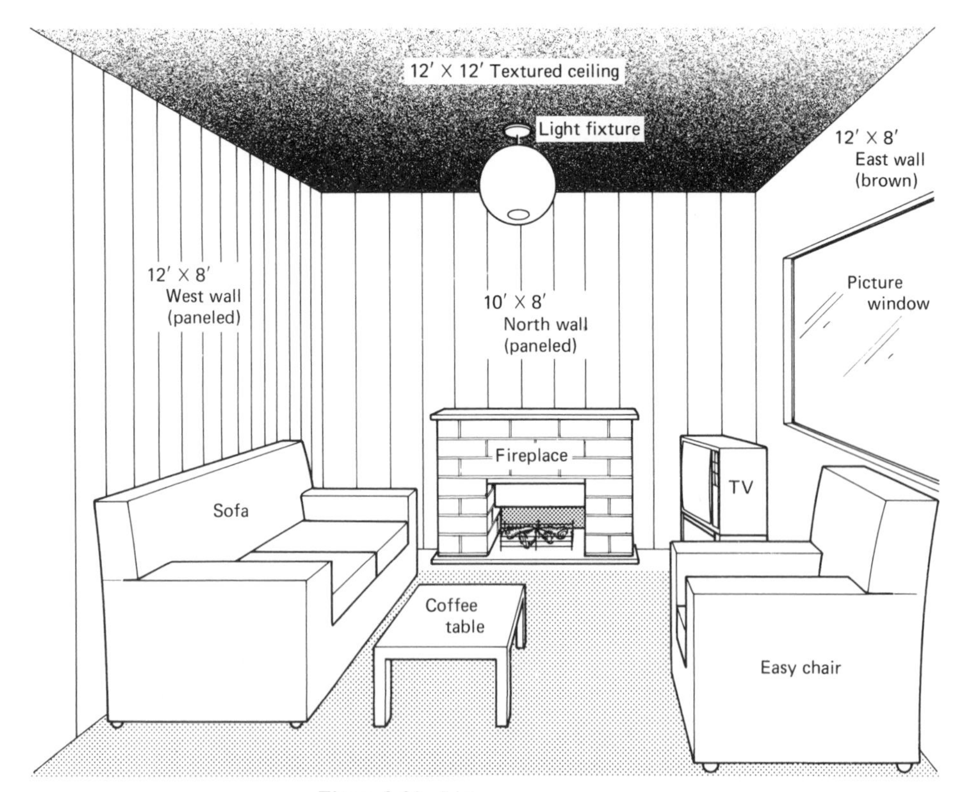

Figure 3-39 Living room scene.

3-19 Construct a family of situational frames that could be used with a robot vision system to represent the living room pictured in Figure 3-39.

3-20 Recall in Chapter 2 that the three-disc robot pick-and-place problem could be divided into three subproblems as follows:

(1) Move discs *A* and *B* from peg 1 to peg 2.

(2) Move disc *C* from peg 1 to peg 3.

(3) Move discs *A* and *B* from peg 2 to peg 3.

Construct a family of situation/action frames that will represent subproblem 2.

3-21 Define an expert system.

3-22 What type of knowledge representation is utilized by most expert systems?

3-23 Briefly explain the three major operating modes of an expert system.

3-24 What questions can be answered by an expert system in its explanation mode?

3-25 List at least three commercial expert systems and their respective application areas.

4 Speech Synthesis

The ultimate user-friendly machine, whether it is a personal computer or a robot, must have complete voice-communication abilities. Voice communication involves two separate, but related, technologies: ***speech synthesis*** and ***speech recognition***. Of the two, speech synthesis is the more developed technology and is the topic of this chapter. Speech recognition will be discussed in the next chapter. As you will see, many of the fundamental ideas behind speech synthesis presented in this chapter also apply to speech recognition. Thus, a sound knowledge of speech synthesis is a prerequisite to understanding speech recognition.

4-1 ELEMENTS OF SPEECH

Before you can begin to design electronic speech synthesizers, you must have a basic understanding of how you and I produce speech. As you will soon discover, some popular electronic speech synthesis techniques are modeled directly from the human vocal tract. Other techniques create speech by putting together the various sounds that are required to produce speech. As a result, it is important that *you* also understand the sound characteristics of speech. The more you know about speech sounds, the easier it is to synthesize speech. So let's begin with a study of human speech production.

Speech Production

To understand how artificial speech is created, or synthesized, it helps to know the fundamental concepts of human speech. The anatomy of the human vocal system is illustrated in Figure 4-1. As you can see, the vocal system can be broken down into three main regions: the lungs, larynx, and vocal cavity.

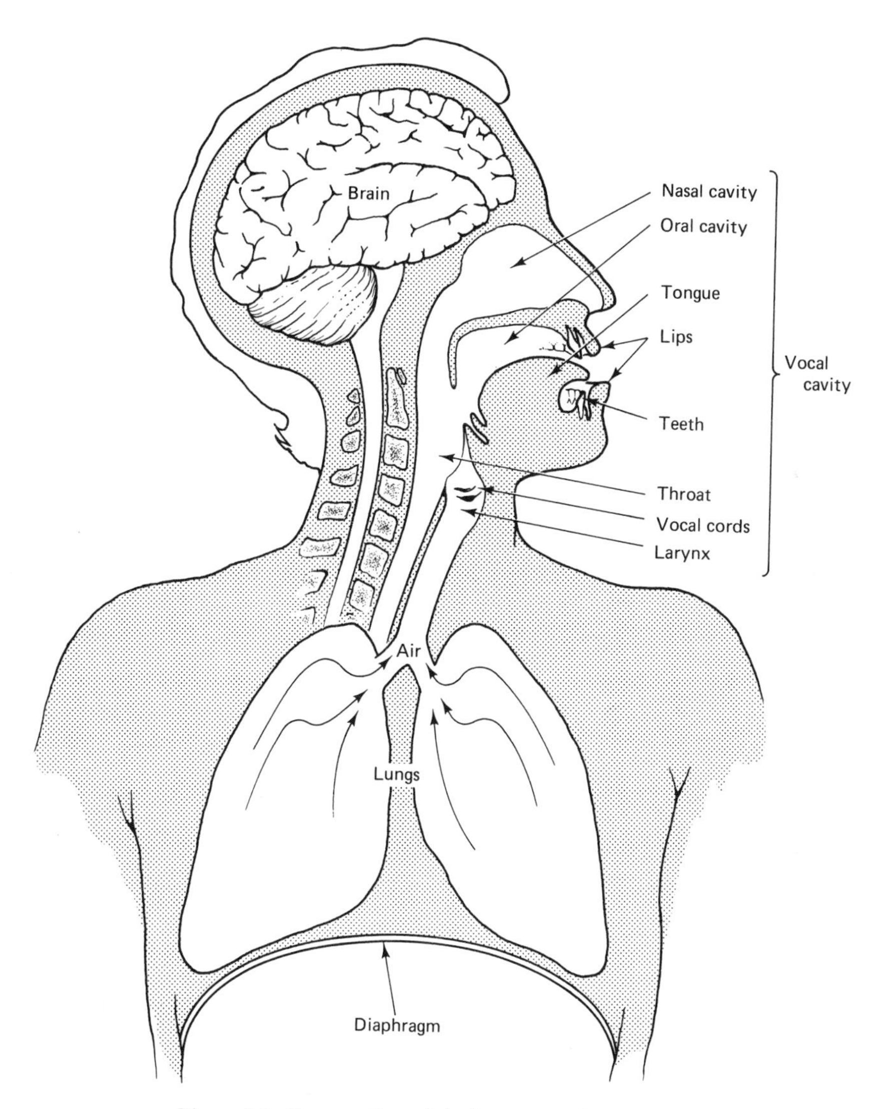

Figure 4-1 Cross section of the human vocal tract.

Let's pretend that you are in the doctor's office with a sore throat and the doctor asks you to say the word "aaaaaaah." Say the word and think about what is happening in your vocal system.

Your lungs are providing the power to the system by forcing air up through the larynx and into the vocal cavity. The sound is made by your vocal cords, which are located in the larynx just behind your Adam's apple. Your vocal cords are made up of skin layers that flap, or vibrate, as the air passes through. This vibrating action generates several resonant frequencies within the vocal cavity. In addition, each resonant frequency has several harmonic frequencies. You make different sounds by changing the shape of the vocal cavity with your throat, tongue, teeth, and lips. This, in turn, changes the resonant frequencies within the vocal tract.

Take a deep breath and say the word "aaaaah" again. While saying this word, change to the word "oooooh." Now, if you still have any breath left, change to the word "eeeeee." What was happening in your vocal cavity when you changed from one word to the next? From "aaaaah" to "oooooh" you changed the open end of your vocal cavity by forming your lips into the shape of an "O." From "oooooh" to "eeeee" you probably smiled and placed your tongue against your lower front teeth. The different sounds were made by changing the shape of your vocal cavity.

A model that is commonly used to illustrate the action of the vocal cavity is the pipe shown in Figure 4-2. This pipe is approximately 17 centimeters (cm) long and is closed at one end and open at the other. Assume that your vocal cords are located at the closed end of the pipe and your lips at the open end.

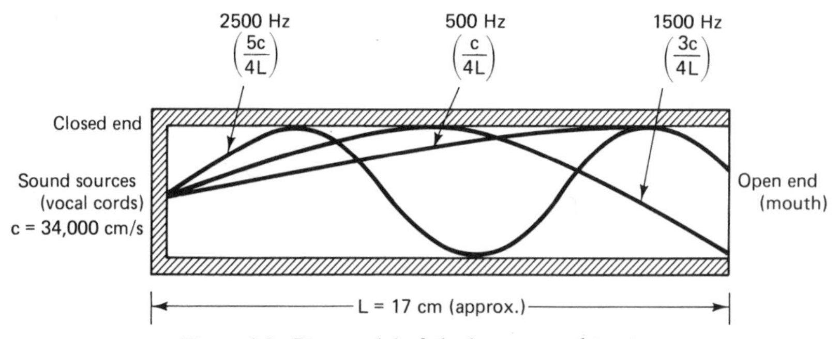

Figure 4-2 Pipe model of the human vocal tract.

You might recall from physics class that sound waves in such a pipe will resonate at odd-quarter wavelength frequencies, or $\frac{1}{4} \times c/L$, $\frac{3}{4} \times c/L$, $\frac{5}{4} \times c/L$, $\frac{7}{4} \times c/L$, and so on. Here, c is the speed of sound, or 34,000 cm/s, and L is the length of the pipe, or 17 cm. If you calculate the resonant frequencies, you get: 500, 1500, 2500, and 3500 hertz (Hz), respectively. If the pipe approximates your vocal cavity with any degree of accuracy, these are the frequencies you generate when you say "aaaaa," since your mouth is open, like the pipe, for that word.

When you change the shape of your mouth to generate different sounds, the frequencies shift according to the amount of physical distortion created by the action of your throat, tongue, teeth, and lips. If you were to graph the resonant frequencies generated in your vocal cavity for the word "aaaaah," you would get the waveform shown in Figure 4-3, which is a graph of relative amplitude in decibels (dB) versus frequency. The graph is actually a frequency spectrum envelope for the "aaaaah" sound. The resonant frequencies form the peaks of the envelope and are called ***formants.*** The lower frequency is labeled formant 1, the next frequency is formant 2, and so forth. Observe that the lower formants have a higher amplitude than the upper formants.

Harmonic frequencies on both sides of the format frequencies form a frequency band. Consequently, a formant frequency combined with its harmonics is called a ***formant band.*** Each sound generated in your vocal cavity produces a unique combination of formant bands. Formant 1 will vary between 200 and 1000 Hz when you are speaking. Formant 2 will vary between 500 and 2500 Hz, and formant 3 between 1500 and 3500 Hz as different sounds are generated during speech.

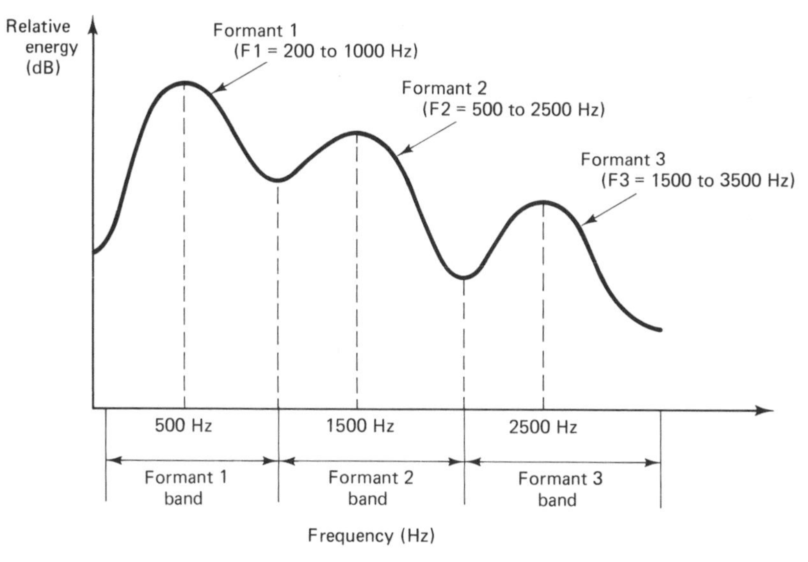

Figure 4-3 Frequency spectrum of the sound "aaah."

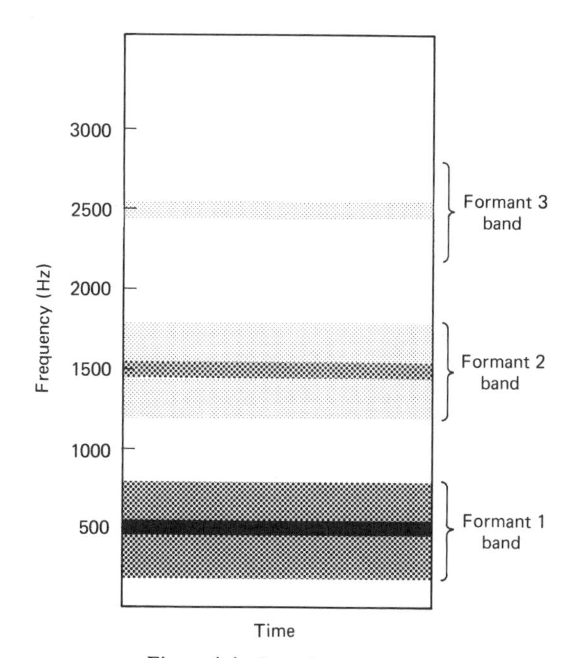

Figure 4-4 Sound spectrogram.

When engineers study speech, they use a frequency analyzer called a sound ***spectrograph.*** The output generated by a spectrograph is a ***spectrogram*** and looks something like that illustrated in Figure 4-4. Notice that frequency is shown on the *y* axis and is graphed against time on the *x* axis. The spectrograph is actually a three-dimensional graph since the relative intensity of each frequency is indicated by the degree of darkness of the graph. Notice that formant 1 is darker than formant 2, which is darker than formant 3. You might think of the spectrograph in Figure 4-4 as a view of the spectral envelope in Figure 4-3 looking down from the top.

Spectrographs have shown that the range of most human speech is from 150 to 3600 Hz. This represents a frequency bandwidth of 3450 Hz. If you consider the range of frequencies between a good bass singer and a soprano, you get a bandwidth of about 15 kilohertz (kHz), from 10 Hz to 15 kHz. Moreover, the corresponding *volume ratio* for human speech is about 16,000 to 1, from a shout to a whisper. Does this mean that, to duplicate human speech, you must design a speech synthesizer to generate frequencies from 10 Hz to 15 kHz with a 16,000:1 volume ratio? No, the idea is to generate a frequency bandwidth and volume ratio that will produce "intelligible" speech. The telephone system, for example, has a bandwidth of 3000 Hz, from 300 to 3300 Hz, with a 1000:1 volume ratio. Of course, anyone sounds a bit different over the telephone than they do in person, but the speech is still acceptable. It is this acceptable speech that allows the speech waveform to be compressed. As you will soon discover, speech compression is highly desirable in computer systems since it reduces the amount of memory required to create, or synthesize, speech.

The Sounds of Speech

Up to this point, you have been dealing with what linguists call ***voiced sounds.*** A voiced sound is a sound created by the vibrating action of the vocal cords. The vowel sounds (*a, e, i, o, u*) are all pure voiced sounds.

Another broad category of sound in human speech is the ***unvoiced sound.*** Unvoiced sounds are created by holding the vocal cords open and forcing air into the oral and/or nasal cavities. The forced air produces turbulence that results in a noisy excitation of the cavity. Sounds made by speaking the consonants such as *s, f, p,* and *t* fall into this category.

Both voiced and unvoiced sounds are subdivided further into several subcategories, as shown in Table 4-1. ***Fricative sounds*** are produced as the result of air friction within the oral cavity. Sounds like *s, f,* and *sh* are called ***unvoiced fricatives*** since they do not involve the vocal cords. All these sounds are produced near the front of the vocal cavity by air friction against the tongue, teeth, and lips.

Voiced fricative sounds utilize the vocal cords in addition to air turbulence. Sounds such as *z* and *v* fall into this subcategory. If you say the words *sip* and *zebra*, you will hear the difference between an unvoiced and a voiced fricative. The *s* in *sip* is an unvoiced fricative, while the *z* in *zebra* is a voiced fricative.

Plosive sounds, sometimes called ***stop sounds***, are the sounds you hear when

TABLE 4-1 **Sounds of Speech**

Voiced Sounds		Unvoiced Sounds	
Pure	a, e, i, o, u uh, aa, ee, er, uu, ar, aw	Fricative	s, sh, f, th
		Plosive (stop)	p, t, k, h
		Affricate	ch
Fricative	z, zh, v, dh	Aspirate	h
Plosive (stop)	b, d, g		
Nasal	m, n, ng		
Affricate	j		
Glides	r, w, l, y		

speaking the consonants like *b, d, p, t,* and *k.* These are called plosive sounds since they sound like small explosions coming from the vocal cavity. These tiny explosions are produced by a sudden release of air pressure that has been built up behind the throat, tongue, or lips. For example, the *p* sound in the word *pen* is generated by releasing air from behind the lips. Try it! Like fricative sounds, plosive sounds can be voiced or unvoiced, as shown in the table.

A ***nasal sound*** is heard when you block your oral tract by closing your mouth and forcing air into your nasal cavity. Nasal sounds are voiced sounds since they utilize the vocal cords. However, the sound-producing cavity is the nasal cavity rather than the oral cavity. Nasal sounds are produced when the consonants *m, n,* or *ng* are spoken, as in *new, man,* and *thing.*

The ***affricate sound*** subcategory is similar to the fricative sounds since they are also produced by air turbulence. There are only two affricate sounds: *j*, which is a voiced sound, and *ch* (as in *change*), which is an unvoiced sound.

A ***glide***, sometimes called a ***liquid***, sound is a voiced sound similar to a vowel sound. The difference is that you change the shape of your vocal cavity while pronouncing the sound. For example, when you pronounce the *w* in the word *watch*, your lips are close together in the beginning and then open at the end of the *w* sound. Another example of a glide is the *r* sound. Some people roll their *r*'s using a tongue motion. As a result, the shape of the vocal cavity is being altered during the sound by the movement of the tongue.

Finally, the *h* sound in the word *head* is called an ***aspirated sound.*** The aspirated h sound is created by forcing air straight through the oral cavity, as you do when you whisper. Aspirated sounds are also heard in connection with other sounds, like the *p* in *pen.* Here aspiration is heard after the air is released from behind the lips.

Phonemes and Allphones. All the sounds listed in Table 4-1 are called phoneme sounds since they are the fundamental sounds of the English language. A ***phoneme*** is the simplest and most basic sound present in any given language. In other words, speech is created by stringing phonemes together. This is called

phoneme concatenation and is one method used for electronic speech synthesis in computers.

The English language consists of about 40 phonemes, which are broken down into the sound categories listed in Table 4-1. It would be nice if speech synthesis simply involved the concatenation of these 40 basic sounds. However, it is not that simple, since a given phoneme may be pronounced many different ways. The way a phoneme is pronounced differs depending on where it is used within a word. For example, say the word *robot.* Now say the word *Robbie.* Notice how you pronounced the *o* phoneme within each of these words. What made you pronounce the phoneme differently? The idea is that you changed the phoneme pronunciation to accommodate the sounds that occurred around it. This is called ***coarticulation.*** In the English language, coarticulation produces approximately 128 sound variations of the 40 basic phonemes. The coarticulated sounds are called ***allophones.*** Each phoneme has several allophones associated with it.

Prosodic Features. You just found out that speech cannot be created by simply connecting the basic phonemes together, since the pronunciation of a given phoneme differs depending on where it is used within a word. The obvious solution to this dilemma is to create speech by connecting allophones rather than phonemes, right?! Look at the punctuation at the end of the last sentence. I didn't accidentally punctuate the end of the sentence with both a question mark and an exclamation point. Is there a difference in the way you say "right?" and "right!"? Of course there is! The allophones required to speak "right?" and "right!" are the same. So what is the difference? The difference in the way they are pronounced is due to a variation in *pitch* at the end of the word. Pitch is one of the prosodic features of speech.

There are three prosodic features in speech: ***pitch, amplitude,*** and ***timing***. Pitch is governed by frequency changes within the voice pattern. Notice that the pitch, or frequency, of your voice usually rises at the end of a question. The term ***intonation*** is used to describe the pattern of pitch changes in speech. The intonation of a spoken phrase often determines the meaning of the phrase. In addition, intonation plays an important part in making computer-generated speech sound natural.

Timing, or rhythm, also plays an important role in speech interpretation and naturalness. In most cases, you adjust the timing of speech by inserting pauses between allophones. For example, consider the following phrases:

"The approximate value is . . . ."
"To approximate the value, you must . . . ."

Think about the way the word *approximate* is pronounced in each phrase. The basic phoneme sounds are the same, but the timing of the sounds within the word is different. In the first phrase, the word *approximate* is used as an adjective, whereas in the second phrase it is used as a verb.

This example illustrates the effect that timing has on the meaning of speech. In addition, without proper timing, computer-generated speech sounds very artificial. As you might suspect, timing and intonation are related. They combine during speech to produce an effect called ***stress.***

Amplitude variations exist within speech, but they do not have the same meaningful effect as timing and intonation. For this reason, amplitude variation is not an important consideration in speech synthesis.

Electronic Speech-synthesis Techniques

Webster defines synthesis as "the putting together of parts or elements so as to form a whole." Thus, speech synthesis is the putting together, or creation, of speech. From the preceding discussion, you should now have some ideas on how speech can be synthesized, or created, with a computer. Basically, two general techniques are used for ***electronic speech synthesis (ESS)***: ***natural speech analysis/synthesis*** and ***artificial constructive/synthesis.***

Natural Speech Analysis/Synthesis. This technique is called natural, since it involves the recording and subsequent playback of human speech. The recording can be in analog form, such as a magnetic tape recording, or digitized and stored in computer memory. The analog record/playback method produces the most natural sounding speech, but it is not very practical for general speech production since the recorded messages must be accessed serially. In addition, it is almost impossible to record all the words and phrases that might be required for general speech synthesis, as in a robot application. This method might be the best choice to produce limited speech as in vending machines, appliances, and automobiles.

The digital record/playback method always involves two separate operations: ***analysis*** and ***synthesis.*** A typical system block diagram is shown in Figure 4-5. During the analysis phase, human speech is analyzed, coded into digital form, and stored. Then, during the synthesis operation, the digitized speech is recalled from memory and converted to analog form to re-create the original speech waveform.

The digital analysis/synthesis method provides more flexibility than the analog method since the stored words and phrases can be randomly accessed from computer memory. However, the vocabulary size is limited by the amount of memory available. For this reason, several different encoding techniques are used

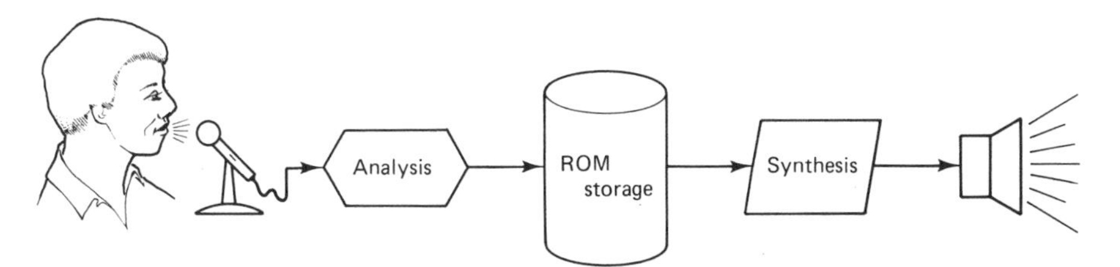

Figure 4-5 Block diagram of a typical analysis/synthesis system.

to analyze and compress the speech waveform. These various encoding processes attempt to discard the unimportant parts of the speech waveform, resulting in fewer bits being required to represent a given speech segment.

One type of digital analysis/synthesis is referred to as ***time-domain analysis/synthesis.*** With time-domain analysis, the speech waveform is digitized in the time domain. In other words, the analog waveform is periodically sampled and converted to a digital code using an A/D converter. The stored samples are then passed through a D/A converter to reproduce the speech during the synthesis operation. Time-domain analysis/synthesis also goes by the name of ***waveform digitization,*** for obvious reasons.

You are probably familiar with waveform digitization of speech from using the telephone system. When you dial directory assistance in most locations, the person's name you wish to call is entered into a computer by the directory assistance operator. The computer then looks up the name and responds verbally with the appropriate number. The computer is simply piecing the number together from digitized speech stored in memory. This type of system works fine for this application, since only ten numbers and a few words need to be stored in memory.

Another type of digital analysis/synthesis is called ***frequency-domain analysis/synthesis.*** With this technique, the frequency spectrum of the analog waveform is analyzed and coded, rather than amplitude variations. In addition, the synthesis operation attempts to emulate the human vocal tract electronically. Briefly, this is accomplished by using stored frequency parameters obtained during the analysis phase to control electronic frequency generators and filters that reproduce the voiced and unvoiced sounds of the human vocal tract.

Artificial Constructive/Synthesis. The second general category of digital speech synthesis is artificial constructive/synthesis. As the name implies, speech is created artificially by putting together, or constructing, the various sounds that are required to produce a given speech segment. The most popular method of doing this is called ***phoneme speech synthesis.*** With your knowledge of phonemes, you can probably imagine how this technique works. A block diagram of a typical system is shown in Figure 4-6. Phoneme and allophone sounds are coded and stored in memory. Software algorithms must then be written to connect the phonemes to produce a given word. Words are then strung together to produce phrases. In some cases, the software consists of a set of production rules that are used to translate written text into the appropriate allophone codes. This is called ***text-to-speech*** translation.

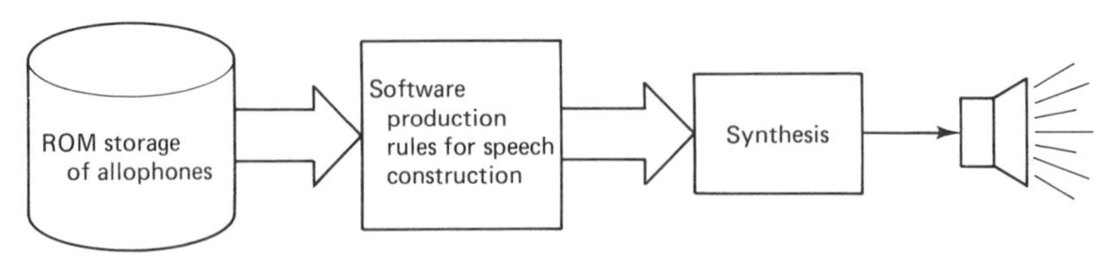

Figure 4-6 Block diagram of a typical constructive/synthesis system.

With the phoneme technique, a computer can produce an unlimited vocabulary using a minimum amount of memory. To sound natural, the software must determine which allophonic variation of a phoneme to use as a result of its location within a word. In other words, the software must anticipate the sound of a phoneme much the same as you do. In addition, the major prosodic features of timing and intonation must be included in the sythesized speech. All this results in a lot of software overhead and produces a very artificial, or computer, sounding speech.

In summary, digital speech synthesis can be accomplished in one of three ways: time-domain analysis/synthesis, frequency-domain analysis/synthesis, or phoneme constructive/synthesis. The remaining sections of this chapter are devoted to the above three digital speech synthesis techniques. Analog speech synthesis will not be considered, since it does not lend itself to robotic applications.

The chart in Figure 4-7 summarizes the various methods that are used for electronic speech synthesis. You will find that none of these is the answer to all the speech synthesis problems you might encounter. A given method might work well for one type of application, while not so well for another. The application must always be considered first, since it will usually dictate the best solution. When con-

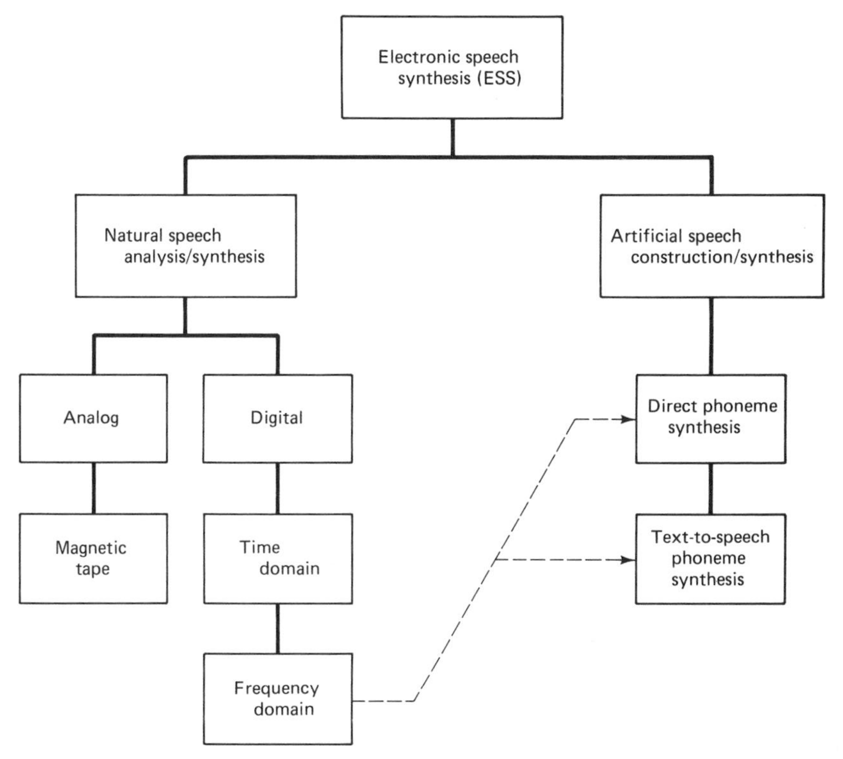

Figure 4-7 Summary of the various methods that are used for electronic speech synthesis (ESS).

sidering an application for speech synthesis, keep in mind the application requirements for speech quality, vocabulary, flexibility, and economy. These application requirements will usually dictate which type of speech-synthesis technology to use.

4-2 TIME-DOMAIN ANALYSIS/SYNTHESIS AND WAVEFORM DIGITIZATION

Any method that converts the amplitude variations of speech to a digital code for subsequent playback can be considered time-domain speech analysis/synthesis. The idea is to digitize human speech and store it in memory, usually ROM, for later playback. The principle is relatively simple and produces good-quality speech. In addition, all the characteristics of the human speaker, such as pitch (male/female), accent, and timing are reproduced during the playback.

Time-domain speech analysis/synthesis involves two fundamental operations: (1) encoding the human speech waveform to digitize and store the speech, and (2) decoding the digitized speech into analog form for playback. Since the human speech waveform is analog, an A/D converter is used for the encoding process. Then a D/A converter is used to decode the digitized speech for playback. A typical synthesizer circuit is shown in Figure 4-8. In addition to the two converters, most circuits usually contain a sample/hold device on the A/D converter input to hold analog input values while they are being converted. Finally, a low-pass filter is connected to the D/A converter output to smooth out the steps in the synthesized waveform.

The problem with waveform digitization is that large vocabularies require large amounts of memory. For this reason, several methods of time-domain encoding attempt to reduce the amount of memory required to store the digitized speech.

Simple Pulse-code Modulation

Pulse-code modulation (PCM) is simply a fancy name for direct waveform digitization using an A/D converter. The analog speech waveform is sampled and converted into a digital code by the A/D converter. Two things control the quality of the digitized speech: the sampling rate and the resolution of the A/D converter.

As you found earlier, human speech can range from 10 to 15,000 Hz. It is a complex waveform with many frequency components that are continuously changing. To "catch" all the subtleties of the waveform, it must be sampled about 30,000 times per second. If each sample were converted to an 8-bit digital code, one second of speech would require 8 X 30,000 or 240,000 bits of memory. Thus, the data-conversion rate would be 240,000 bits per second (bps). Obviously, this is highly impractical for extended continuous speech. To reduce the data rate, you must reduce the sampling rate. The effect of sampling rate on an analog waveform is illustrated by Figure 4-9. Notice that the quality of the synthesized output increases as the sampling rate is increased.

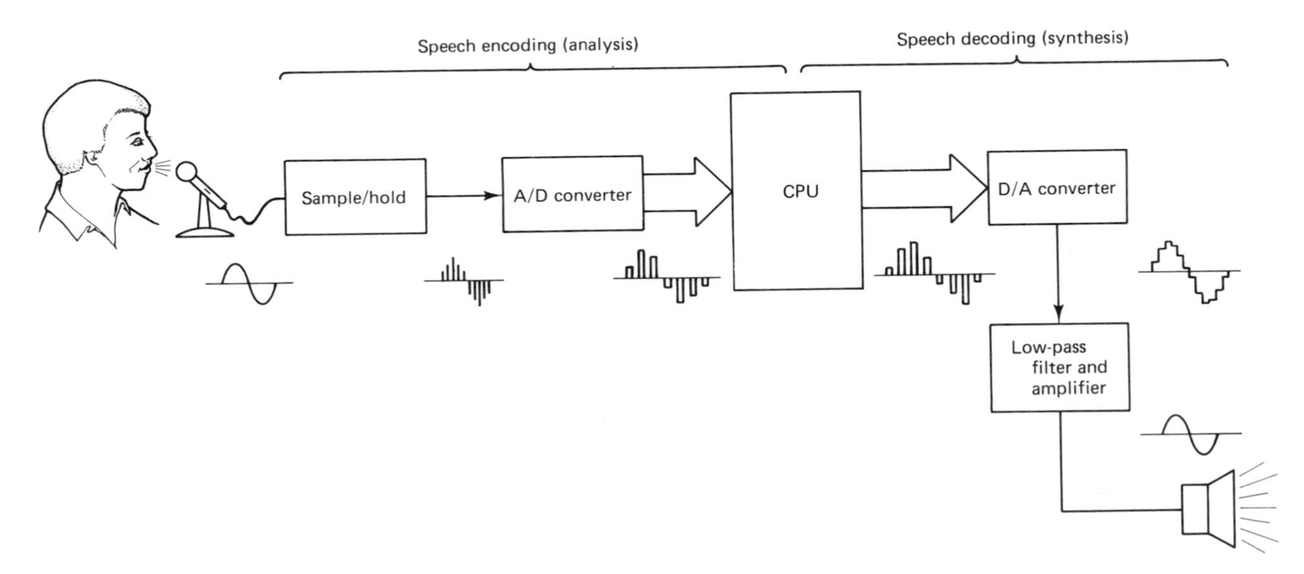

Figure 4-8 Typical time-domain system.

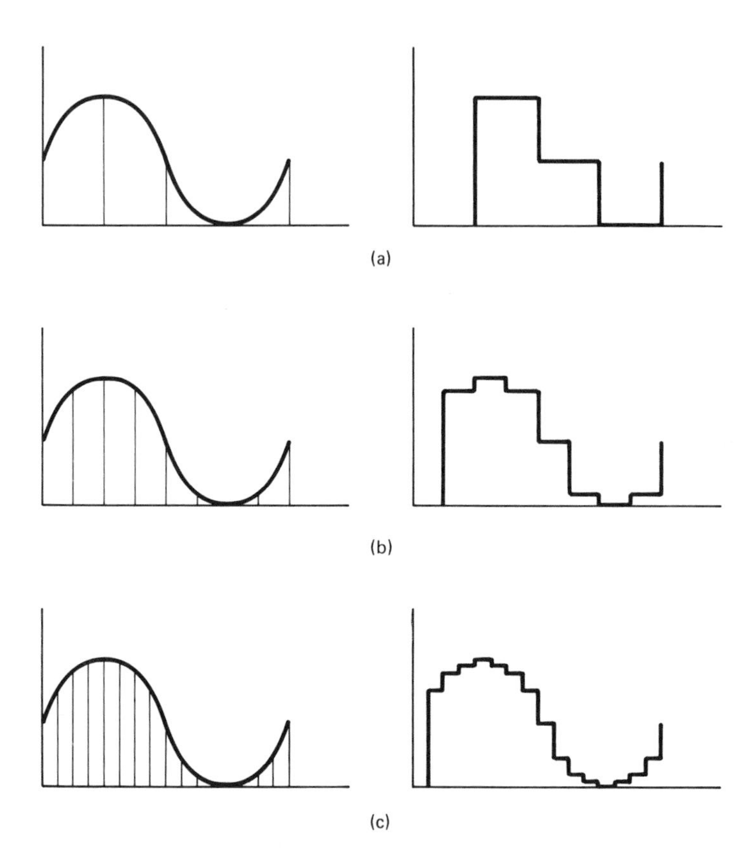

Figure 4-9 Effect of sampling rate on an analog waveform: (a) low sampling rate; (b) medium sampling; (c) high sampling rate.

Experimentation has shown that acceptable speech can be created using a sampling rate of at least two times the highest frequency component in the speech waveform. Recall from earlier discussions that most speech falls in the 300- to 3000-Hz range. As a result, 6000 conversions per second is the minimum sampling rate required to produce acceptable speech.

The second factor controlling the quality of the synthesized speech is the resolution of the A/D converter. The A/D converter resolution determines the smallest analog increments that will be detected by the system. Acceptable speech can be synthesized using an 8-bit A/D converter. With 6000 samples per second, this translates to a data rate of 8 $\times$ 6000, or 48,000 bps. As a result, 10 seconds of speech would require 60,000 bytes of memory, or roughly 64k bytes. Better-quality speech can be produced with a 12-bit converter. However, the same 10 seconds of speech would require 96k bytes of memory.

You are probably beginning to see the problem with simple pulse-code modulation. It just requires too much memory to produce acceptable speech for any length of time. One answer to this dilemma is to reduce the amount of bits that are required to represent a given speech segment. This reduces the data rate, thereby reducing the amount of memory required. Three popular methods are employed: ***delta modulation, differential pulse-code modulation***, and ***adaptive differential pulse code modulation.***

Delta Modulation

Delta modulation is the simplest form of data reduction. With delta modulation, only a single bit is stored for each sample of the speech waveform, rather than 8 or 12 bits. The A/D converter still converts a sample to an 8- or 12-bit value. However, this value is not stored for subsequent playback. Rather, it is compared to the last sample value. If the present value is greater than the last value, the computer stores a single logic 1-bit value. If the present value is less than the last value, a logic 0 is stored. In this way, a single bit is used to represent each sample. The analog speech is then stored in memory as a series of single bit values. An integrator is used on the circuit output to convert the serial bit stream to an analog waveform resembling the original speech waveform. The delta modulation method is illustrated in Figure 4-10.

On the surface, delta modulation might sound like the perfect solution to the data-rate problem. However, since only a single bit is being used to represent each sample, the sampling rate must be very high to catch all the details of the speech signal. A typical sampling rate is 32,000 samples per second. This translates to a 32,000-bps data rate. Thus, 10 seconds of speech would require 39k bytes of memory. By comparison, this represents a 40 percent data reduction from simple 8-bit pulse-code modulation.

Another problem with delta modulation is ***compliance error***, which results when the speech waveform changes too rapidly for a given sampling rate. This is sometimes called ***slope overload,*** since the slope of the input speech waveform changes too fast. The resulting digitization does not truly represent the analog waveform and produces audible distortion in the output. The only solution to slope overload is to increase the sampling rate. Of course, this results in an increased data rate and more memory.

Differential Pulse-code Modulation (DPCM)

Differential pulse-code modulation (DPCM) uses the same idea as delta modulation. However, several bits are used to represent the *actual* difference between two successive samples, rather than using a single bit. Since the speech waveform contains many duplicate sounds and pauses, or silence, the change in amplitude from one sample to the next is relatively small as compared to the actual amplitude of a sample. As a result, less bits are required to store the difference value than the absolute sample value. A typical DPCM circuit is shown in Figure 4-11.

In many cases the difference between two successive samples can be represented with a 6- or 7-bit value. This multibit value consists of a sign bit plus 5 or 6

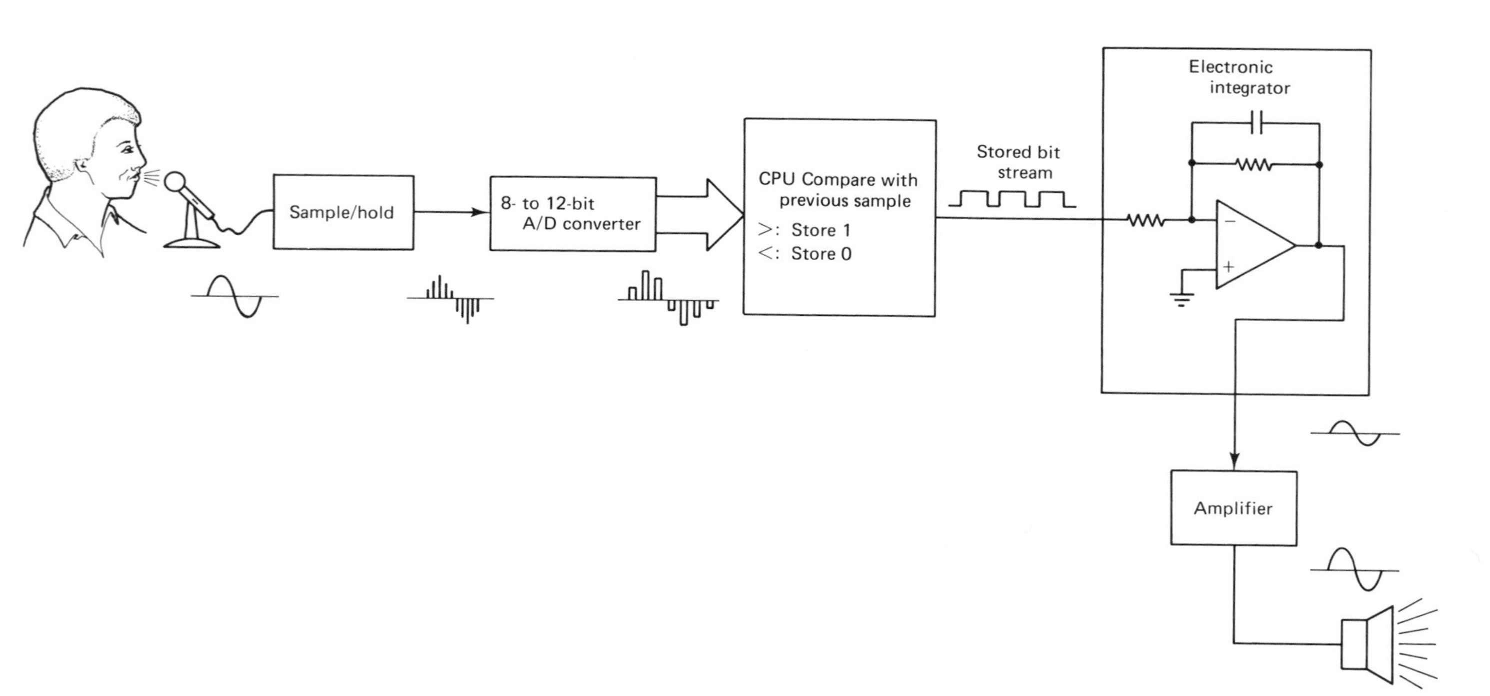

Figure 4-10 Delta modulation.

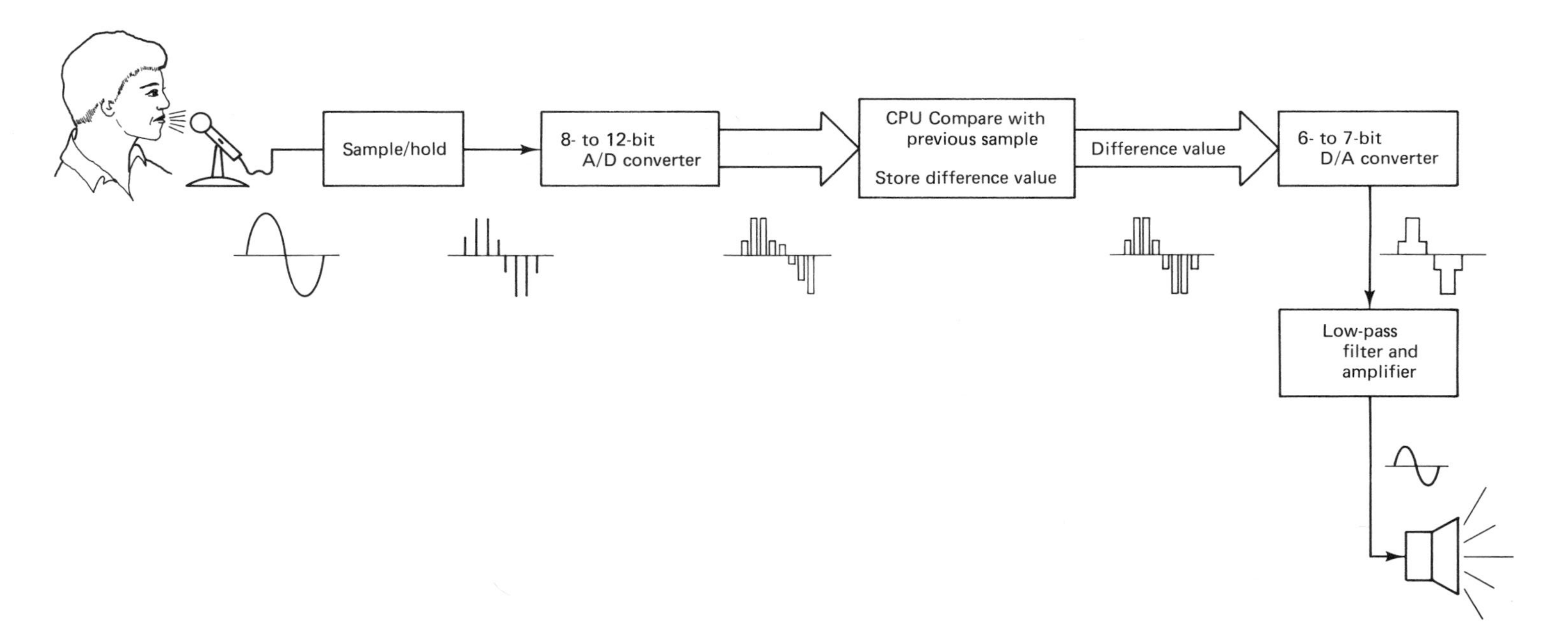

Figure 4-11 Differential pulse-code modulation (DPCM)

bits for the difference value. The sign bit is used to represent the slope of the input waveform. A bipolar D/A converter is used for playback to convert the successive difference values to a continuous analog waveform that is proportional to the original speech waveform.

A 7-bit DPCM system using a sampling rate of 6000 samples per second would require about 51k bytes of memory. This represents a 12 percent savings as compared to a simple 8-bit PCM system. A 25 percent savings would be realized using a 6-bit DPCM system.

Like delta modulation, DPCM suffers from slope overload. If the amplitude of the speech waveform changes too abruptly, the 6 or 7 bits might not be enough to represent the large difference value. The only solution is to increase the number of bits or increase the sampling rate, both of which defeat the data-reduction purpose of the system.

Adaptive Differential Pulse-code Modulation (ADPCM)

Adaptive differential pulse-code modulation is a variation of DPCM that eliminates the slope-overload problem. In addition, only 3 or 4 bits are required to represent each sample using ADPCM. As a result, the required storage is only 25 to 50 percent of that needed for simple PCM.

ADPCM uses a relatively complicated procedure to digitize the speech waveform. The waveform is sampled at about 6000 samples per second with an 8- or 12-bit A/D converter. The computer must then subtract the current sample value from the previous sample value to get a differential value, as with DPCM. However, the differential value is then adjusted to compensate for slope, using a calculated value, the ***quantization factor.*** The quantization factor adjusts the differential value dynamically, according to the rate of change, or slope, of the input waveform. In other words, the quantization factor is simply a scaling factor that is used to adjust the differential value as the input changes. The adjusted differential value can then be represented using only 3 or 4 bits. Consequently, slope overload is minimized and fewer bits are required to represent the speech signal.

In addition to requiring fewer bits, the sampling rate of an ADPCM system can be reduced, since slope overload is minimized. Sampling rates as low as 4000 samples per second are possible. With this sampling rate and a 3-bit code, the data rate becomes 3 bits × 4000 samples per second, or 12,000 bps. Thus, 10 seconds of speech would only occupy about 15k bytes of memory. However, 8000 samples per second and 4-bit codes are more common. Even so, such a system would only require 39k bytes of memory for 10 seconds of speech. This represents quite an improvement over simple PCM, delta modulation, and DPCM. The price you must pay is more sophisticated software.

The ADPCM method is currently employed in several commercial speech-synthesis chips. These chips perform the complete ADPCM waveform digitization process, producing a code that can be easily stored by a microprocessor for subsequent playback via a D/A converter.

One final point: since the first step in speech recognition is to digitize the speech waveform, ADPCM is an attractive approach to the "front end" of a speech-recognition system. More about this later.

4-3 FREQUENCY-DOMAIN ANALYSIS/SYNTHESIS

The time-domain methods discussed in Section 4-2 all involved some sort of record/playback scheme to synthesize speech. In particular, time-domain analysis/synthesis is concerned with recording and playing back the amplitude variations within the speech waveform. On the other hand, frequency-domain analysis/synthesis is concerned with the reproduction of the frequency spectrum within the speech waveform, while less concerned with amplitude variations. Moreover, frequency-domain analysis does not attempt to digitize the speech waveform on a one-to-one analog-to-digital conversion basis. Instead, a mathematical model of the frequency spectrum is stored and used to control an electronic model of the human vocal tract.

During the analysis phase, the frequency characteristics of the human voice are analyzed to produce a series of mathematical parameters that is stored and subsequently recalled to control an electronic speech synthesizer. The electronic synthesizer emulates the human vocal tract using frequency generators and filters that are controlled by the stored speech parameters.

Basically, two methods are employed for frequency domain analysis/synthesis: ***linear predictive coding (LPC)*** and ***formant analysis/synthesis.*** Of the two, linear predictive coding produces the most natural sounding speech. However, formant synthesis is usually less complicated and thus more economical. Let's explore each method, keeping in mind the application requirements for speech quality, vocabulary, flexibility, and economy.

Linear Predictive Coding (LPC)

Linear predictive coding, commonly known as LPC, was introduced in the early 1970s. The first commercial application of LPC was by Texas Instruments as part of their popular Speak & Spell(TM) child's educational toy. Other LPC-based products, such as language tutors, have since been introduced by TI and other manufacturers. If you have ever used any of these products, you are aware of the superior-quality speech produced by LPC.

A typical LPC system is illustrated in Figure 4-12. The first step in linear predictive coding is to digitize the speech waveform with an A/D converter using simple pulse-code modulation. Once in digital form, the waveform is analyzed to extract the frequency, intensity, and other vocal-tract-type variables needed to mathematically reconstruct the waveform. The extracted speech data are then coded into a series of linear equation parameters, called LPC codes, that model the frequency characteristics of the spoken waveform.

Once the speech waveform has been encoded into the LPC format, the stored speech parameters are used to control a synthesizer circuit like the one shown in Figure 4-13. The synthesizer circuit is designed as a model of the human

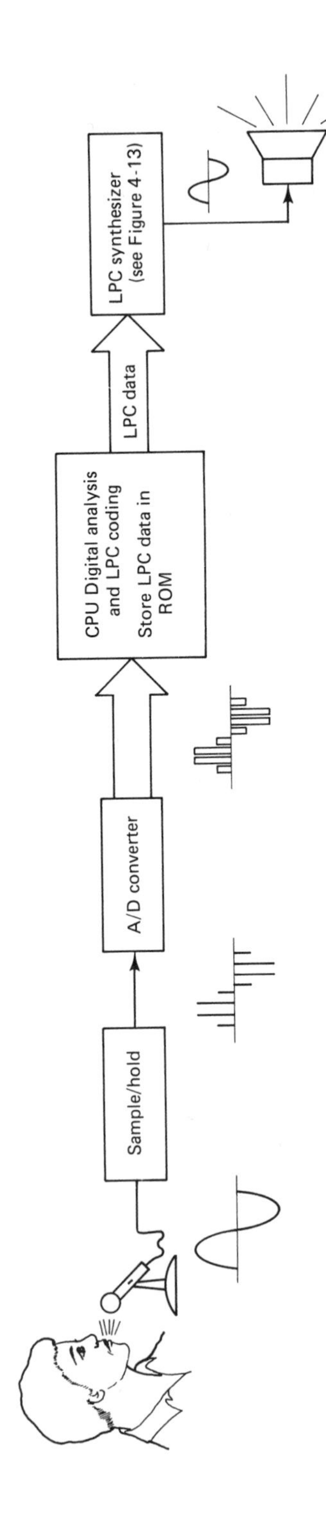

Figure 4-12 Typical linear predictive coding (LPC) system.

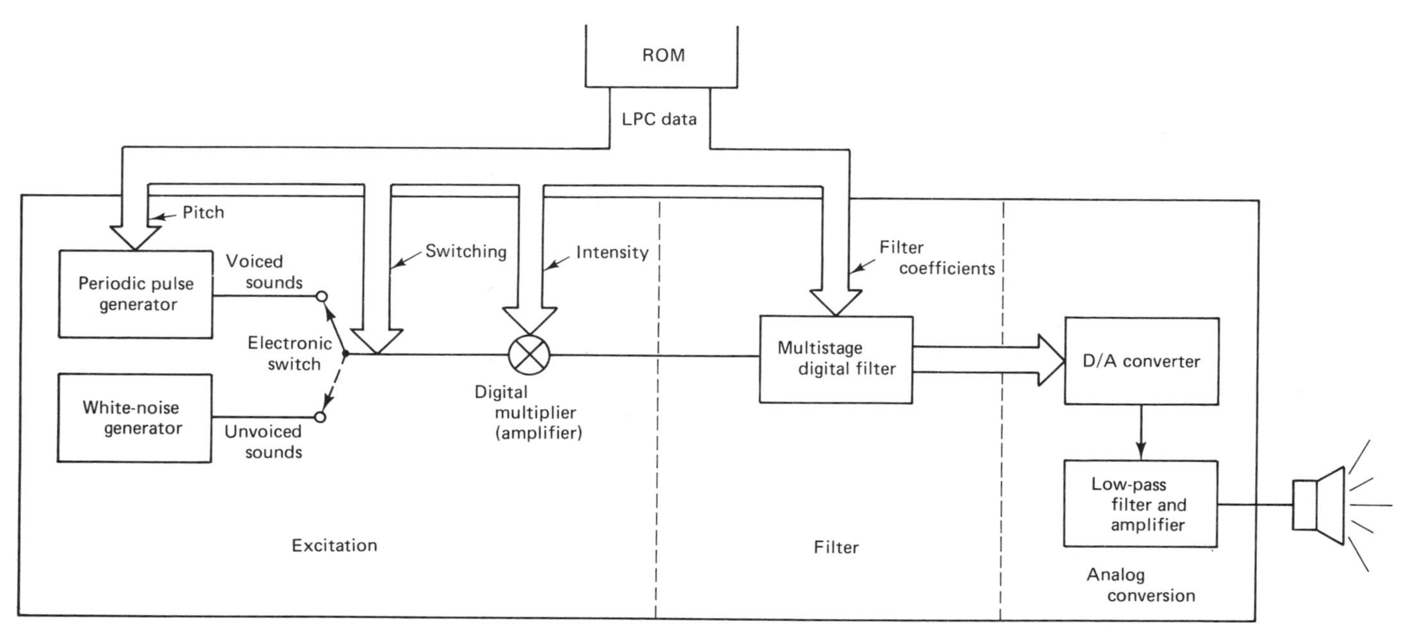

Figure 4-13 An LPC speech-synthesizer circuit models the human vocal tract.

vocal tract. It consists of three major sections: an ***excitation source***, a ***multiple-stage digital filter***, and a ***D/A converter.***

As you can see from Figure 4-13, the excitation source includes a periodic pulse generator, a white-noise generator, an electronic switch, and an amplifier. The periodic pulse generator emulates the action of your vocal cords by producing the periodic voiced sound frequencies. Recall that the rate at which your vocal cords vibrate determines the pitch of the sound. Likewise, the frequencies generated by the periodic pulse generator determine the pitch of the synthesized sound.

The white-noise generator produces the unvoiced sounds. Recall that these sounds are not created by your vocal cords. Rather, unvoiced white-noise sounds are produced as the result of air turbulence in the vocal cavity. The white-noise generator produces unvoiced sounds by generating a random frequency pattern that results in a hissing type of noise.

The combination of voiced and unvoiced sounds produces speech. This is the job of the electronic switch. The voiced and unvoiced sounds are combined by electronically switching between the two sound generators. The selected sound is then amplified and passed through a multistage digital filter circuit.

The purpose of the digital filter is to shape, or modulate, the excitation signal in much the same way that your throat, tongue, teeth, and lips modulate your vocal cavity sounds. Thus, the digital filter acts as a model of your vocal tract. The excitation signal is modulated to produce the desired formant frequency spectrum by varying the characteristics of the filter. LPC data stored in ROM are used to control the filter characteristics. The digital filter output is then converted to an analog speech signal by a D/A converter.

In most cases, the LPC data are stored in ROM. To reproduce the stored speech, the data are applied to the synthesizer circuit and control the following four circuit functions:

1. Pitch of the voiced sounds
2. Selection between the voiced or unvoiced sounds
3. Amplitude of the excitation signal for sound intensity
4. Control of the digital filter by providing the filter coefficients required to modulate the excitation signal and produce the digital speech waveform

In summary, linear predictive coding involves the conversion of PCM data to frequency parameters that are stored and subsequently used to control a synthesizer circuit. The synthesizer circuit is an electronic model of the human vocal tract that is controlled by the stored LPC parameters. These parameters control the pitch, selection between voiced and unvoiced sound, intensity, and modulation of the speech signal.

Because of the computational complexity involved, it can take several minutes with a large computer to convert just a few seconds of speech to the required LPC format. But once coded, the LPC data rate required to reproduce speech is less than

2400 bps. This means that 10 seconds of speech can be stored in less than 2.9k bytes of memory. In fact, Speak & Spell(TM) uses a 16k byte ROM to store about 170 words, resulting in 120 seconds of speech. This translates to a data rate of only 1200 bps. In addition, the synthesized speech retains all the pitch and accent characteristics of the original speaker and is language-independent. This superior speech quality, coupled with the extremely efficient storage of speech, has made LPC a very popular speech synthesis technique.

Formant Analysis/Synthesis

Formant analysis/synthesis is similar to LPC since it is based on the frequency spectrum found in natural speech and utilizes the same type of synthesizer circuit to model the human vocal tract. As the name implies, formant analysis/synthesis attempts to generate speech by reconstructing the formants that exist in the speech waveform.

Recall that voiced sound consists of several resonant frequencies called formants. As you speak, the formant frequencies are constantly shifting to produce different sounds. The formant frequency characteristics of a spoken waveform can be digitally coded and used to control frequency generators and filters in an electronic synthesizer to reproduce the original speech waveform.

The digitized speech in formant analysis/synthesis can take on two different forms. The original speech formants can be coded and synthesized one word at a time as required. In this way, individual words are stored and played back to produce connected speech. This is called the ***stored-word***, or ***dictionary***, method of formant analysis/synthesis and is similar to the LPC technique. The obvious disadvantage of this method is that the entire vocabulary is fixed and limited by the amount of memory available. Typical data rates are comparable to LPC and range from 1000 to 2000 bps.

Since the formant frequencies produce the fundamental sounds of speech, suppose you code all the formants that create the unique speech sounds, store them in memory, and then put them together as needed to produce any given speech segment. This way the speech vocabulary would be unlimited, and only a small amount of memory would be required to store the limited number of speech sounds. Recall that this is the idea behind artificial constructive/synthesis.

As you are aware, the sounds of the English language can be broken down into about 40 fundamental units called phonemes. In addition, there are about 130 phoneme variations called allophones. Each allophone has a unique frequency spectrum made up of several formant bands. Thus, to produce an allophone electronically, all you need to do is generate the unique formant frequencies for that particular sound. The generated allophonic sounds can then be strung together with software to produce connected speech. This is the basic idea behind phoneme speech synthesis, which, as you can see, is clearly a form of formant synthesis.

As stated earlier, the advantages of storing phonemes and allophones rather than words are memory conservation and the flexibility of an unlimited vocabulary.

This is particularly important for robot speech synthesis. Phoneme speech synthesis is capable of producing connected speech with a data rate of only 400 to 600 bps, as compared to 1000 to 2000 bps for other frequency-domain analysis/synthesis techniques.

The disadvantage of phoneme synthesis is speech quality. Software must be written to connect the allophonic sounds and provide the prosodic features required to make the speech sound natural. Even then, phoneme speech synthesis "sounds" machinelike and artificial. In addition, phoneme synthesis is language-dependent, since a different set of allophones must be used for different languages. Nevertheless, for many applications, including robotics, the flexibility and memory conservation benefits of phoneme speech synthesis outweigh its relatively poor speech quality and language dependency. This leads us to the next section of this chapter, which is devoted entirely to a discussion of phoneme synthesis.

4-4 PHONEME SPEECH SYNTHESIS

With phoneme synthesis, all the various speech sounds, or allophones, of a given language are coded in memory. The allophonic sound codes are usually coded in the LPC format and stored in a ROM lookup table. Address codes, called phoneme codes, are then applied to the ROM, which translates them to the LPC parameters using the lookup table. The LPC parameters are subsequently applied to the synthesizer circuit to generate the respective allophonic sounds.

Most phoneme synthesizers are really LPC synthesizers since they operate using LPC parameters. In fact, many times the same synthesizer circuit can be used with either a phoneme constructive/synthesis system or an LPC analysis/synthesis system since both operate off the LPC code. The synthesizer consists of an excitation source and digital filter structure that emulates the human vocal tract. As a result, they are sometimes called ***source-filter*** models of the human vocal tract. A block diagram of a typical phoneme synthesizer is shown in Figure 4-14. Notice that it consists of three major sections: a ***lookup ROM***, an ***excitation source***, and a ***multistage digital filter***. When a phoneme code is applied to the synthesizer, the lookup ROM translates the code into a set of LPC parameters that is applied to the excitation sources and the digital filter. The LPC parameters control which excitation source is selected, its pitch, and the filter settings that are required to produce the given phoneme.

A phoneme speech synthesizer can be used in one of two ways: for ***direct*** speech synthesis or ***text-to-speech*** synthesis. In direct speech synthesis a computer is programmed to generate a series of phoneme codes to the synthesizer ROM when a given event occurs. For example, a mobile robot might be programmed to say "low voltage" when its batteries need recharging. Such a program would most likely be a subroutine consisting of the phoneme code string required to speak the phrase. The subroutine could be stored in RAM or ROM. It would be executed when voltage-sensing circuits detected the low-voltage condition.

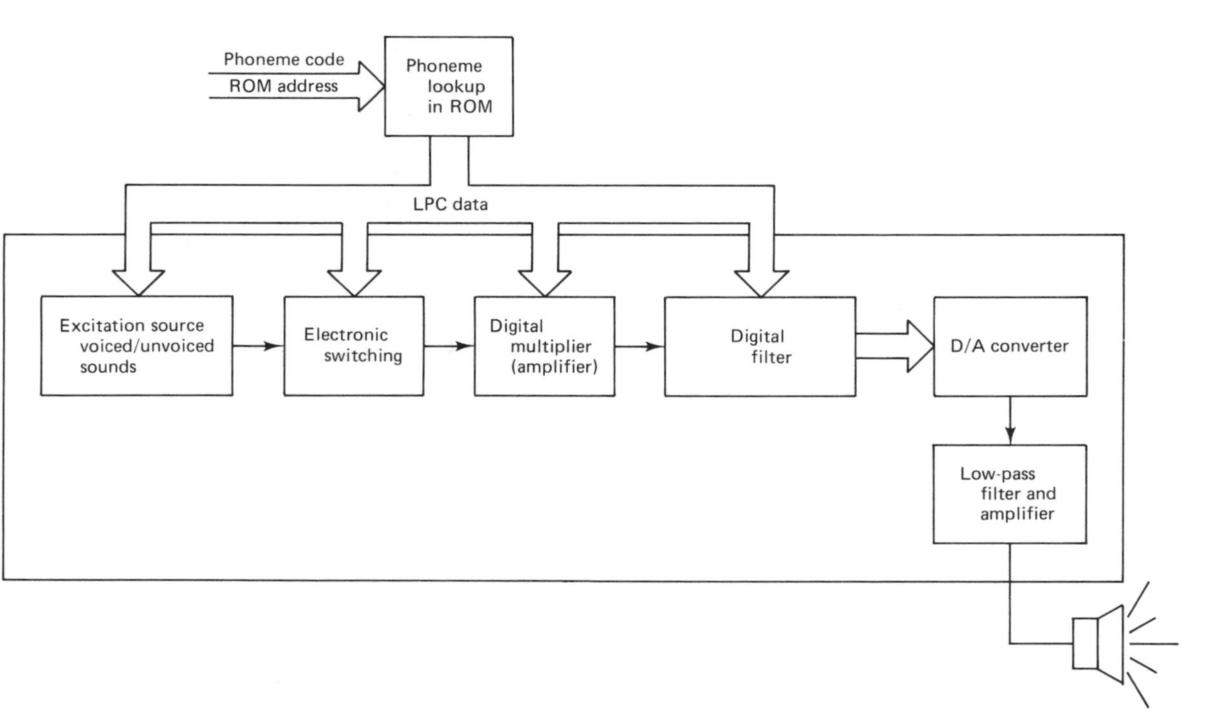

Figure 4-14 A phoneme synthesizer circuit is the same as an LPC synthesizer circuit since both operate off the LPC code.

Another use for phoneme synthesis is for text-to-speech conversion. As the name implies, written text is presented to a computer, which must translate the written symbols into the respective phoneme codes. The phoneme codes are then applied to a phoneme synthesizer circuit for audible reproduction of the written text. A popular application for this type of phoneme synthesis allows the blind to read books without using braille. Let's explore these two types of phoneme synthesis in more detail.

Direct Phoneme Synthesis

When programming a computer or robot for direct phoneme synthesis, you must determine the phoneme codes required for a given phrase. These codes are called a ***phoneme string*** and are usually stored as part of a speech subroutine in RAM or ROM. The subroutine is executed when the programmed phrase must be spoken. Many times, the speech subroutine is part of an interrupt service routine that is activated as the result of a given event. Let's now learn how to program a typical phoneme synthesizer so that you get an idea of what is involved in the programming task. Since phoneme synthesizers are similar, the following discussion will teach you the basics of writing phoneme strings for most commercial phoneme synthesizers.

A typical list of phonemes for a commercial synthesizer is provided in Tables 4-2 and 4-3. The phonemes in Table 4-2 are listed in numerical order, according to their respective phoneme codes. In Table 4-3, the same phonemes are divided into the various sound classifications discussed earlier in this chapter.

There are 64 phonemes/allophones in the list, each with an associated 6-bit hexadecimal phoneme code, which is shown in the left column of Table 4-2. These are the codes that you must program into your speech subroutine to form the phoneme strings. When the subroutine is executed, the phoneme code is passed to the synthesizer ROM, where a lookup table produces the associated LPC code for the synthesizer circuit. This process is illustrated in Figure 4-15.

The phoneme symbols are listed next to the phoneme codes in Table 4-2. The symbols are a mnemonic type of code for you, the programmer, that provide an alphabetical representation of the phonemes and allophones. When writing phoneme strings, you will first write a string of phoneme symbols and then convert the symbols to a phoneme code string for entry into a computer.

Observe also from Table 4-2 that each phoneme has an associated duration specified in milliseconds. This is the time it takes the synthesizer to pronounce the given phoneme sound. Finally, Table 4-2 provides an example word for each phoneme. This word shows you how the phoneme will be pronounced by the synthesizer.

Developing Phoneme Strings. To create a phoneme string, you must perform the following five tasks:

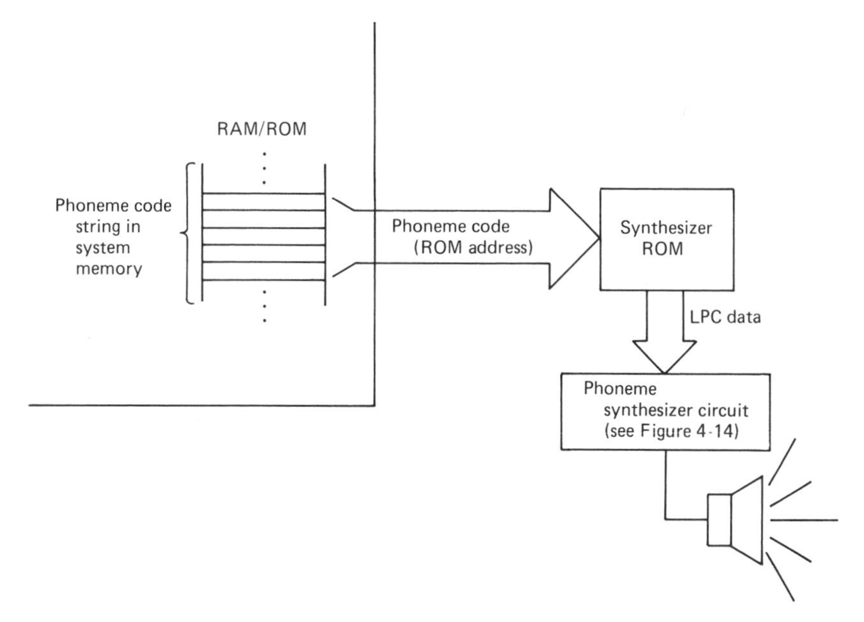

Figure 4-15 The required phoneme code string is stored in system memory and then passed to the synthesizer ROM where a lookup table generates the associated LPC code for the phoneme synthesizer.

1. Determine the phoneme symbol string required for the given words within a phrase.
2. Provide pauses between syllables and words as needed for timing and rhythm.
3. Provide intonation for the individual words as well as the entire phrase.
4. Convert the phoneme symbol string to a phoneme code string.
5. Execute the phoneme string, listen to the results, and modify accordingly.

Phoneme symbol string. The first thing you must do is to determine the phoneme symbol string. As an example, consider the following phrase: "This text is great!?" Take each word in the phrase separately and look it up in a general word dictionary. A good dictionary will provide a pronunciation of the word in parentheses directly to the right of the word. *Webster's New World Dictionary of the American Language* shows the pronunciation of the preceding words as follows:

Word	Pronunciation
this	t̲his
text	tekst
is	iz
great	grāt

TABLE 4-2 Typical Phoneme/Allophone Chart (*Courtesy VOTRAX*)

Hexadecimal Phoneme Code	Phoneme Symbol	Duration (ms)	Example Word	Hexadecimal Phoneme Code	Phoneme Symbol	Duration (ms)	Example Word
ØØ	EH3	59	jacket	20	A	185	day
Ø1	EH2	71	enlist	21	AY	65	day
Ø2	EH1	121	heavy	22	Y1	80	yard
Ø3	PA0	47	no sound	23	UH3	47	mission
Ø4	DT	47	butter	24	AH	250	mop
Ø5	A2	71	made	25	P	103	past
Ø6	A1	103	made	26	O	185	cold
Ø7	ZH	90	azure	27	I	185	pin
Ø8	AH2	71	honest	28	U	185	move
Ø9	I3	55	inhibit	29	Y	103	any
ØA	I2	80	inhibit	2A	T	71	tap
ØB	I1	121	inhibit	2B	R	90	red
ØC	M	103	mat	2C	E	185	meet
ØD	N	80	sun	2D	W	80	win
ØE	B	71	bag	2E	AE	185	dad
ØF	V	71	van	2F	AE1	103	after
1Ø	CH*	71	chip	30	AW2	90	salty
11	SH	121	shop	31	UH2	71	about
12	Z	71	zoo	32	UH1	103	uncle
13	AW1	146	lawful	33	UH	185	cup
14	NG	121	thing	34	O2	80	for
15	AH1	146	father	35	O1	121	aboard
16	001	103	looking	36	IU	59	you
17	00	185	book	37	U1	90	you
18	L	103	land	38	THV	80	the
19	K	80	trick	39	TH	71	thin
1A	J*	47	judge	3A	ER	146	bird
1B	H	71	hello	3B	EH	185	get
1C	G	71	get	3C	E1	121	be
1D	F	103	fast	3D	AW	250	call
1E	D	55	paid	3E	PA1	185	no sound
1F	S	90	pass	3F	STOP	47	no sound

TABLE 4-3 Phonemes/Allophones by Sound Categories (*Courtesy VOTRAX*)

Pure Voiced					Voiced Fricative	Voiced Plosive	Unvoiced Plosive	Unvoiced Fricative	Nasal	No Sound
E	EH	AE	UH	OO1	Z	B	T	S	M	PA0
E1	EH1	AE1	UH1	R	ZH	D	DT	SH	N	PA1
Y	EH2	AH	UH2	ER	J	G	K	CH	NG	STOP
Y1	EH3	AH1	UH3	L	V		P	TH		
I	A	AH2	O	IU	THV			F		
I1	A1	AW	O1	U				H		
I2	A2	AW1	O2	U1						
I3	AY	AW2	OO	W						

You will want to check the Key to Pronunciation at the beginning of the dictionary to interpret the pronunciation symbols.

Next, using the dictionary pronunciation as a guide, look up the phoneme symbols in Table 4-2 for each word sound. According to Table 4-2, the phoneme symbols for the preceding words are as follows:

Word	Pronunciation	Phoneme Symbols
this	this	TH, I2, S
text	tekst	T, EH1, K, S, T
is	iz	I, Z
great	grāt	G, R, A1, T

The phoneme string for the phrase is therefore

$$\underbrace{\text{TH, I2, S,}}_{\text{This}}\ \underbrace{\text{T, EH1, K, S, T,}}_{\text{text}}\ \underbrace{\text{I, Z,}}_{\text{is}}\ \underbrace{\text{G, R, A1, T}}_{\text{great}}$$

Notice how the dictionary pronunciation symbols help you to determine which phoneme to use (particularly the *k* in *tekst* and the *ā* in *grāt*). Also, Table 4-2 provides several timing options for some of the phonemes. The *i* sound in *this* and the *e* sound in *tekst* both have three timing options. The *i* sound could be created with the phonemes I, I1, I2, or I3. Each has a different duration, with I3 the shortest and I the longest. Likewise, the options for the *e* sound in *tekst* are EH, EH1, EH2, EH3.

Example 4-1

Determine the phoneme string required to synthesize the following phrase: "Voice synthesis is technical."

Solution

According to the dictionary, these words must be pronounced as follows:

Word	Pronunciation
voice	vois
synthesis	sin'thə-sis
is	iz
technical	tek'ni-k'l

From Table 4-2, you find the respective phonemes to be as follows:

Word	Pronunciation	Phoneme
voice	vois	V, O1, UH3, E1, S
synthesis	sin'thə-sis	S, I, N, TH, I1, S, I1, S
is	iz	I, Z
technical	tek'ni-k'l	T, EH1, K, N, I3, K, UH3, L

As a result, the phoneme symbol string is

V, 01, UH3, E1, S, S, I, N, TH, I1, S, I1, S, I, Z, T, EH1, K, N, I3, K, UH3, L

voice — synthesis — is — technical

Pauses. From the list of phonemes in Table 4-3, you find two pauses: PA0 and PA1. The PA0 pause produces no sound for 47 ms; PA1 produces a 185-ms pause. The short pause might be required between the syllables of some words, whereas the long pause is usually inserted between words within a phrase. Since our example phrase does not contain any multisyllable words, the following phoneme string only shows pauses inserted between the individual words.

PA1, TH, I2, S, PA1, T, EH1, K, S, T, PA1, I, Z, PA1, G, R, A, T, PA1

^ this ^ text ^ is ^ great ^

For clarity, a ^ symbol is used to show the pause insertions.

Additional pauses are required when phrases are strung together. Punctuation such as commas, periods, and colons will usually dictate more pauses. To create longer pauses, as for a period at the end of a sentence, you simply string pauses together by inserting several pause phonemes.

Another use for the pause is to emphasize words or sounds within a phrase. When you and I speak, we emphasize a word by increasing the volume of that word so that it stands out from the other words. Unfortunately, most simple phoneme synthesizers do not accommodate programmed volume control. As a result, word emphasis must be accomplished using pauses. To make a word stand out, simply increase the number of pauses around the word. In our example, the word *great* could be emphasized by using two or three pauses before and after the word, rather than just one.

Other uses of pauses are less defined, but still important for natural sounding speech. In general, pauses must be used to control the timing, or rhythm, of the synthesized speech.

Example 4-2

Modify the phoneme string in Example 4-1 to create pauses between the words and to emphasize the word *is*.

Solution

Use a PA1 phoneme between each individual word to create the word pauses. To emphasize the word *is*, place three PA1 pauses on each side of the word. The resulting phoneme symbol string would be as follows:

PA1, V, O1, UH3, E1, S, PA1, S, I, N, TH, I1, S, I1, S, PA1, PA1, PA1, I, Z, PA1,

^ voice ^ synthesis ^ ^ ^ is ^

PA1, PA1, T, EH1, K, N, I3, K, UH3, L, PA1

^ ^ technical ^

Intonation. Recall that intonation has to do with the pattern of pitch changes within a spoken phrase. Proper intonation not only helps the synthesized speech sound natural, but, more importantly, it often determines the meaning of a phrase. For instance, our example phrase can be understood three ways, depending on its intonation. The phrase can be a statement, exclamation, or question, as follows:

This text is great.
This text is great! (Author's note: I like this one!)
This text is great?

What determines the meaning of the phrase is its intonation.

Commercial phoneme synthesizers are available with programmable intonation. The intonation is programmed by providing an inflection, or pitch, level code with each phoneme within the phoneme string. Simple synthesizers provide four levels of inflection, while advanced devices provide up to 64 levels of programmable inflection. Let's assume that our synthesizer is relatively simple and has only four levels of inflection, or pitch. Thus, the four inflection levels can be designated as levels 0 through 3, using a 2-bit binary code of 00 through 11. This idea is illustrated by Figure 4-16.

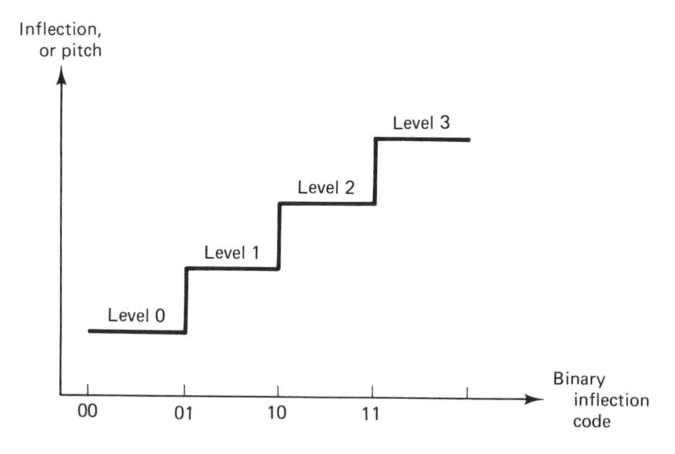

Figure 4-16 Simple phoneme synthesizers use four levels of inflection and a corresponding 2-bit inflection code.

To help program the inflection levels, it is a good idea to construct a graphical device known as an ***intonation bar graph.*** This graph is used something like a music score to illustrate the pitch changes within a phrase. The graphs in Figure 4-17 plot the pitch changes that are necessary for the three different meanings of our example phrase. In addition, the pitch changes for each individual word are plotted. Notice that each graph is divided into four levels, representing the four levels of programmable inflection.

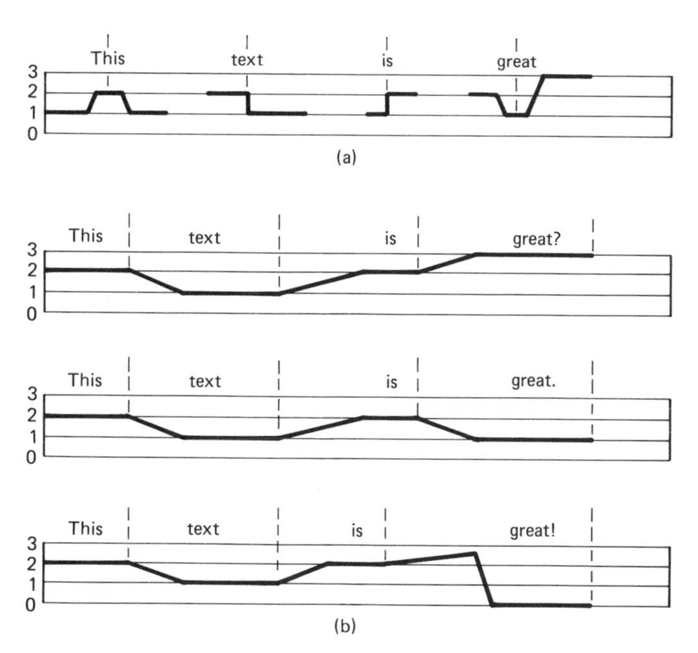

Figure 4-17 Intonation bar graphs illustrate pitch changes within (a) individual words and (b) phrases.

To designate the pitch, or inflection, levels within a phoneme string, place the inflection level obtained from the bar graph before the phoneme symbol, followed by a slash (/). The slash is used to separate the inflection level from the symbol. For instance, to ask the question "This text is great?", the phoneme string would be

$$\underset{\wedge}{\text{PA1}},\ \underbrace{\text{1/TH, 2/I2, 1/S}}_{\text{This}},\ \underset{\wedge}{\text{PA1}},\ \underbrace{\text{2/T, 1/EH1, 1/K, 1/S, 1/T}}_{\text{text}},\ \underset{\wedge}{\text{PA1}},\ \underbrace{\text{1/I, 2/Z}}_{\text{is}},\ \underset{\wedge}{\text{PA1}},$$

$$\underbrace{\text{2/G, 1/R, 3/A, 3/T}}_{\text{great?}},\ \underset{\wedge}{\text{PA1}}$$

Observe that the individual word inflection levels are taken directly from Figure 4-17(a). Furthermore, because the phrase is a question, the pitch is raised at the end of the phrase, as shown by the graph in Figure 4-17(b). If the phrase were an exclamation, the pitch must be lowered at the end of the phrase.

Example 4-3

Construct an intonation bar graph for the following question: "Is voice synthesis technical?"

Solution

First, construct a bar graph for each individual word within the phrase. To do this, say the word aloud and listen to the pitch changes within the word. Plot these changes on the bar graph as if you were writing music. The resulting word bar graph is shown in Figure 4-18(a).

Next construct a bar graph for the entire phrase. Say the entire phrase aloud and plot the relative pitch changes. Since the phrase is a question, the overall pitch increases at the end of the phrase, as shown in Figure 4-18(b).

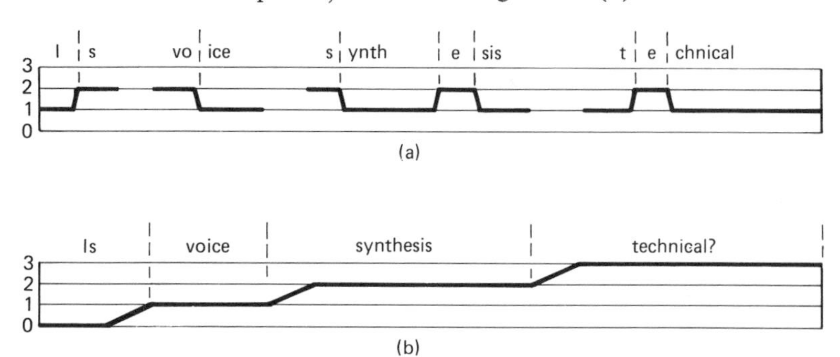

Figure 4-18 Intonation bar graphs for Example 4-3: (a) individual word graph; (b) phrase graph.

Example 4-4

Using the bar graphs you developed in Example 4-3, write a phoneme symbol string for the given phrase.

Solution

You determined the phoneme symbols for each of the words within the phrase in Example 4-1. Using these symbols, adding pauses, and using the bar graphs in Example 4-3 results in the following string:

$$\underset{\wedge}{\text{PA1}},\ \underbrace{\text{1/I, 2/Z}}_{\text{Is}},\ \underset{\wedge}{\text{PA1}},\ \underbrace{\text{2/V, 1/01, 1/UH3, 1/E1, 1/S}}_{\text{voice}},\ \underset{\wedge}{\text{PA1}},$$

$$\underbrace{\text{2/S, 1/I, 1/N, 1/TH, 2/I1, 1/S, 1/I1, 1/S}}_{\text{synthesis}},\ \underset{\wedge}{\text{PA1}},\ \underbrace{\text{1/T, 2/EH1}}_{\text{te}},$$

$$\underbrace{\text{2/K, 2/N, 2/I3, 3/K, 3/UH3, 3/L}}_{\text{chnical?}},\ \underset{\wedge}{\text{PA1}}$$

Code conversion. Now that you have developed the complete phoneme string, it is time to convert the phoneme symbols into their corresponding phoneme codes. Recall that the phoneme code in Table 4-2 is a 6-bit code. To provide intonation, you must add the 2-bit inflection code to produce an 8-bit value. The required bit pattern definition is shown in Figure 4-19. Here the lower six bits are the

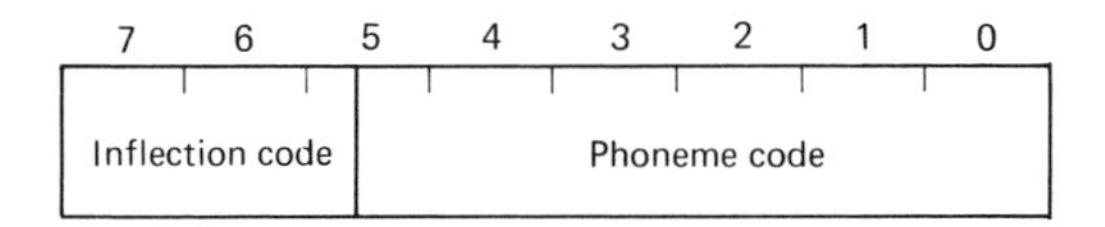

Figure 4-19 An 8-bit phoneme code consists of a 6-bit code from Table 4-2 and a 2-bit inflection code.

phoneme codes from Table 4-2, and the two most significant bits are the inflection code.

Earlier we developed the phoneme symbol string for the question "This text is great?" Converting the symbol string into a string of phoneme codes results in the following:

	PA1	= 00 111110 = 3E
	1/TH	= 01 111001 = 79
This	2/12	= 10 001010 = 8A
	1/S	= 01 011111 = 5F
	PA1	= 00 111110 = 3E
	2/T	= 10 101010 = AA
	1/EH1	= 01 000010 = 42
text	1/K	= 01 011001 = 59
	1/S	= 01 011111 = 5F
	1/T	= 01 101010 = 6A
	PA1	= 00 111110 = 3E
	1/I	= 01 100111 = 67
is	2/Z	= 10 010010 = 92
	PA1	= 00 111110 = 3E
	2/G	= 10 011100 = 9C
	1/R	= 01 011011 = 5B
great?	3/A	= 11 100000 = E0
	3/T	= 11 101010 = EA
	PA1	= 00 111110 = 3E

Execution and modification. The next step is to enter the phoneme code string into a computer for execution. You will literally hear the results. However, don't be disappointed if the programmed speech does not sound quite the way you expected it to sound. I call this the debugging stage, since the phoneme string must usually be modified several times to get acceptable and natural sounding speech.

As you can see, programming a computer for direct phoneme synthesis is quite a tedious task. Wouldn't it be much easier to simply type in your phrases and have the computer assemble the required phoneme code? As you can well imagine, such a computer would have to possess a certain degree of intelligence to interpret the written text. This is the topic of our next discussion.

Text-to-Speech Conversion

Let's review, for a minute, the process that you just went through to write a phoneme string. Looking at a written phrase, you decided which phoneme codes were required to pronounce the words within the phrase. Then you added the appropriate pauses and intonation to make the synthesized phrase sound natural. The fundamental intelligent task was to convert the written English symbols into the appropriate phoneme code string. Thus, the entire process reduces to a code conversion.

As you are aware, computers can be programmed to provide code conversions rather efficiently. So why not enter phrases into a computer by means of a keyboard and let the computer perform the code conversion? Since most computers represent letters and symbols using the ASCII code, the program task reduces to converting ASCII code to phoneme code. This is called ***text-to-speech conversion***. The idea sounds simple, but it is complicated by the various ways in which words must be pronounced within a phrase to convey the intended meaning. Interpretation, intonation, and emphasis all provide a challenge to the software. For these reasons, many text-to-speech converters incorporate the principles of AI production systems to produce acceptable translations.

There are basically three ways that written text can be converted to phoneme code strings: ***word lookup, morpheme lookup,*** or ***phoneme lookup.*** As you will soon discover, each has its own advantages and disadvantages. As a result, most commercial systems incorporate some combination of these three techniques to produce accurate conversions. Let's take a brief look at each of these conversion techniques.

Word Lookup. This technique is sometimes called the ***dictionary*** method. It is the simplest text-to-speech conversion technique, but it is the least flexible and requires the most memory. From its name, you have probably guessed that whole words are stored in computer memory, and you are right. When text is input to the computer, the software simply looks for the ASCII representation of a space to divide up the phrase into individual words. Each individual word is then compared to the list of stored dictionary words until a match is found. Once a match is made (if it ever is), a lookup table produces the phoneme code string required to pronounce the word. The phoneme code strings are sequentially passed to a phoneme synthesizer for immediate speech reproduction or temporarily stored in a phoneme memory buffer for subsequent playback.

The obvious disadvantage of word lookup is the size of the stored dictionary. As with other dictionary-type systems, vocabulary size is a direct function of dictionary size. You can increase the vocabulary by increasing the size of the dictionary, but this creates another problem: large dictionaries require too much time to search for a given word. In addition, an abbreviation, misspelled word, or unusual word might never be found, resulting in an error condition.

Morpheme Lookup. A ***morpheme***, or simply ***morph***, is any word or word fragment that conveys meaning. In other words, a morph is the smallest combination of letters or symbols that means something. Examples of morphs include the *sun* in sundown, *ortho* in orthopedic, and *blue* in blueberry. It has been shown that about 8000 morphs are required to form most of the words in the English language.

A morph lookup system works like a word lookup system in that the morphs are stored in memory. The input text is analyzed and divided into morphs for subsequent matching to the dictionary. Here is where the problem arises: the input text must be dissected and analyzed to produce the appropriate morph string. To do this, the software must look at all the possible ways that a given word can be broken up in order to find the respective morphs. Once the proper morph string is determined, the morph dictionary must be searched to produce the corresponding phoneme code string. This analysis and searching process generally requires a relatively large amount of computer time and is inefficient, especially in microprocessor-based systems.

The advantage of a morph system over a word lookup system is flexibility. Only 8000 or so morphs need to be stored to obtain a very large vocabulary. New and unusual words rarely need to be added to the dictionary, since in most cases they will consist of existing morphs.

Phoneme Lookup. Phoneme lookup is probably the most efficient and flexible text-to-speech conversion technique available. This technique is sometimes referred to as letter-to-phoneme lookup, since the software attempts to convert each individual text letter or symbol to its corresponding phoneme.

A system developed by the Naval Research Laboratory (NRL) uses approximately 400 production rules to convert written text into phonemes. Each production rule has the following format:

IF ⟨left context (text character) right context⟩ THEN ⟨phonemes⟩

Notice that the conditional part of the rule consists of a text character(s) along with a left and right context. The character portion of the rule must match the current character(s) being translated by the rule. The left and right contexts specify what must appear to the left and right of the current text character(s) in order for the rule to fire. For instance, consider the following rule:

IF #:(AL)! THEN UH,L

Here the characters to be matched are the letters AL. The left context is represented in symbols by a colon and a pounds sign (#:). The right context is represented by a single exclamation mark (!). These symbols refer to subroutines that are called during the matching process to examine the context on either respective side of the current character. The preceding context symbols are defined as follows:

\# Context must be one or more vowels.

: Context must be zero or more consonants.

! Context must be a nonalphanumeric character.

Thus, the left context symbol, #:, means that the character(s) immediately to the left of the current character, AL, must be zero or more consonants, followed by one or more vowels, from right to left. The right context symbol, !, means that the character immediately to the right of the current character must be a nonalphabetic character such as a space.

If you apply this rule to the word FICTIONAL you will see that all the rule conditions are met for the AL characters at the end of the word. Notice that the context to the left of AL is ION, which satisfies the left context conditions (#:). The context to the right of AL is a space, which satisfies the right context condition (!).

The action part of the production rule provides the phonemes required by the phoneme synthesizer. Once all the rule conditions are met, the rule is fired and the respective phoneme codes are temporarily stored in a phoneme buffer for subsequent transmittal to a phoneme synthesizer. In our example, the phoneme codes for the phonemes UH,L would be placed in the phoneme buffer.

Example 4-5

Given the following text-to-phoneme production rule:

IF !:(E)! THEN E1

For which letter or group of letters in the following sentence will the rule be fired?

THIS TEXT IS THE GREATEST!

Assume the following context symbol definitions:

! Context must be a nonalphabetic character.
: Context must be zero or more consonants.

Solution

The rule is for the letter E. As a result, the candidates for this rule in the sentence are the E's in the words T<u>E</u>XT, TH<u>E</u>, and GR<u>E</u>AT<u>E</u>ST. However, before the rule can fire, the context conditions must be met.

The E's in the words TEXT and GREATEST all fail, since the right context is not a nonalphabetic character as required by the right context symbol (!). However, the E in the word THE satisfies both the right and left context conditions. Notice that the right context for this word is a space, satisfying the right context condition (!). The left context symbols, !:, require the left context to be zero or more consonants, followed by a nonalphabetic character. The TH in the word THE satisfies this condition. Consequently, the only letter in the sentence that will cause the given rule to be fired is the E in the word THE.

The rules listed in Table 4-4 are used with a commercial text-to-phoneme synthesizer. These rules were derived from the set of 400 rules developed by the NRL. The conditional part of the rule contains the character(s) and context requirements, while the action part of the rule provides the corresponding phoneme symbols. Notice that these are the same phoneme symbols that appear in Table 4-2.

TABLE 4-4 Typical Set of Text-to-Phoneme Production Rules

A	B	C

IF (A)! THEN UH2
IF (AR) THEN AH1, R
IF #:(AL)! THEN UH,L
IF (AL)$ THEN AW1,UH3,L
IF (A) THEN AE1
	IF (B) THEN B	IF (CH) THEN T,CH
		IF (C)+ THEN S
		IF (C) THEN K

D	D	F

IF (D) THEN D	IF #:(E)! THEN ;
	IF !:(E)! THEN E1
	IF #:(E)D! THEN ;
	IF (ER) THEN ER
	IF #:(E)S! THEN ;
	IF (EE) THEN E
	IF (EA) THEN E
	IF (E) THEN EH1
		IF (F) THEN F

G	H	I

IF (G) THEN G	IF (H)# THEN H
	IF (H) THEN ;
		IF !(IN) THEN I1,N
		IF (I)$+:# THEN I
		IF (I)$+ THEN AH2,I2
		IF (I) THEN I1

J	K	L
IF (J) THEN D,J	IF (K) THEN K	IF (L)L THEN ; IF (L) THEN L
M	N	O

IF (M) THEN M	IF (NG) THEN NG
	IF (N) THEN N
		IF (OF)! THEN UH2,V
		IF (OR) THEN AW,R
		IF I(ON) THEN UH2,N

P	Q	R
IF (P) THEN P	IF (QU) THEN K,W IF (Q) THEN K	IF (R) THEN R
S	T	U

IF (SH) THEN SH
IF #(SED)! THEN Z,D
IF S(S) THEN ;
IF (S) THEN S
	IF !(THE)! THEN THV,UH2
	IF (TO)! THEN T,IU,U1
	IF (THAT)! THEN THV,AE,T
	IF (TH) THEN TH
	IF (TI)O THEN SH
	IF (T) THEN T
		IF (U)$$! THEN UH1
		IF (U) THEN Y1,IU,U1

V	W	X
IF (V) THEN V	IF !(WAS)! THEN W,AH1, IF (W) THEN W	IF (X) THEN K,S
Y	Z	PAUSE
IF #:$(Y)! THEN 2E1 IF (Y) THEN 3I1	IF (Z) THEN Z	IF ( ) THEN PA1

Context Symbols:

- ! Context must be a nonalphabetic character.
- : Context must be zero or more consonants.
- $ Context must be one consonant.
- # Context must be one or more vowels.
- \+ Context must be E, I, or Y.
- . Context must be a voiced consonant.
- ; Terminate rule; go to next rule or next text character.

From Table 4-4 you can see that there are one or more rules for each letter in the alphabet. Some of the rules refer to whole words, such as THE, TO, THAT, and WAS, which are very common in written text. As a result, Table 4-4 is a combination word and phoneme lookup table. When a text letter such as T is encountered, the T-word rules are tested first, followed by (TH), (TI)O, and (T), until a match is made. Of course, the last rule (T) will always assure a match.

Notice also that Table 4-4 makes provision for pauses with the ( ) = PA1 rule. This means that a PA1 pause will be inserted in the phoneme string for each space encountered in the written text.

Additional rules can be added to recognize punctuation marks such as a comma, period, exclamation mark, and question mark. Punctuation mark rules are sometimes used to raise or lower the pitch of the synthesized phrase to provide more realistic intonation. Another way to handle punctuation is to develop rules that cause the system to actually speak the punctuation. This method is less efficient, but assures that the proper meaning is conveyed.

Abbreviations and acronyms can also be handled with special production rules. One way is for the system to speak the individual letters in an abbreviation containing all consonants and/or numbers. Abbreviations containing one or more vowels are pronounced as individual words. Using this idea, the abbreviation RS-232C would be pronounced as R, S, dash, 2, 3, 2, C, while the abbreviation ASCII would be pronounced as a whole word, *asky*. But what about abbreviations such as USA, CPU, ALU, or even AI? Such abbreviations would require special rules if they occur frequently within the written text.

SUMMARY

Human speech is created in the vocal cavity by generating a complex waveform of resonant frequencies called formants. These frequencies range from 10 Hz to over 15 kHz with a volume ratio of 16,000 : 1. However, acceptable speech can be reproduced electronically using a frequency range from 300 to 3300 Hz and a volume ratio of 1000 : 1.

The sounds of speech include voiced sounds and unvoiced sounds. Voiced sounds are generated by the vocal cords and subsequently modulated by the vocal cavity. Unvoiced sounds, or white-noise sounds, are created by holding the vocal cords open and forcing air into the vocal cavity. Both voiced and unvoiced sounds are subdivided into several sound subcategories according to the way in which the sound is modulated by the vocal cavity (see Table 4-1). During speech, voiced and unvoiced sounds are strung together, and prosodic features such as pitch, amplitude, and timing are added to convey meaning to the basic sounds.

The two fundamental techniques used for electronic speech synthesis are natural analysis/synthesis and artificial constructive/synthesis. Natural analysis/synthesis involves the analog or digital recording and subsequent playback of natural human speech, whereas constructive/synthesis attempts to artificially piece together speech using speech sounds called phonemes.

Digital analysis/synthesis is accomplished using either time-domain or frequency-domain techniques. Time-domain analysis attempts to digitize the amplitude variations of the speech waveform using simple pulse-code modulation (PCM), delta modulation, differential pulse-code modulation (DPCM), or adaptive differential pulse-code modulation (ADPCM). Frequency-analysis techniques attempt to code the frequency spectrum of the spoken waveform using linear predictive coding (LPC) or formant analysis/synthesis.

Phoneme speech synthesis is also a frequency technique that uses the principles common to both LPC and formant synthesis. The sounds of the English language can be broken down into about 40 fundamental units called phonemes. In addition, there are about 130 phoneme variations called allophones. With phoneme speech synthesis, these speech sounds are coded in memory using the LPC format and subsequently recalled to construct words and phrases. Thus, phoneme synthesis is a form of constructive analysis/synthesis.

Phoneme synthesis is done one of two ways: direct or text-to-speech. With direct phoneme synthesis, a phoneme code string is programmed for a given phrase, stored in RAM or ROM, and then passed to a phoneme synthesizer circuit when the phrase is to be spoken. The difficulty with this is that you, the programmer, must add the proper intonation and timing for the synthesized speech to sound natural.

As the name implies, text-to-speech conversion translates written text, represented by ASCII code, into phoneme strings. The three ways that this can be done are by word lookup, morpheme lookup, or phoneme lookup. Most commercial systems incorporate some combination of these three techniques and utilize the principles of AI production systems to produce acceptable translations.

Digital techniques that involve the use of LPC produce the most natural sounding speech, since the electronic synthesizer models the human vocal tract.

PROBLEMS

4-1 Explain how voiced sounds are created by a human speaker.

4-2 What frequency range and volume ratio will produce acceptable speech with a speech synthesizer?

4-3 What is the purpose of a sound spectrograph?

4-4 Which speech sound produces a tiny explosion within the vocal cavity?

4-5 Describe the difference between a voiced fricative and an unvoiced fricative. List at least three examples of each sound.

4-6 List the features of speech that convey meaning to a word or phrase.

4-7 Explain the fundamental difference between natural speech analysis/synthesis and artificial constructive/synthesis.

4-8 Another name for time-domain analysis/synthesis is ______________ .

4-9 Describe the fundamental differences between PCM, delta modulation, DPCM, and ADPCM.

4-10 Suppose an 8-bit A/D converter is used to sample a speech waveform at a rate of 5000 samples per second.

(a) Calculate the data rate.

(b) How many bytes of memory are required to store 10 seconds of speech with simple PCM?

(c) How many bytes of memory are required to store 10 seconds of speech using delta modulation?

(d) How many bytes of memory are required to store 10 seconds of speech using a 6-bit DPCM system and the same sampling rate?

(e) How many bytes of memory are required to store 10 seconds of speech using a 3-bit ADPCM system and the same sampling rate?

4-11 Describe the major functional regions of a speech synthesizer that is designed to operate using the LPC code.

4-12 What circuit functions in an electronic speech synthesizer are controlled by the LPC code?

4-13 Name the two types of phoneme synthesis.

4-14 List the steps that must be performed to create a phoneme string for a given phrase.

4-15 Using Table 4-2, determine the fundamental phoneme string required to synthesize the following phrase: "The United States of America."

4-16 Modify the phoneme string in Problem 4-15 to create pauses between the words and to emphasize the word *United.*

4-17 Construct intonation bar graphs for the phrase in Problem 4-15, assuming that a period is present at the end of the phrase.

4-18 Add intonation to your phoneme string using the bar graphs of Problem 4-17.

4-19 List the three fundamental types of text-to-speech conversion and describe the difference between them.

4-20 Using Table 4-4, determine which rules will fire to synthesize the word "fictional."

5 Speech Recognition and Understanding

Now that you have learned about speech synthesis, it is time to discuss the second half of voice communication: speech recognition and understanding. Imagine a computer world with no keyboards, where all programming and operational commands are provided by voice entry. In other words, rather than a keyboard interface, all computers are equipped with a voice interface. In such a world, the product inspector can enter data verbally, leaving his or her hands free to perform production operations. An industrial robot can be ''told'' what to do. An office manager can dictate memos directly into a computer for immediate printout and distribution. In such a world the blind or physically handicapped would no longer be handicapped in the field of computer science. In fact, when you really think about it, the keyboard is one of the biggest handicaps to computer users and the sole obstacle for many potential computer applications. For this reason, a large amount of research and development is presently taking place in the field of speech recognition and understanding.

As the years go by, you will see keyboards gradually disappear and be replaced with voice interfaces. Many predict that speech will be the primary I/O channel used by the next generation of computers. In fact, the telephone will eventually become the universal computer terminal. This transition will not take place overnight. Rather, voice interfaces will gradually replace traditional keyboards. As you will discover in this chapter, voice interface technology is an evolutionary, not a revolutionary, technology. In other words, no sudden invention or scientific breakthrough will produce computers capable of voice communication. Instead, voice interfaces will evolve with other high-technology fields, in particular the field of artificial intelligence.

In the last chapter, you were introduced to the elements of both human and computer speech. You found that a fundamental understanding of the sounds of human speech was prerequisite to learning computer speech-synthesis technology. In particular, with phoneme speech synthesis, the programming task required you to construct artificial computer speech by putting together the various sounds of speech, called phonemes. Now, for computer speech recognition, you must reverse the process. Here the task will be to break down a spoken word or phrase into the fundamental sounds of speech in order for a computer to verify and/or understand what was said. Let's begin this chapter with an overview of several different types of speech-recognition systems.

5-1 TYPES OF SPEECH-RECOGNITION SYSTEMS

Earlier you found that there were several design approaches to speech synthesis. Moreover, the application requirements dictated which type of speech-synthesis technology to use. The same is true of speech recognition and understanding. When designing a speech-recognition or understanding system, you must keep the following application requirements in mind:

- *Speaker dependence/independence:* Who is going to use the system, an individual or many individuals?
- *Vocabulary:* What is the maximum vocabulary size that the system will ever be required to recognize or understand?
- *Isolated word versus connected speech:* Will the system be required to recognize single isolated words, or must it understand connected words and phrases?
- *Recognition versus understanding:* Must the system simply recognize a given word(s) when it is spoken, or must it interpret the meaning and thus understand the spoken word(s)?

As with any type of engineering, the application must always be considered first, since it usually dictates the best solution. With speech-recognition systems, careful consideration of the preceding design criteria as related to the application will usually dictate the most efficient and economical system.

Speaker Dependence/Independence

Our first criterion requires you to consider *who* is going to use the system. The reason for this is that a speech-recognition system is either ***speaker-dependent*** or ***speaker-independent***. A speaker-dependent system must be trained by the user. The training session requires that the user speak each word to be recognized into the system. In most cases, each word must be repeated at least five times for the training session to be successful. The system must be retrained for different users.

These training sessions are obviously time consuming. However, for many applications a speaker-dependent system is the most practical and economical choice, since it is very accurate in recognizing a given user's voice. Speaker-dependent systems are particularly suited to industrial applications. But what about those times when the user has a cold or must be replaced while sick or on vacation?

The obvious solution to the problems of a speaker-dependent system is a speaker-independent system. Such systems do not require training and will respond to the voice input of any user. However, the trade-off is usually a smaller vocabulary and increased software complexity. In addition, the accuracy of today's speaker-independent systems is from 85 to 95 percent, compared to 91 to 99 percent for speaker-dependent systems. Nevertheless, the ultimate voice interface must be a speaker-independent system. This is particularly true for consumer applications.

Vocabulary

Here is where the application indeed dictates the best system. Why design a system capable of recognizing several hundred words when just a few might do the job? Consider, for example, the words *yes, no, start, stop,* and the numerals 0 through 9. Many industrial and consumer applications exist that could utilize this simple vocabulary.

Both speaker-dependent and speaker-independent systems are possible with small vocabularies. However, as the vocabulary increases, speaker independence becomes less of a reality with today's technology. Speaker-dependent systems with vocabularies of over 100 words are common. However, most existing commercial speaker-independent systems have vocabularies of roughly only 10 to 25 words. With today's technology, good speech recognition is directly related to small vocabularies. Remember, as the system vocabulary increases, the system memory, complexity, response time, and cost all increase in direct proportion.

Isolated Word Versus Connected Speech

Aside from speaker dependence/independence, speech-recognition and understanding systems fall into one of two *functional* categories: ***isolated-word recognition*** or ***connected-speech understanding*** systems. As the name implies, isolated-word recognition systems are designed to recognize, or verify, single-word utterances. Such systems can be speaker-dependent or speaker-independent. The technology for these systems is well developed and, as a result, many commercial products employ isolated-word recognition.

A speech signal that contains whole phrases and sentences is called connected, or continuous, speech. Such speech is said to be "natural" or "normal," since it is the way you and I normally talk. Since connected speech is made up of individual words, you might suggest that, to interpret connected speech, the speech signal can be broken down into individual word signals. Then isolated-word recognition can be used to recognize each individual word. From this, the meaning of the entire phrase or sentence can be determined.

Here is where the problems begin. First, it is difficult to determine the precise boundaries of an individual word within a connected speech signal. Some so-called connected-speech recognition systems are really nothing more than isolated-word recognition systems, since they impose speaking-rate restrictions on the user. The speaker must make a 200- to 300-ms pause between words. This allows the system to accurately determine the boundaries, or end points, of individual words. Second, the signal for a given word in a connected phrase does not closely resemble the signal for the same word if spoken separately. Finally, the pronunciation of a given word is affected by the words and punctuation around it. All these problems make connected-speech recognition or understanding a difficult task indeed. Consequently, the principles of AI must be employed to achieve any satisfactory degree of accuracy for connected-speech systems.

Speech Recognition Versus Speech Understanding

The preceding discussion leads us directly into distinguishing between speech recognition and speech understanding. You have probably noticed the use of both of these terms in our discussion (note the chapter title). The terms recognition and understanding cannot be used interchangeably, since they suggest two different approaches to voice interface technology. Speech recognition is usually associated with the signal-matching, or template, techniques used in isolated-word recognition systems. Speech understanding, on the other hand, is usually associated with the AI techniques used in connected-speech understanding systems. Such techniques attempt to interpret the meaning of speech signals using *knowledge* about speech, rather than simply matching patterns as in speech-recognition systems. The basic difference is that speech-understanding systems employ the principles of AI, while speech-recognition systems do not.

In the remainder of this chapter, you will explore both types of systems: first, speech-recognition by means of isolated-word recognition systems; then, speech understanding by means of connected-speech understanding systems. I should emphasize that the technology used for isolated-word speech recognition is relatively proved, as compared to connected-speech understanding technology. Since speech understanding employs the principles of AI, the associated technology will evolve naturally with the field of artificial intelligence.

5-2 ISOLATED-WORD RECOGNITION

Recall from the previous discussion that both speaker-dependent and speaker-independent isolated-word recognition systems are possible. Speaker-dependent systems have large vocabularies (over 100 words), while speaker-independent systems have relatively small vocabularies (less than 50 words). In addition, speaker-dependent systems are more accurate than speaker-independent systems (99 percent as compared to 95 percent at best).

Speaker-dependent isolated-word recognition systems operate by attempting to match individual input word signals against stored patterns, or ***templates***, of the vocabulary words. For this reason, isolated-word recognition systems are sometimes referred to as ***signal-matching, template-matching,*** or ***pattern-recognition*** systems.

Speaker-independent systems using template-matching techniques are not as reliable as speaker-dependent systems because of the signal variations from speaker to speaker. To solve this problem, some speaker-independent systems operate using a ***phonetic analysis*** technique. With phonetic analysis, the speech sounds are divided into several broad phonetic categories. This information is then used to determine a small set of possible word candidates, from which a more detailed phonetic analysis produces the final word selection. Let's take a closer look at both the template-matching and phonetic-analysis techniques.

Template Matching

Template matching is the most proven speech-recognition technique and has resulted in many low-cost, highly accurate commercial systems. A block diagram of a typical system is shown in Figure 5-1. Notice the simplicity of the system. The hardware consists of a microphone, amplifier, bandpass filter, sample/hold circuit, and A/D converter.

Although a simple part of the system, the microphone is an important consideration. A microphone with a high signal-to-noise ratio must be used to reduce the possibility of recognition error due to environmental noise. The signal-to-noise ratio should be at least 25 decibels (dB).

From the microphone, the speech signal is amplified and bandpass-filtered to produce a 200- to 3200-hertz (Hz) frequency range. This is done to increase the signal level and eliminate as much environmental noise as possible. The signal is then sampled and digitized by the sample/hold and A/D converter circuits. The sampling rate must be at least twice the highest frequency component, or 2 X 3200 Hz = 6400 samples per second. In practice, sampling rates of 8000 to 12,000 samples per second are common. An 8-bit A/D converter is adequate for most applications. Once the speech signal is in digital form, the rest of the process is handled by the system software.

The most complex portion of any speech-recognition system is the software. The major software operations are illustrated by Figure 5-2. There are two operational modes: ***training*** and ***recognition.***

Training Mode. As you are now aware, a template-matching system must be trained by the system user. The associated software performs two functions during training: ***acoustic analysis*** and ***template generation.*** The user must speak each vocabulary word into the system several times for the training to be successful. The system software analyzes the speech signal and generates word templates in memory, usually RAM, which are later used to match unknown utterances.

During the training session, the A/D converter samples the analog speech

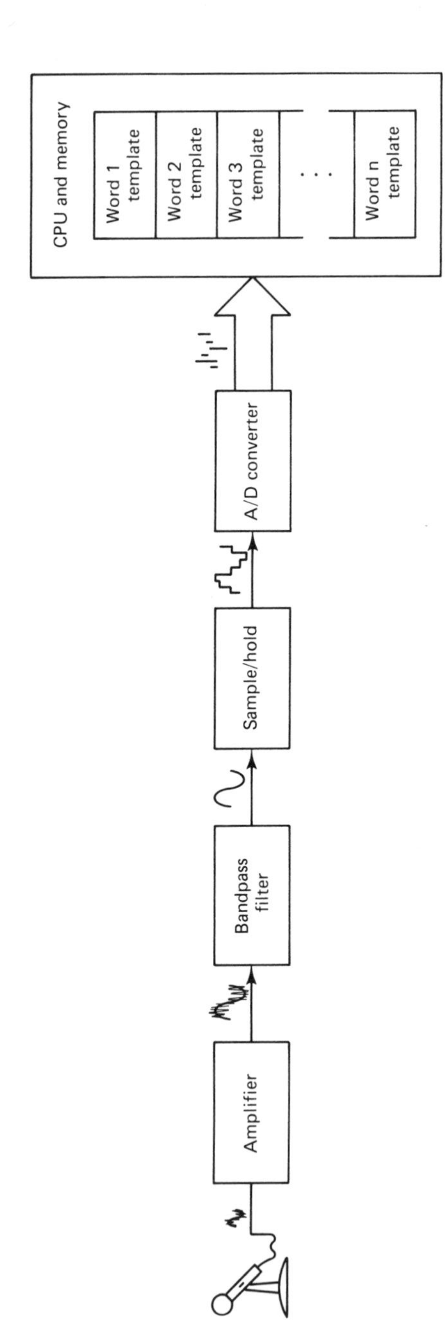

Figure 5-1 Typical template matching system.

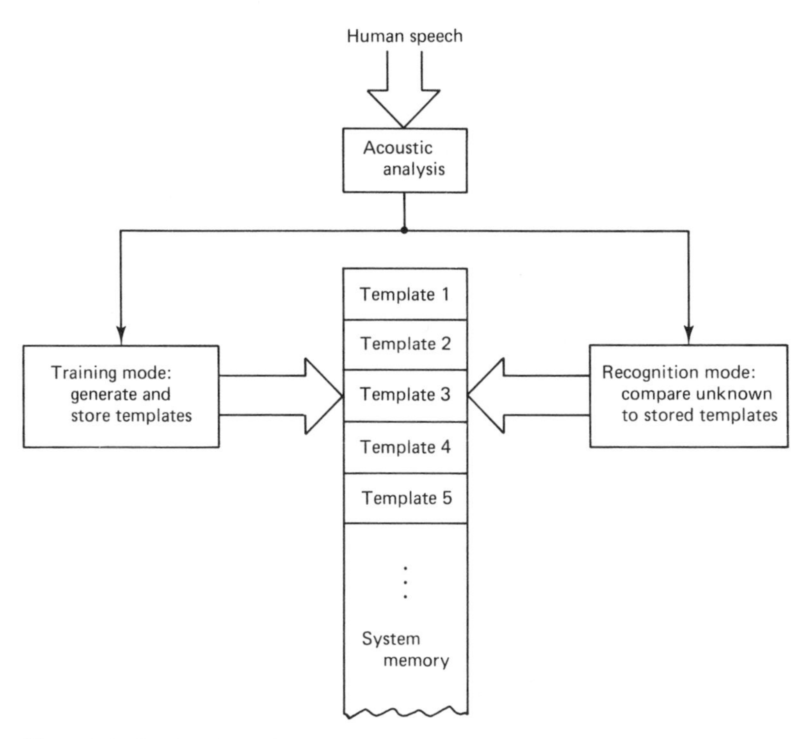

Figure 5-2 Speaker-dependent isolated-word recognition involves two major software operations: training and recognition.

signal at a fixed rate. Since direct wave-form digitization (PCM) requires too much template memory, the system software must acoustically analyze and convert the digitized samples into a more efficient storage code. Any of the speech-compression techniques discussed in Chapter 4 can be used. However, linear predictive coding (LPC) is the most common in commercial systems.

Once coded, the word templates are stored in RAM. Depending on the type of coding used, a single-word template will require between 10 and 1000 bytes of memory. A typical LPC system requires about 100 bytes of memory for each word template in the vocabulary. The LPC data are stored in matrix form as reference patterns for subsequent comparison to unknown signals during the recognition mode.

Recognition Mode. In the recognition mode, the input signal is converted and acoustically analyzed as in the training mode. The coded (LPC) data are then temporarily stored in matrix form for comparison to the reference templates that were generated during the training mode. The *unknown template* can then be compared directly to the *reference templates*, element by element. Direct comparison is accomplished by computing a distance factor between the reference and unknown matrices. This is done by taking the sum of the square of the differences between corresponding elements in the two matrices. The reference template that has the smallest distance factor from the unknown template is picked as the match to the input signal. Ideally, the unknown matrix must be compared to all the reference matrices to find the best match. However, in some cases special search algorithms are used to reduce the comparison task and increase the system response time. Of course, this increases the possibility of recognition error.

A more accurate way to compare templates is to use a mathematical algorithm called ***dynamic programming.*** Dynamic programming and a related technique, ***time warping***, are used to reduce recognition errors due to improper time alignment. Improper time alignment, or synchronization, accounts for most of the errors in isolated-word recognition systems.

Figure 5-3 illustrates the time-alignment problem. Here, for example purposes,

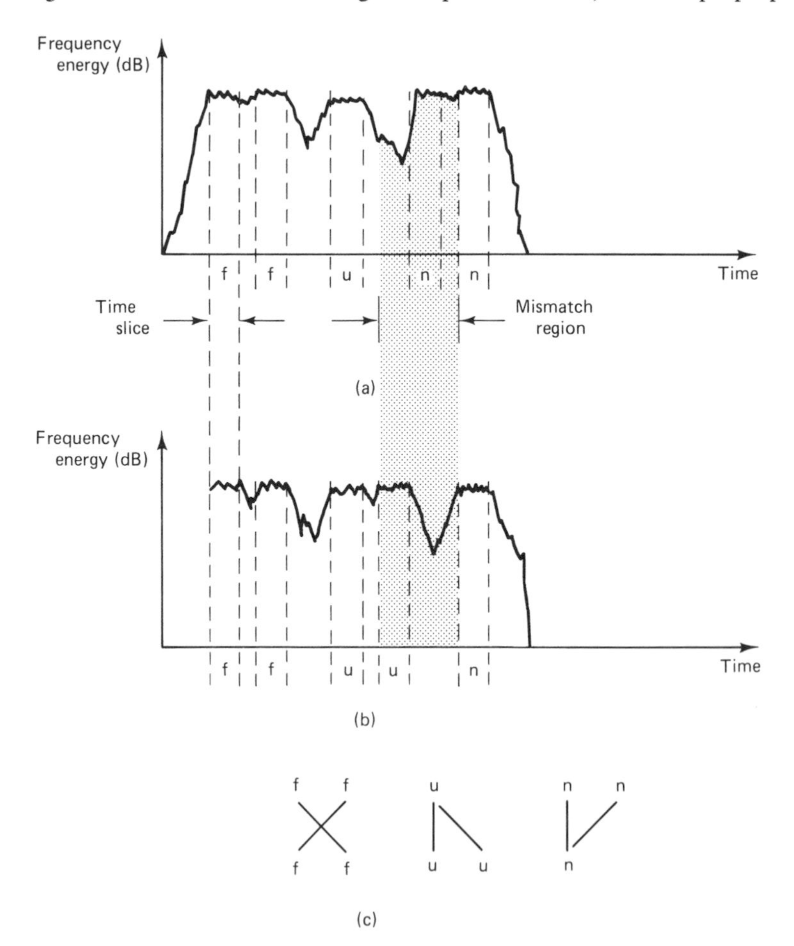

Figure 5-3 Time alignment for (a) reference and (b) unknown templates is a problem during template matching. (c) Dynamic programming time aligns the two templates.

the isolated word to be recognized is the word *fun.* Suppose that the analog signal shown in Figure 5-3(a) was used to make the reference template, and the signal shown in Figure 5-3(b) is the current user's utterance of the word *fun.* When the reference template was made, the signal in Figure 5-3(a) generated a template of *ffunn.* The current unknown signal in Figure 5-3(b) generates the template *ffuun.* Notice that the sounds within the two waveforms do not time align, resulting in two unique templates for the same word and the same speaker.

Dynamic programming attempts to produce the best possible fit between the two templates so that the system does not reject the current utterance simply because the speaker did not say the word exactly the same as when the reference template was made. The results of dynamic programming can be seen by the time-alignment lines in Figure 5-3(c). Notice how dynamic programming compensates for the variation in the two templates.

Dynamic programming is a pattern-matching algorithm that is used for both speech recognition and visual pattern recognition (Chapter 6). It is a matrix analysis technique that computes all the possible combinations of time alignments between the reference and unknown templates, the result being the best match between the two templates. A complete discussion of the algorithm is beyond the scope of this text. A good text in matrix theory will provide the details.

One of the biggest advantages of dynamic programming is that end-point detection is not as critical as it is with direct matrix comparison. In most speech-recognition systems, the end points of a spoken word are determined by the energy of the incoming signal, as illustrated in Figure 5-4. A pause produces an idle line with little or no energy. At the beginning of an utterance the sound energy increases, triggering the system to begin analyzing a word. The signal is analyzed until the sound energy drops below a certain ***threshold level***, indicating the end of the word.

Since dynamic programming compensates for time alignment, or signal mis-

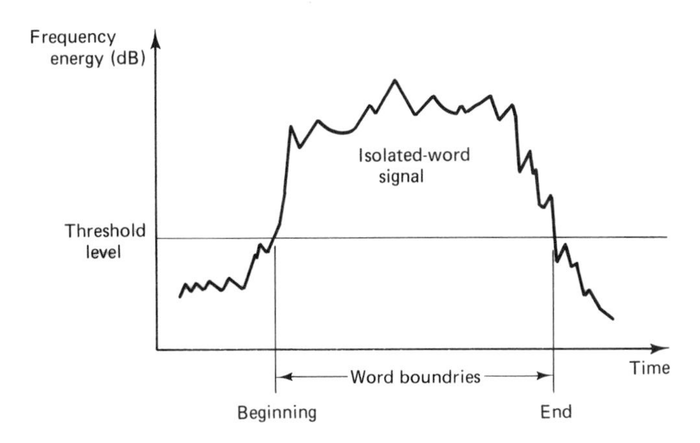

Figure 5-4 End-point detection of words in a speech signal is determined using a threshold level.

match, precise detection of the word end points is not so critical. This becomes an extreme advantage in connected-speech recognition, where words often run together, making end-point detection difficult at best. However, dynamic programming does not come without a price. As you might guess, the trade-off is slower response time due to more complex calculations during the template-matching operation. Dynamic programming increases the response time by a factor of 100 over direct matrix comparison. Several search strategies have been developed to reduce this factor, thereby making dynamic programming a realistic alternative to matrix comparison.

Isolated-word Recognition Devices. Single-chip devices called ***speech chips*** are commercially available that perform *both* the speech-synthesis and speech-recognition task. Recall that both speech synthesis and isolated-word recognition utilize linear predictive coding (LPC) for data compression. So why not integrate an LPC synthesizer and LPC acoustic analyzer on a single chip? This is precisely what is done to manufacture speech chips.

A typical microprocessor-based speech-chip system is shown in Figure 5-5. The function of the speech chip is twofold: (1) to accept LPC code from the microprocessor for speech synthesis and (2) to accept speech signals from a microphone and generate LPC code to the microprocessor for speech recognition.

For speech synthesis, the microprocessor generates LPC code from a memory lookup table to the speech chip. The chip contains the excitation sources, digital filter, and D/A converter required to synthesize speech from the LPC code. The analog speech signal is passed from the chip to an external audio amplifier, a low-pass filter, and finally to a speaker for sound reproduction.

In the speech-recognition mode, the speech chip performs the acoustic-analysis function. Speech signals from an external microphone and amplifier are sampled and converted to digital by an internal A/D converter. The digitized speech is then compressed using linear predictive coding and read by the microprocessor at periodic intervals. Notice, however, that the speech chip only performs the acoustic-analysis task. Template generation, storage, and matching must be performed by the microprocessor system software.

Phonetic Analysis

Unfortunately, the template-matching techniques just discussed are not very reliable for speaker-independent systems due to signal variations from one speaker to the next. In addition, even speaker-dependent systems with vocabularies of more than a few hundred words become highly impractical. This is because of the large amounts of template memory required and the amount of time it takes to search the memory during the matching process, not to mention the time it would take to train a system with a vocabulary of 1000 or more words.

A speech-recognition technique called ***phonetic***, or ***feature, analysis*** has been

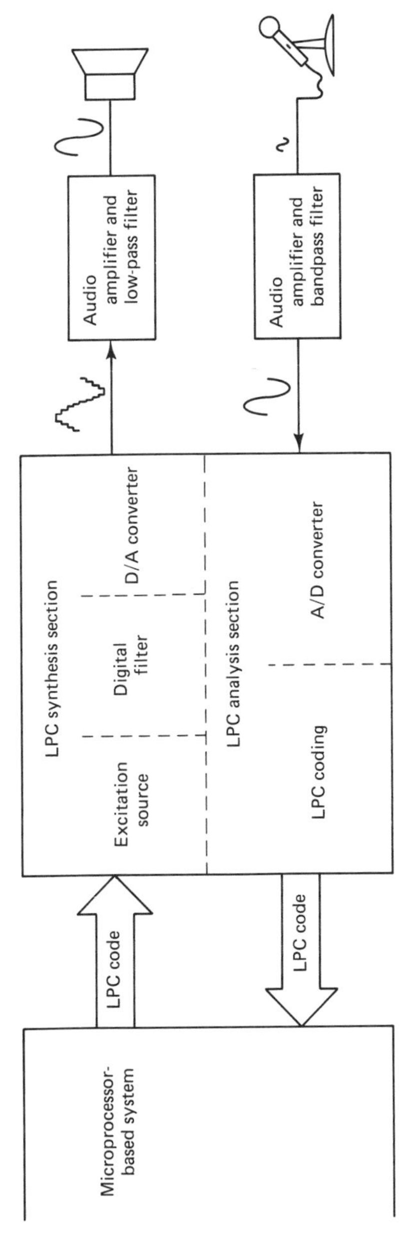

Figure 5-5 Typical speech-chip interface.

developed to handle large-vocabulary, speaker-independent, isolated-word recognition. As you might guess from its name, the technique is based on the detection and analysis of the phonetic features of the speech signal.

Early research in speech recognition concentrated on a *detailed* phonetic analysis of the speech signal. The idea was to analyze the speech signal and separate the phoneme sounds to produce a detailed phoneme representation of the utterance. From this phonetic representation, a memory lookup operation would produce the corresponding word. This idea sounds good in theory, but it has several serious drawbacks.

First, it is very difficult to make fine phonetic distinctions from the speech signal. For instance, consider the two words *boy* and *buoy.* A phonetic transcription of these words would most likely result in the same phonetic representation for both words. Second, allophonic variations such as the *p* in *pin, pen, pan, pick,* and *place* create problems, especially when pronounced by different speakers. Finally, depending on who is doing the talking, phonemes or whole syllables may be dropped from a word. The word *international* is a prime example, since the first *t* is dropped by many speakers. In fact, it was the early difficulties with phonetic-based systems that led to the development of template-matching systems.

However, in the early 1980s researchers began to look at speech signals in terms of broad, rather than fine, phonetic categories. Broad phonetic classifications such as vowel, nasal, and fricative are less senstive to fine phonetic variances from speaker to speaker. The idea is to interpret the spoken word into a series of broad phonetic categories. This information is then used to determine a small set of possible word candidates, from which a more detailed phonetic analysis produces the final word selection. You might say that broad phonetic classification is a backward, or top-down, reasoning strategy as opposed to the forward, or bottom-up, reasoning employed by fine phonetic analysis.

Let's take a close look at a typical feature-analysis system. Suppose you divide all the sounds of speech into the six broad phonetic categories listed in Table 5-1.

TABLE 5-1 Phonetic Categories Used for a Typical Feature Analysis System

Category	Sounds
Pure voiced vowel (V):	a, e, i, o, u, uh, aa, ee, er, uu, ar, aw
Nasal (N):	m, n, ng
Voiced fricative (VF):	z, zh, v, dh
Unvoiced fricative (UF):	s, sh, f, th
Plosive (P):	b, d, g, p, t, k, h
Glide (G):	r, w, l, y

Now consider the ten digits 0 through 9. Using the six categories of Table 5-1, the sounds of each digit can be expressed as a sequence of sound categories as shown in Table 5-2.

TABLE 5-2 Sound Category Sequences for the Digits 0 to 9

Digit		Dictionary Pronunciation	Phonetic Sound Sequence
0	Zero	zêr'ō	VF-V-G-V
1	One	wun	G-V-N
2	Two	tōō	P-V
3	Three	thrē	UF-G-V
4	Four	fôr	UF-V-G
5	Five	fīv	UF-V-VF
6	Six	siks	UF-V-P-UF
7	Seven	sev''n	UF-V-VF-N
8	Eight	āt	V-P
9	Nine	nīn	N-V-N

For example purposes, let's assume that the vocabulary of our speaker-independent system consists solely of the ten digits. A speaker utters one of the digits into the system, and an analysis algorithm translates the speech signal into one of the phonetic sequences in Table 5-2. Next, a decision-making algorithm must be used to determine which of the ten digits was actually spoken. The algorithm shown in Figure 5-6 should do the job. Notice that the algorithm initially looks at the first phonetic classification of the utterance. From this, any one of the digits 0, 1, 2, 8, or 9 can be uniquely determined. If none of these digits was spoken, the algorithm assumes that the first phonetic classification is an unvoiced fricative (UF)

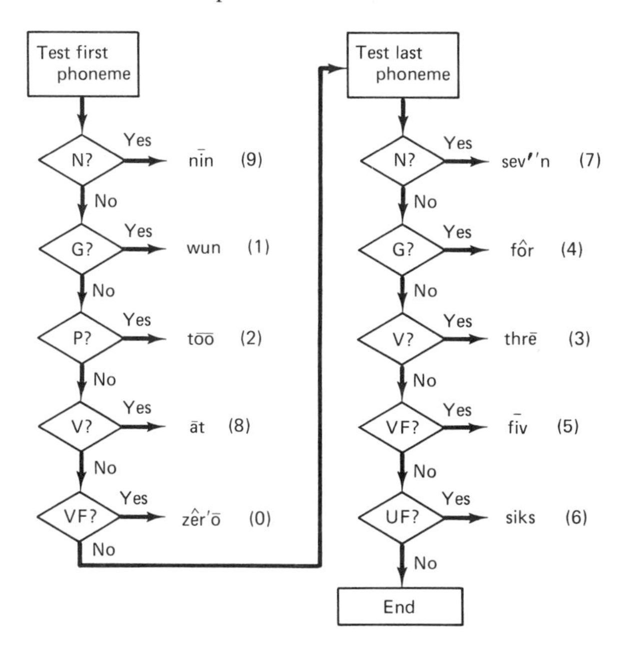

Figure 5-6 Decision algorithm flowchart for the numbers 0 through 9.

and then examines the last phonetic classification. This leads the system to selecting one of the digits 3, 4, 5, 6, or 7.

Example 5-1

Using Table 5-1 and a dictionary, construct a table of phonetic sound sequences for the words *yes, no, start,* and *stop.*

Solution

First, you must go to a good dictionary and look up the pronunciation of each word. I used *Webster's New World Dictionary of the American Language*, College Edition. Then from the dictionary pronunciations you must use Table 5-1 to form the phonetic sound sequences of each word. The resulting sequences are shown in Table 5-3.

TABLE 5-3 Solution Sequences for Example 5-1

Word	Dictionary Pronunciation	Phonetic Sound Sequence
Yes	yes	G-V-UF
No	nō	N-V
Start	stärt	UF-P-V-P
Stop	stop	UF-P-V-P

Example 5-2

Add decision-making steps to the digit flowchart in Figure 5-6 that will allow a speaker-independent system to recognize the words in Example 5-1, in addition to the digits 0 through 9.

Solution

The resulting flowchart is shown in Figure 5-7. Notice that the system can uniquely determine the digits 0 through 9 and the words *yes* and *no*. However, a problem arises between the words *start* and *stop* because they are so phonetically close to each other. Here the broad classifications have resulted in two possible word candidates.

When the broad phonetic classifications result in several word candidates, a more detailed phonetic analysis must be performed on the signal to select between the words. In Example 5-2, a fine phonetic analysis on the last phonemes in each word (the *t* in *start* and *p* in *stop*) should distinguish between the two. An even simpler, and more commonsense solution, would be to change the word *start* to the word *begin.*

Of course, as the vocabulary increases in such a system, the number of word candidates for a given phonetic sequence also increases. It has been shown that about 7000 words of a 20,000-word dictionary can be uniquely identified using the preceding broad phonetic classifications. Of the remaining words, over 10,000 fit into a phonetic pattern containing 5 word candidates or less. The largest pattern contains about 200 word candidates. Even so, this is only 1 percent of the entire dictionary. The point is this: the memory and processing requirements for large-

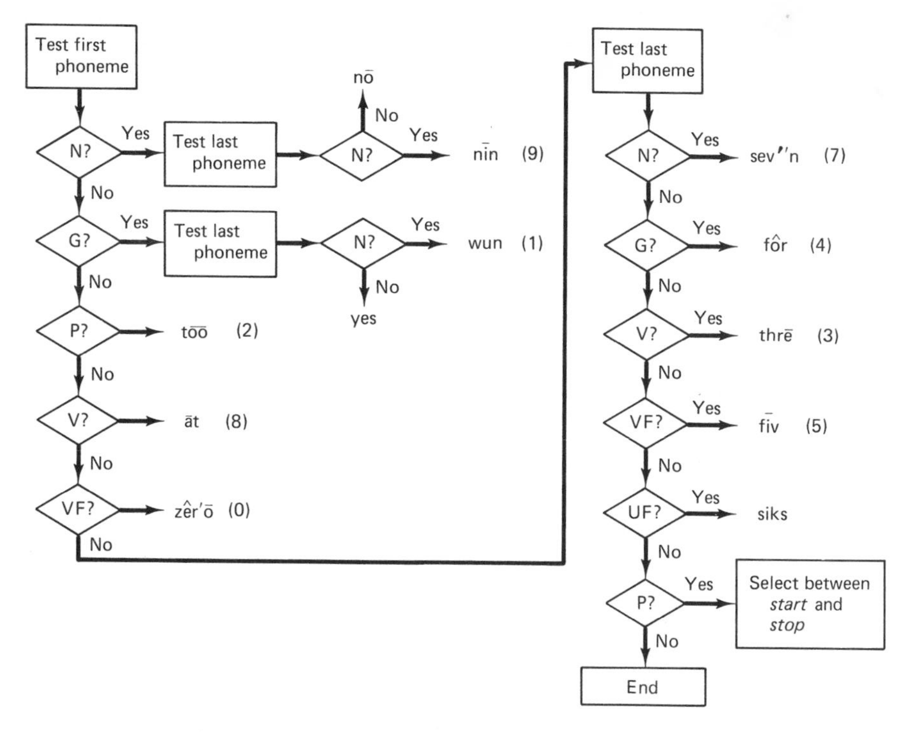

Figure 5-7 Design algorithm flowchart for Example 5-2.

vocabulary, speaker-independent, isolated-word recognition systems can be drastically reduced by using the broad-classification phonetic analysis technique.

Acoustic Analysis. The analysis of the speech signal to determine the phonetic sound classes usually involves linear predictive coding. In addition, a zero-crossing analysis of the speech wave form is sometimes performed.

Zero-crossing analysis involves counting the number of times the speech signal crosses zero in a fixed time period, usually 10 ms. A simple op-amp comparator circuit, like the one shown in Figure 5-8, can be used to generate a logic transition

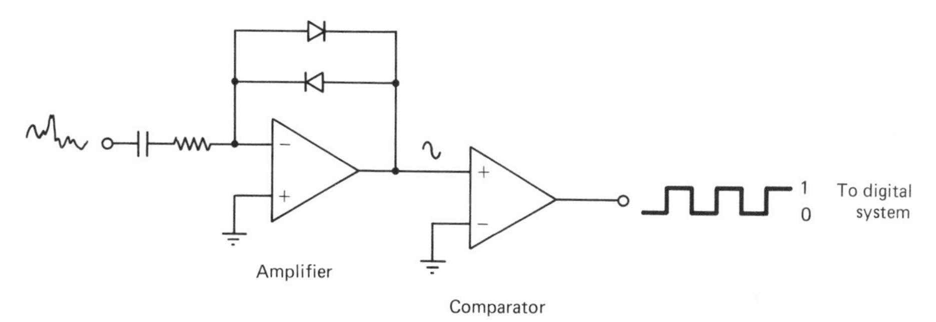

Figure 5-8 Zero-crossing detector circuit.

each time the amplified speech signal crosses zero. The zero-crossing detector output is then applied to an input port line, where a software counter counts the zero crossings. Zero-crossing analysis is used along with linear predictive coding to determine the frequency spectrum, and thus the phoneme sounds, within the speech waveform.

5-3 CONNECTED-SPEECH UNDERSTANDING

The interpretation of connected, or continuous, speech is a difficult task indeed. To do so requires the direct application of artificial intelligence. As of this writing, connected-speech understanding is in the research and development stage. However, as the field of AI evolves, so will speech-understanding systems.

Since speech understanding requires intelligence, a workable system must possess several degrees of knowledge about speech. This knowledge must include the phonetic knowledge of speech sounds as well as knowledge about how words are used in phrases and general knowledge about human conversation.

Knowledge Sources

The fundamental difference between speech recognition and speech understanding is found in the use of speech-knowledge sources to interpret, or understand, speech. Seven fundamental sources of knowledge are used in speech-understanding systems:

1. Pragmatic knowledge
2. Semantic knowledge
3. Syntactic knowledge
4. Lexical knowledge
5. Prosodic knowledge
6. Phonological knowledge
7. Phonetic knowledge

The preceding listing is a top-down listing. In other words, pragmatic knowledge is the highest level since it involves knowledge about conversation, while phonetic knowledge is the lowest level since it deals directly with raw signal data. In general, the lower-level knowledge sources are derived from the speech signal data, whereas higher-level sources are comprised of rules about how we talk.

The combination of these knowledge sources using an effective AI control strategy is the basis for connected-speech understanding. Let's take a closer look at each of these speech-knowledge sources in preparation for a discussion of speech-understanding control strategies. Our discussion will begin at the bottom level with phonetic knowledge.

Phonetic Knowledge. Of the seven speech-knowledge sources, phonetic knowledge is the one with which you are most familiar. This knowledge consists of the fundamental sounds of speech, which include phonemes and phoneme variations, or allophones. As a result, the phonetic knowledge source in a speech-understanding system is derived directly from the speech signal during the acoustic-analysis phase. Consequently, the front-end software of a speech-understanding system is a phonetic analysis algorithm like those used for isolated-word recognition.

Phonological Knowledge. Phonological knowledge must be used to translate the allophones that can be identified in the speech signal into the actual phonemes. Such knowledge involves the understanding of how allophones are determined during conversational speech. Rules that govern the way allophones are pronounced within words, as well as when words are spoken together in phrases, are included as part of this knowledge source. In addition, phonological knowledge includes variations in pronunciation due to dialect.

For instance, most people say *car*, but a speaker from Boston says *caw*. Likewise, a speaker from the Northeast says *you all*, while a speaker from the south says *y'all*. The first is an example of phonological variation *within a word* due to dialect; the second is an example of variation *between words* due to dialect.

Prosodic Knowledge. As you are aware, the prosodic features of speech include amplitude, intonation, and timing. The latter two combine to form a prosodic feature called ***stress***. Prosodic knowledge is extremely important in speech-understanding systems, since it is the prosodic features that often determine the meaning of a spoken phrase. Furthermore, the prosodic features within a speech signal can help to identify word and phrase boundaries.

The speech signal must be analyzed to extract the prosodic features of an utterance. Prosodic knowledge rules must then be used to map the identified features to possible word and phrase candidates.

Lexical Knowledge. Lexical knowledge is knowledge about different words in a given vocabulary. Such knowledge includes all the various pronunciations of words that can result from different context, punctuation, and phraseology. In addition, the lexical knowledge base includes rules that determine plurals, tense, possession, and so on.

Syntactic Knowledge. Syntax is used to describe sentence structure. In other words, the syntax of a sentence has to do with the arrangement of words within the sentence. Syntactic knowledge rules in a speech-understanding system are used to describe all the "legal" combinations of words that can be used to form phrases and sentences. Such knowledge is called ***constraining knowledge*** since it restricts the number of word combination possibilities. In turn, this reduces the number of sentence possibilities, since not all word combinations are legal for a given application. A sentence such as "Attack the operator!" would definitely not be a legal combination of words in a robotic application. Again, the application always dictates the system. Such a command might be perfectly legal in a security or military application.

Syntactic knowledge can also be applied to anticipate the meaning of an utterance. As an example, consider the phrase "Pick and ____." In a robotic application, syntactic knowledge might anticipate that the command is to *pick and place* a part before the *place* command is even uttered. In addition, any other interpretation would be judged illegal since the robot knows that a *pick* command must always be followed by a *place* command.

Semantic Knowledge. Semantics has to do with the actual meaning of words and phrases. As a result, semantic knowledge is a relatively high level knowledge source that is used to interpret the final meaning of an utterance.

Semantic knowledge is also used as a constraining knowledge. Phrase and sentence possibilities can be eliminated because they don't make sense, even though they might be syntactically correct for a given application. In our robot application, a command of "Pick the part and place it on the cart" makes sense, whereas a command of "Pick the cart and place it on the part" does not make sense. Consequently, the robot semantic knowledge source would not allow the second interpretation of the command.

Pragmatic Knowledge. Finally, the highest level of knowledge available to a speaker-understanding system is pragmatic knowledge. Pragmatic knowledge contains the rules of ordinary conversation, as well as information about a current dialogue. The pragmatic knowledge source must keep track of such things as what *it* is within a conversation. In addition, pragmatic knowledge must dictate that the response to a question such as "How old are you?" requires a numerical response rather than a simple *yes* or *no*.

Knowledge Organization and Reasoning Strategies

All the knowledge in the world does not do you any good unless you organize and control it in an efficient manner. For this reason, the organization and control of the preceding knowledge sources is an important part of any speech-understanding system.

Assuming a speech-understanding system has the preceding knowledge sources available to it, the next task for the system designer is to decide how to organize the knowledge base. Then, once the knowledge organization is established, a reasoning and control strategy must be implemented that results in the most accurate and efficient use of the knowledge base. Let's begin with a discussion of different knowledge organizations and their associated reasoning strategies.

Three fundamental knowledge-organization models have experienced various levels of success in speech-understanding systems: the ***hierarchical model, blackboard model,*** and ***compiled-network model.*** You will find that all these organization models are familiar to you from your learning experience with knowledge representation in Chapter 3.

Hierarchical Model. The hierarchical model shown in Figure 5-9 is the most natural and probably the simplest organization of the seven knowledge sources. It is a natural organization scheme since the seven knowledge sources are inherently hierarchical by definition. Phonetic knowledge is the lowest form of speech knowledge, followed by phonological knowledge, and so on, up to the highest level, pragmatic knowledge.

The next question is, what type of reasoning strategies must be used to control the flow of information within the knowledge base. As you are aware, any hierarchical knowledge base can employ one of two fundamental reasoning strategies: ***forward*** or ***backward*** reasoning.

Forward reasoning. A forward, or bottom-up, reasoning strategy would dictate that the system acoustically analyze the speech signal to identify its phonetic characteristics for the phonetic knowledge source. Phonetic knowledge is then

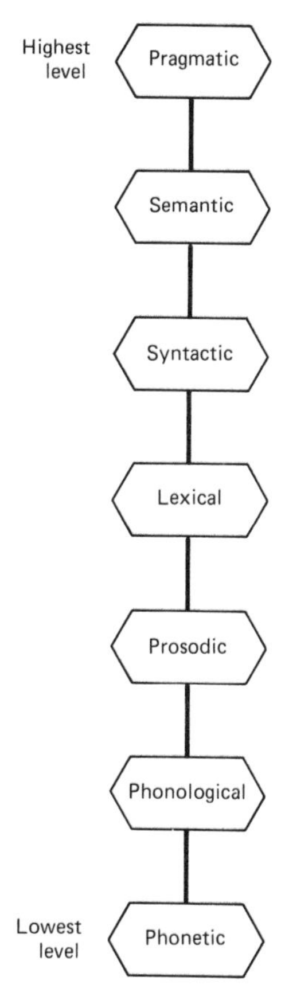

Figure 5-9 Hierarchical knowledge-organization model.

passed up to the phonological source for further analysis and interpretation. The upward flow of knowledge continues in serial fashion until a final interpretation of the utterance is made by the higher-level knowledge sources. The flow of information in a forward-reasoning, speech-understanding system is illustrated in Figure 5-10.

Forward reasoning is the simplest to implement because of the natural hierarchical structure of the knowledge sources. However, it has several inherent disadvantages. First, it encourages combinatorial explosion, since an unrecognizable sound might result in an exhaustive search at the lower knowledge levels. Second, syntactic and semantic knowledge cannot be used to constrain the reasoning process, thereby making the system less efficient. Third, errors made at the lower levels will be passed on to higher knowledge sources, resulting in an inaccurate interpretation of the utterance. Finally, forward reasoning does not lend itself to verification and explanation of a conclusion, since the higher-level knowledge sources cannot interact with the low-level sources.

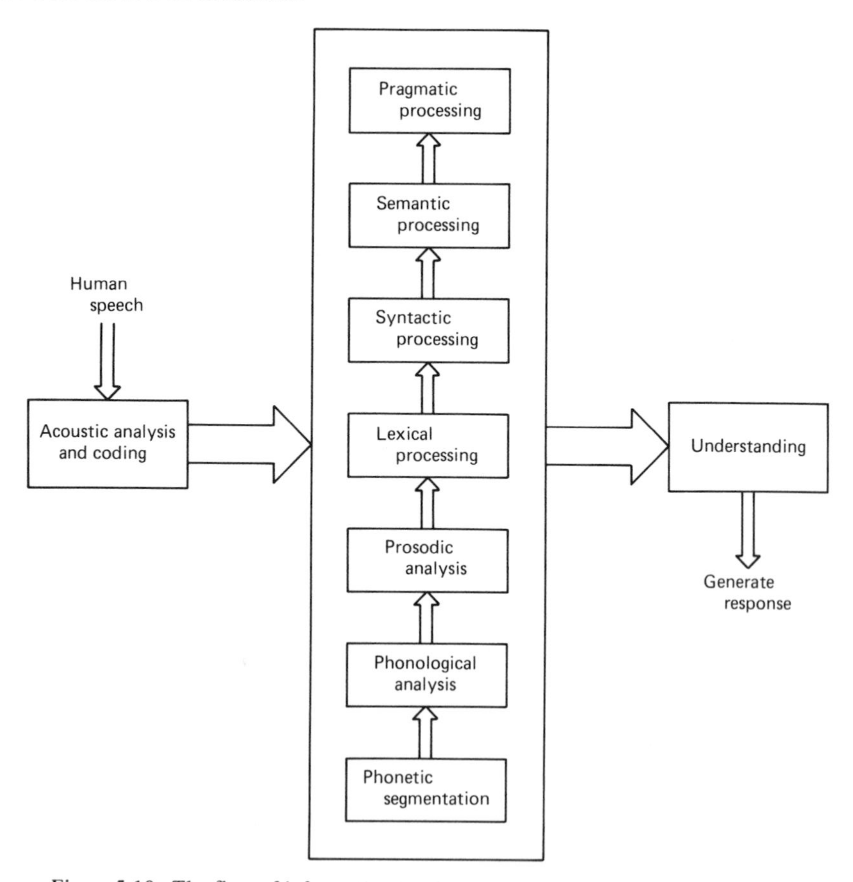

Figure 5-10 The flow of information in a forward-reasoning speech-understanding system is from the bottom up.

Backward reasoning. Backward reasoning results when the direction of the knowledge source arrows in Figure 5-10 is reversed. Reversing the direction of reasoning provides a solution to many of the inherent problems of forward reasoning.

Backward, or top-down, reasoning in speech-understanding systems is sometimes called ***expectation-driven reasoning.*** The idea is that the higher knowledge sources generate expectations of what is contained in the utterance. An expectation is simply an educated guess at the interpretation of an utterance based on heuristic knowledge of speech contained in the knowledge base. The expectations generated by the higher knowledge sources are then used to direct the analysis of the speech signal by the lower knowledge sources. In other words, the lower knowledge sources attempt to support the expectations through analysis of the speech signal. If unsuccessful, additional expectations are generated until an accurate interpretation is produced.

In a sense, you and I do the same thing when listening to someone talk. We tend to interpret an utterance without really recognizing every single word that was said. Sometimes, too much expectation gets us into listening trouble. The same is true of the expectation-driven reasoning process in speech understanding. The system uses any words that have already been recognized and makes an interpretation based on pragmatic, semantic, and syntactic rules contained in the higher-level knowledge sources. It has been shown that a speech-understanding system can accurately interpret an utterance based on just pragmatic, semantic, and syntactic knowledge without performing a complete and accurate acoustic analysis of the speech signal.

Expectation processing not only streamlines the reasoning process by applying high-level constraint knowledge; it also reduces the possibility of interpretation error, since the expectations are usually supported by acoustic analysis at the lower knowledge levels.

A combination of both forward and backward reasoning often produces the optimum reasoning strategy in an AI system. The same is true of speech-understanding systems that employ the hierarchical model of knowledge organization.

Blackboard Model. The blackboard model of knowledge organization shown in Figure 5-11 has been extremely successful in speech-understanding applications. Recall from Chapter 3 that the blackboard idea was first developed for a speech-understanding system called HEARSAY. The reason it works so well for speech-understanding systems is that speech knowledge is easily modularized into independent chunks of knowledge.

Using the blackboard idea in speech understanding requires that each knowledge source become a system specialist. The independent knowledge sources monitor the blackboard for the information that they need to analyze their respective levels of the utterance. Conclusions generated by the individual knowledge sources are "written" onto the blackboard for use by other knowledge sources. In this way, the knowledge sources interact by sharing information.

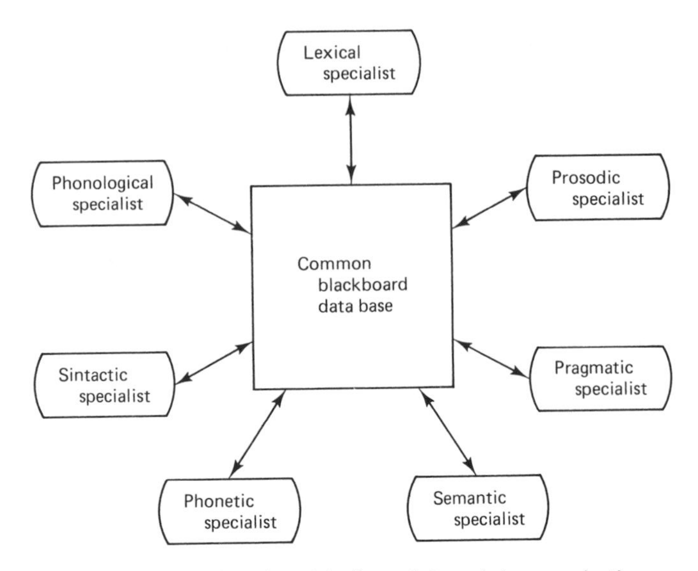

Figure 5-11 Blackboard model of speech-knowledge organization.

Constraining knowledge and expectations generated by the pragmatic, syntactic, and semantic knowledge sources are used by the lower-level knowledge sources to streamline the reasoning process. Likewise, phonetic and syllable information is used by the higher-level knowledge sources to verify conclusions and generate new expectations.

As new information is written onto the blackboard, it is used as required by the independent sources. Each knowledge source processes the data independently and asynchronously. In other words, the speech signal is processed in a parallel rather than a serial fashion, resulting in improved overall system efficiency. This parallel processing continues until no new information can be generated by the knowledge sources.

The major advantage of the blackboard model is its modularized structure. The knowledge sources are completely independent and "know" nothing about each other. In fact, they don't even know each other exists! This allows the system knowledge base to be easily changed and updated. It is interesting that HEARSAY can operate without all its knowledge sources "plugged in." However, the system performance level changes as different knowledge sources are plugged in and out. The overall performance of the system improves by 25 percent for the addition of the syntactic knowledge source and another 25 percent for the semantic source. This shows the power of these constraining knowledge sources.

Compiled-network Model. This last form of speech-knowledge organization compiles, or combines, all the speech-knowledge sources into one large search-tree type of network that I call a ***speech tree.*** For a given application, all possible word pronunciations and legal sentence structures are within the network. To interpret an utterance, the network is searched using a comparison technique until the closest match to the unknown speech signal is found.

A small portion of a typical speech tree is provided in Figure 5-12. Here the nodes represent allophonic templates that are used to match the coded speech signal. The branches from each node connect to all the legal sound extensions of that node. Once a given node is matched, the system attempts to match its associated branch nodes. The search process continues through the tree until the best match is found. Thus, the final solution path is the sound sequence that provides the best match to the utterance. Notice from Figure 5-12 how the tree structure is used to connect speech sounds that, in combination along a given path, produce intelligible words and phrases. All the different search strategies discussed in Chapter 2 can be applied to finding the best solution path within the network.

The problems associated with compiled-network, speech-understanding systems are the same as those of any treelike network used in AI. First, the whole network must be reconstructed to add new speech knowledge or to make any kind of change in the knowledge base. Second, a large vocabulary means a large network, which, in turn, requires a large computer system for processing. Finally, a compiled network can easily get off on the wrong track, since it is not expectation driven.

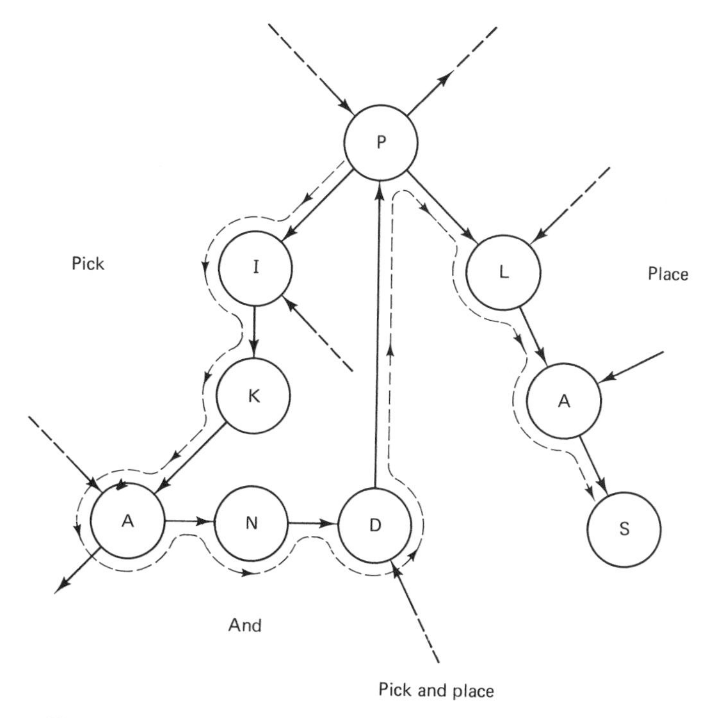

Figure 5-12 Small portion of a robot speech-understanding tree.

Control Strategies

Regardless of the type of knowledge organization used, an overall strategy must be developed for control of the interpretation process. Let's take a brief look at several popular control strategies.

The control strategy for a speech-understanding system must dictate how the system is to interpret the utterance. It must decide where to start, what to do next, and when to conclude the interpretation process. In addition, the control program must score the accuracy of knowledge-source conclusions and decide on the best possible interpretation of the utterance.

Left-to-Right Control. You might suggest that the logical place for a system to start the interpretation process is with the first word of the utterance and then to proceed from left-to-right, interpreting each successive word in the utterance. This is called a ***left-to-right control strategy***, for obvious reasons. As it turns out, the first word is not always the most efficient place to start. What happens if the first word is interpreted wrong or cannot be identified at all?

Middle-out or Island-driving Control. To literally get around the problems of left-to-right control, the system could start in the middle of the utterance and pick out those words that are easily identified. The interpretation process could then proceed left or right from the identified words. This control strategy goes by two names: ***middle-out*** or ***island-driving*** strategy.

The term island driving is used since the system creates islands of words within an utterance that have been identified. The system then generates expectations based on the constraining knowledge sources of possible word extensions of the islands. The island-driving idea is illustrated in Figure 5-13. The problem with this strategy is that there can be too many extensions to investigate, leading to combinatorial explosion.

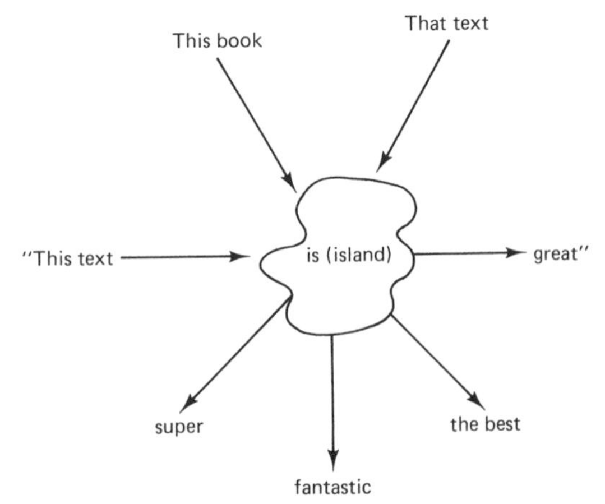

Figure 5-13 Island-driving control strategy.

Hybrid Control. The most efficient control strategy is a combination, or hybrid, of the left-to-right and island-driving techniques. With a hybrid control strategy, the system attempts to interpret any of the first few words of an utterance. Any identifiable words then become islands for expansion. Chances are that at least one of the first two or three words of an utterance will be identifiable, thereby eliminating the major problem with the left-to-right strategy.

To reduce the combinatorial explosion problem of island driving, the hybrid system first attempts to interpret the words to the left of the island back to the beginning of the utterance. Then the words to the right of the island are interpreted to the end of the utterance.

As you can see from Figure 5-13, once an island is created, many possible expectations, or extension candidates, can be generated from the island. The control strategy must decide which extension is the most promising candidate and should be used as the interpretation.

In Chapter 2 you learned that a way to make control decisions is to assign a cost factor to each operation. The control strategy was to perform the operation with the minimum accumulated cost. This idea is also used in speech-understanding control strategies. Once an island is created, the system begins calculating a cost factor for each island extension. The cost factor is calculated as the extension is analyzed based on how close the speech signal matches allophonic templates in memory. A threshold, or maximum-cost, level can be set such that the system is directed to another extension if the threshold is exceeded.

An opposite approach to the minimum-cost control strategy is a scoring strategy. Here the idea is to score each island extension as it is analyzed. The interpretation provided by the extension with the highest score is used. Again, a threshold can be established to discard low-scoring extensions.

In a blackboard system, the individual knowledge sources each score the expectation. The scoring can be weighted according to the level of the knowledge source, or each knowledge source can have an equal opportunity for scoring. In any event, the control strategy is to explore the expectation with the highest score.

You now are acquainted with the fundamental concepts used in speech recognition and understanding technology. As these technologies evolve, the basic ideas presented in this chapter will be expanded on to produce more accurate, efficient, and intelligent voice interfaces.

SUMMARY

Speech systems fall into several different functional categories. There are speaker-dependent and speaker-independent systems, isolated-word and connected-speech systems, and speech-recognition and speech-understanding systems. The application will usually dictate the type of speech system required.

Speaker-dependent, isolated-word recognition systems are most common and operate using a template-matching technique. These systems generally have vocabularies between 100 and 200 words. They are relatively accurate, but require training by the user. During the training session, the system generates reference template matrices in memory, which are subsequently compared to unknown templates during the recognition process. The unknown and reference templates are compared using direct matrix comparison or a pattern-matching algorithm called dynamic programming.

Speaker-independent, isolated- and connected-word recognition systems use a phonetic analysis technique rather than template matching because of signal variations from speaker to speaker. With this technique, the speech signal is analyzed and divided into several broad phonetic categories. The unknown utterance is converted into phonetic sequences from which a decision algorithm recognizes the spoken word(s). Phonetic analysis systems do not require as much memory as template systems for large vocabularies, since only the basic phonetic sounds need to be matched during the recognition process.

Connected-speech understanding requires the direct application of artificial intelligence. Speech-knowledge sources are used to interpret rather than simply recognize spoken phrases. These knowledge sources range from low-level phonetic knowledge to high-level syntactic, semantic, and pragmatic knowledge. The higher-level knowledge sources are used to provide expectational and constraining knowledge to the system. Knowledge sources in a speech-understanding system are organized using one of several common AI knowledge representation schemes: hierarchical, blackboard, or compiled-network knowledge representation. Each has its inherent advantages and disadvantages.

Regardless of the knowledge organization used, a control strategy must be developed that dictates how the system is to interpret the utterance. Such a strategy must decide where to start, what to do next, and when to conclude the interpretation process. Typical control strategies include left-to-right control, island-driving control, and hybrid control. The most efficient knowledge organization and control schemes are those that allow expectational and constraining knowledge to guide the interpretation process.

PROBLEMS

5-1 What design criteria must be considered when designing a speech-recognition or speech-understanding system?

5-2 What types of applications are particularly suited for speaker-dependent systems?

5-3 Why is it difficult to use an isolated-word recognition system for connected speech?

5-4 Describe the difference between speech recognition and speech understanding.

5-5 A typical speaker-dependent, isolated-word recognition system requires a vocabulary of 50 words. About how much memory is required to store all the word templates using an LPC coding format?

5-6 Describe the hardware structure of a typical speech-recognition system.

5-7 What two methods are employed for template matching in isolated-word recognition systems?

5-8 Why is dynamic programming superior to direct matrix comparison?

5-9 Describe a typical speech chip and explain its operation in a microprocessor-based system.

5-10 Name the six broad phonetic categories used in a typical phonetic analysis system.

5-11 You want to design a speaker-independent system to recognize the following robot control words: *pick, place, left, right, up,* and *down.* Construct a table of phonetic sound sequences for these words using the six broad phonetic sound categories discussed in this chapter.

5-12 Develop a flowchart for a decision-making algorithm that could be used to determine which of the words in Problem 5-11 was actually spoken.

5-13 List the seven fundamental knowledge sources that are used for speech understanding.

5-14 Explain why the higher-level knowledge sources are so important to the speech-understanding task.

5-15 Describe how knowledge can be organized in a speech-understanding system.

5-16 Develop a compiled network of phonemes for the robot commands in Problem 5-11.

5-17 Explain the function of a control strategy in a speech-understanding system.

5-18 Describe the island-driving control strategy for speech understanding.

6 Vision

Vision is the single most important sense that you and I possess. Likewise, vision provides the most enhancement to an intelligent machine's abilities. Truly intelligent machines, like robots, must be able to see if they are to perform humanlike tasks such as assembly, inspection, and navigation. In addition, vision provides a means whereby an intelligent machine can acquire information and learn from its own environment, rather than being limited to a knowledge base provided by a programmer.

In this chapter, you will explore the exciting world of machine vision. You will soon discover that analyzing and understanding visual information is probably the most complex task that an intelligent machine can perform. However, once analyzed and understood, visual information provides the most useful knowledge that an intelligent machine can possess about its environment.

Vision involves the transformation, analysis, and understanding of light images. As a result, the science of machine vision can be reduced to three fundamental tasks: ***image transformation, image analysis,*** and ***image understanding.***

Image transformation involves the conversion of light images to electrical signals that can be used by a computer. Once a light image is transformed into an electronic image, it must be analyzed to extract such image information as object edges, regions, boundaries, color, and texture. This is called image analysis. Finally, once the image is analyzed, a vision system must interpret, or understand, what the image represents in terms of knowledge about its environment. This is by far the most difficult of all machine vision tasks and requires the direct application of artificial intelligence.

The next three sections of this chapter are devoted to a detailed discussion of these three areas of image technology. I should caution you, however, that this technology is in a state of continuous change and improvement. The intent of the following discussion is to provide you with those fundamental concepts that have been successfully applied in vision systems. This should give you a feel for the technology and provide you with the basics required for further study in this field. One other point: keep in mind that existing machine vision systems are relatively crude when compared to human vision. You will realize this in the last section of the chapter. Further developments in image technology, especially as related to image understanding, are sure to narrow the gap between human and machine vision. Such improvements will be the direct result of AI research and devleopment.

6-1 IMAGE TRANSFORMATION

As you are now aware, image transformation is the process of transforming a light image to an electronic image for use by a computer. Consequently, image transformation involves the electronic digitization of the light image. Once an image has been digitized, it can be analyzed using software techniques.

Any vision system, human or machine, can be broken down into two functional parts: the ***imaging device*** and ***image analysis/understanding*** system. This idea is illustrated by Figure 6-1. Image transformation is performed by an imaging device. The imaging device is the front end of the system, which acts as an image transducer to convert light energy to electrical energy. In humans, the imaging device is the eye. In a machine, the imaging device is a ***TV camera, photodiode array, charge-coupled device (CCD) array,*** or ***charge-injection device (CID) array.***

The function of the image analysis and understanding system should be self-explanatory. The brain performs this function in humans, while computer software is used in machines. Let's take a look at different imaging devices in preparation for a discussion of both image analysis and understanding. We will begin with a very natural imaging device, the human eye. A brief discussion of the eye will provide a foundation for a discussion of electronic imaging devices, like the TV camera, which model the human eye.

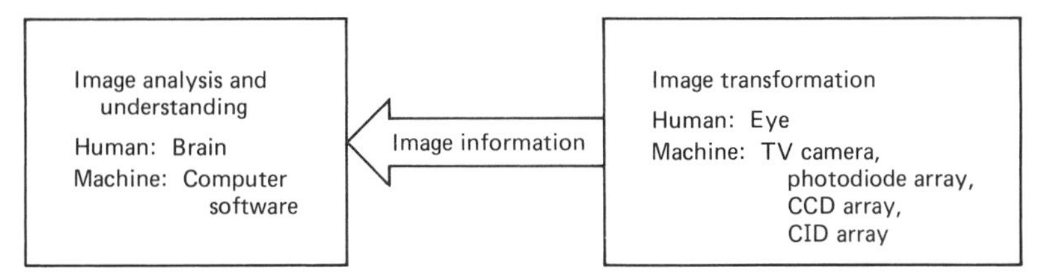

Figure 6-1 A computer vision system must perform three tasks: image transformation, analysis, and understanding.

Human Eye

A cross section of the human eye is shown in Figure 6-2. Light energy enters the eye via a transparent tissue called the ***cornea.*** The cornea begins to focus the image by bending the light and directing it through an opening called the ***pupil.*** The amount of light entering the pupil is controlled by the ***iris***, which increases or decreases the size of the pupil opening. Light energy is then passed through the ***lens***, where the image is focused onto the back layer of the eye, called the ***retina.*** Image focusing is achieved by changing the shape of the lens through muscular action.

The actual conversion of light energy to electrochemical energy is performed in the retina. The retina consists of a layer of cells called ***rods*** and ***cones.*** The rods react to the brightness levels within an image, while the cones react to color. Electrochemical signals generated by the rods and cones are then transmitted via the ***optic nerve*** to the brain for analysis and understanding.

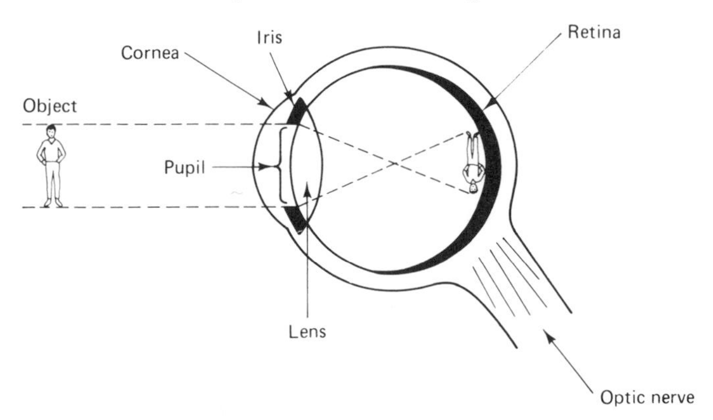

Figure 6-2 Cross-section of a natural imaging device, the human eye.

Television Camera

Standard mechanical cameras, as well as TV cameras, are modeled after the human eye. A simple mechanical camera is shown in Figure 6-3. Here the aperture setting governs the amount of light entering the camera, as the iris does in the human eye. The camera lens is moved back and forth to focus the image on the film. Light striking the camera film causes a chemical change within the film, which can be subsequently developed to reproduce the image.

In a TV camera, an electronic vacuum tube device, called a ***vidicon***, takes the place of the film in a mechanical camera and the retina of the human eye. The vidicon converts light energy to an analog signal, called a ***video signal***, that is used to represent the image.

Here's how it works. The vidicon operates using a photoconduction principle. Light enters the vidicon tube in Figure 6-4(a) through a focusing lens and strikes a

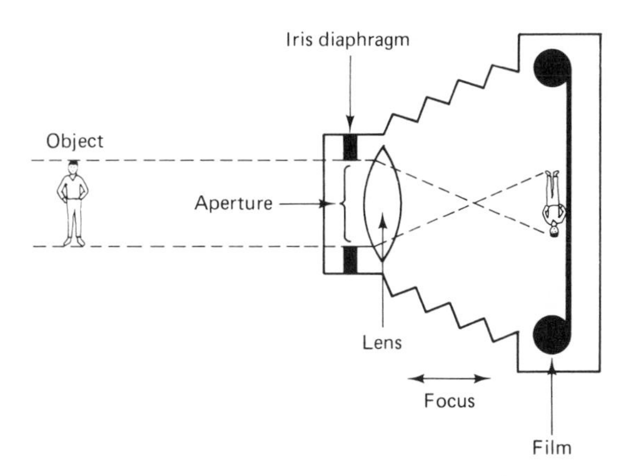

Figure 6-3 A typical mechanical camera is a model of the human eye.

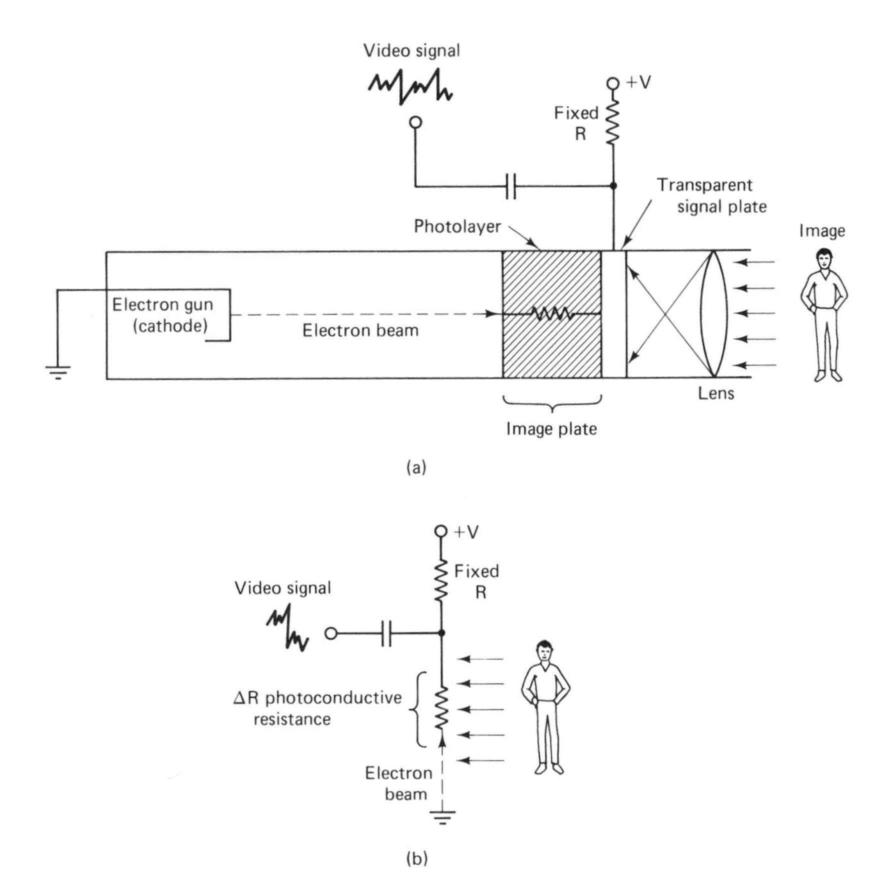

Figure 6-4 (a) Vidicon structure; (b) a simple vidicon model.

photoconductive ***image plate.*** The image plate has two layers. On the lens side, the plate is coated with a transparent conductive film called the ***signal plate*** of the vidicon, since the video signal is taken from this plate.

Light passes through the transparent signal plate into a photolayer made of a photoconductive material. This photolayer is an insulator whose resistance changes inversely to the light intensity. For instance, the resistance of the photolayer may be 20 megohms (MΩ) for a black image and decrease to 2 MΩ for a white image.

An electron beam completes the circuit to ground via the cathode of an electron gun. The electron beam scans the entire photolayer from left to right and top to bottom. At any given point within the scan, a voltage drop appears between the signal plate and ground, whose potential is directly related to the resistance of the photolayer at that point. In other words, the photolayer and electron beam form a resistor to ground whose voltage drop is proportional to the light intensity at that point within the electron beam scan.

Thus, the photoconductive resistance is part of a simple voltage-divider circuit, as shown in Figure 6-4(b). More light produces less resistance, which generates a smaller voltage from the circuit by Ohm's law. The resulting analog signal is capacitively coupled to the external circuit as shown. Thus, the image is represented by a series of voltages, or charges, across a coupling capacitor.

In summary, as the electron beam scans the image plate, the vidicon generates a time-varying voltage, or video signal. The amplitude of the video signal represents various shades of gray between 0 and 1.5 volts (V), as shown in Figure 6-5. Since

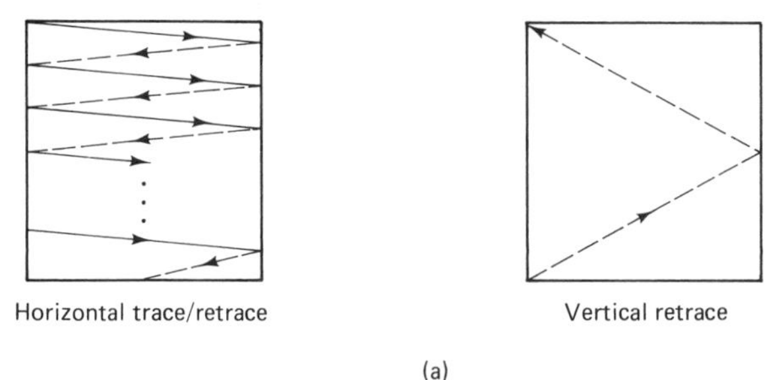

(a)

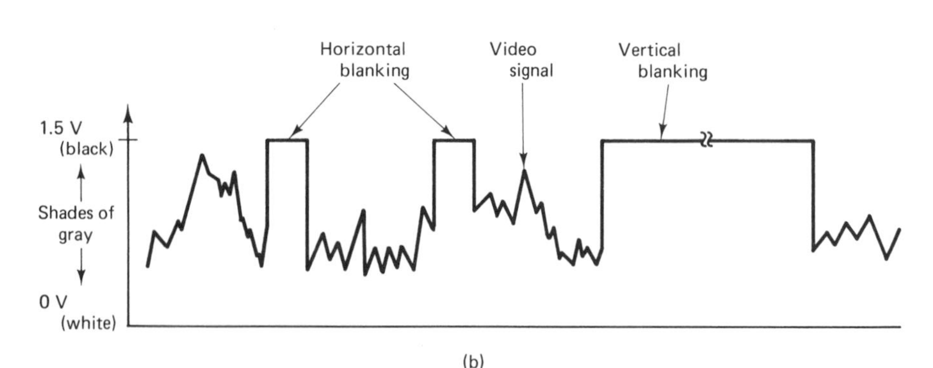

(b)

Figure 6-5 (a) Electron beam scan, and (b) resulting video signal.

the electron beam scans horizontal rows from top to bottom, it must retrace at the end of a given row to get to the beginning of the next row. In addition, when it reaches the bottom of the scan, it must retrace to get to the top of the scan. These retrace periods are called ***horizontal*** and ***vertical blanking***, respectively. Because of this, there are gaps in the video signal called ***blanking periods*** when horizontal and vertical retrace occurs.

Since the vidicon is a vacuum-tube device, a TV camera has several major disadvantages when used as an imaging device in vision systems. First, it is obviously bulky and fragile. A typical vidicon is from five to eight inches long, about one inch in diameter, and is enclosed in glass. Second, vidicons require relatively high voltage sources to operate (600 V) and must be calibrated frequently. Finally, like most vacuum tubes, vidicons have a relatively short life (5000 to 20,000 hours). The obvious solution to many of these inherent problems is found in a solid-state camera made from photodiodes, charge-coupled devices (CCDs), or charge-injection devices (CIDs).

In the preceding discussion, only a black and white TV camera has been considered. If a vision system is to make use of color information, a color TV camera must be used. I will defer a discussion of these cameras until the next section, where you will learn how color can be used to interpret visual information.

Photodiodes and Photodiode Arrays

A photodiode is a solid-state, light-sensitive device that uses a *PN* junction as shown in Figure 6-6. As you can see, the diode consists of a *P*-type island diffused into an *N*-type substrate. Metal electrodes are attached to the *P* and *N* materials to provide for external connection to the device.

When used as a light-detection device, the photodiode is reverse biased as shown. This produces a wide depletion region around the *PN* junction, resulting in a relatively large reverse bias resistance and a very small reverse current. When light strikes the *PN* junction, the light energy decreases the reverse resistance and allows more current to flow in the circuit. Consequently, the photodiode is a photoconductive device that acts like a light-variable resistor. As the light intensity in-

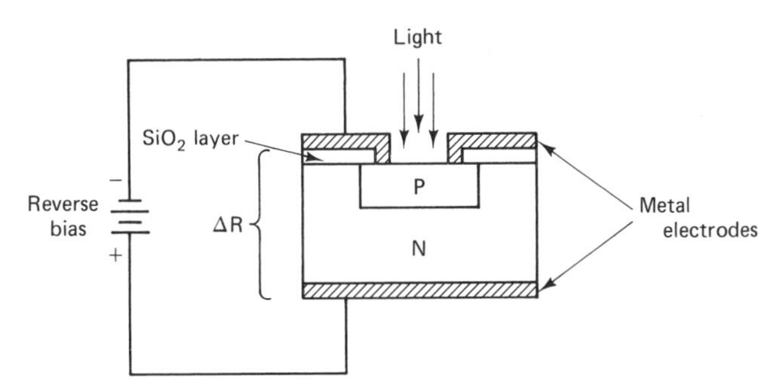

Figure 6-6 A photodiode acts like a variable resistor when reverse biased.

creases, the reverse resistance decreases, allowing more current to flow in the bias circuit.

Another type of photodiode is the *PIN* photodiode shown in Figure 6-7. Here, an undoped intrinsic (*I*) layer is sandwiched between the *P* and *N* materials. The intrinsic layer has a very high resistance, since it contains few impurities. As a result, a reverse-biased *PIN* diode has a higher reverse resistance than a standard *PN* diode. In a *PIN* photodiode, this higher reverse resistance allows it to be more sensitive to lower light frequencies, thereby making it useful over a broader spectrum of light. In addition, the higher reverse resistance of a *PIN* diode decreases the junction capacitance, resulting in a faster response to changes in light intensity.

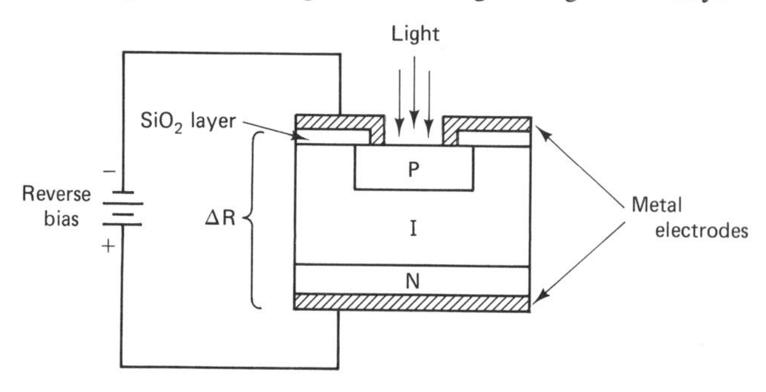

Figure 6-7 A *PIN* photodiode increases the light-frequency spectrum and has a faster response time.

Individual photodiodes can be used to detect the presence or absence of an object. However, they must be connected in an array fashion to be used as imaging devices for vision systems. Photodiodes are sometimes used to form the imaging plate of a vidicon, as shown in Figure 6-8. Here, an array of *P*-type islands is diffused into an *N*-type substrate. This produces an image plate made from an array of individual photodiodes. A 540 × 540 diode array is typical.

The diodes are reverse biased by applying a positive potential to the *N*-type substrate via the transparent signal plate. A ground potential is applied individually to the *P*-type islands by the scanning electron beam. Thus, when the electron beam scans a given photodiode, its reverse resistance is proportional to the light striking its junction. As a result, the photodiode acts as a variable resistor in a voltage-divider circuit to generate a video signal, as shown in Figure 6-8(b). As light strikes the array, the reverse resistance of the individual photodiodes changes according to the light variations within the image to produce a composite analog video signal of the image.

The advantage to using a photodiode array as the image plate in a vidicon is faster response to changes in light levels and increased sensitivity, especially in the infrared light region. Furthermore, the photodiode array is more resistant to burns from excess light intensity than the photoconductive image plate used in the vidicon described earlier. However, such an imaging device still has many of the inherent disadvantages associated with vacuum tubes, since it is enclosed in a glass envelope.

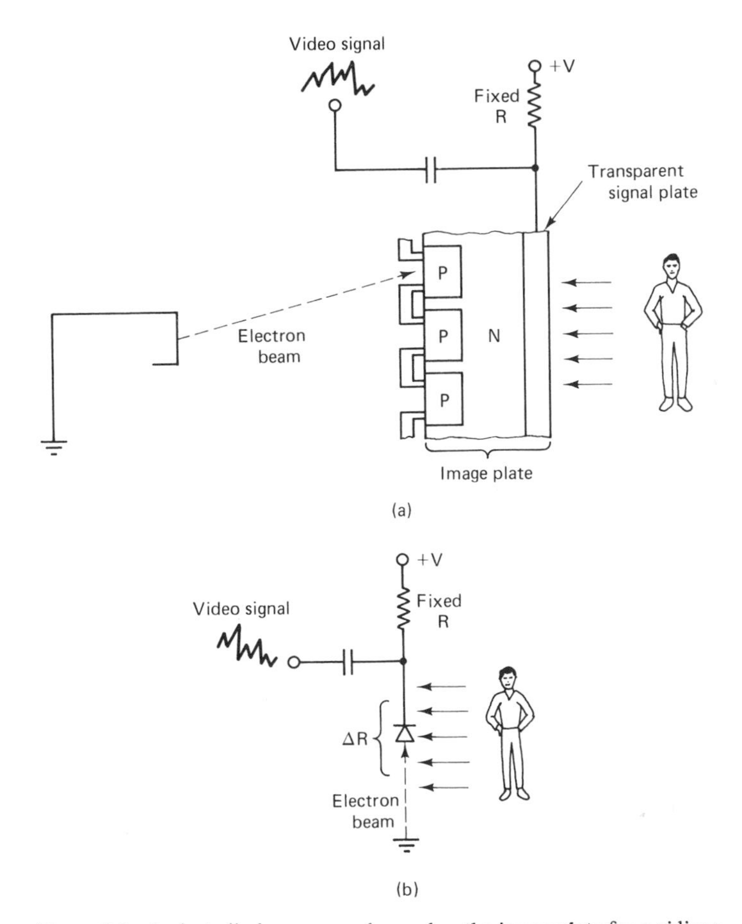

Figure 6-8 A photodiode array can be used as the image plate for a vidicon.

An all solid-state imaging device can be constructed using photodiodes as shown in Figure 6-9. Here the photodiodes have been integrated into a single chip and connected in a linear array. A series of integrated MOS transistors acts as switches to scan the array. An externally supplied clock pulse sequentially closes the MOS switches, reverse biasing the diodes one at a time. The current through a given diode is proportional to the light intensity striking its junction. The result is a continuous analog output signal that is proportional to the light striking the array.

Imaging devices using a linear array of photodiodes are called ***line-scan cameras.*** They are useful for determining size and shape and detecting the edge of an object. Line-scan cameras used for industrial applications require up to 2048 photodiodes in the linear array.

The linear array idea in Figure 6-9 can be expanded into a rectangular array by integrating a matrix of photodiodes within a single chip. Such a rectangular array requires two-phase clocking to control the scanning logic. Rectangular diode

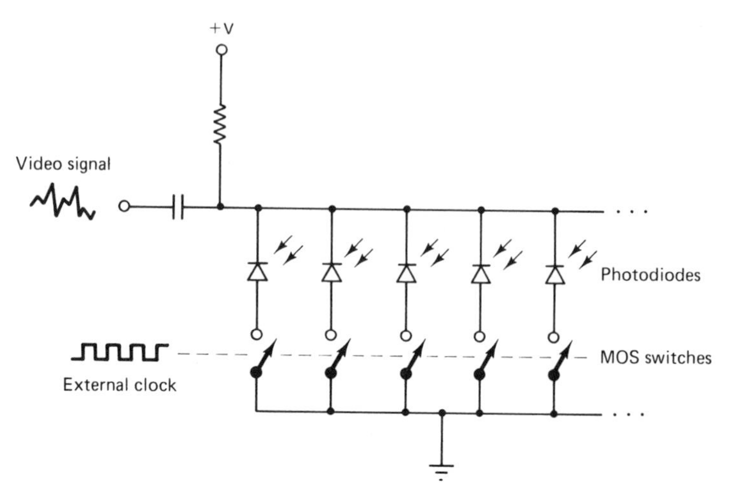

Figure 6-9 Linear photodiode array.

arrays add another dimension to the image over linear arrays. This allows them to be used like a vidicon as a two-dimensional imaging device, without the inherent disadvantages associated with vacuum tubes. Imaging devices that use a rectangular diode array are sometimes called ***diode-matrix cameras.***

Charge-coupled Device Arrays

Charge-coupled device (CCD) cameras are becoming extremely popular as imaging devices. The reason is their high sensitivity over a wide light spectrum, low power consumption, small size, and light weight.

A CCD array is constructed by integrating an array of MOSFET (metal oxide semiconductor field-effect transistor) devices. A partial cross section of a typical CCD cell row is shown in Figure 6-10(a). You can think of the individual cells as a series of integrated capacitors that temporarily store a series of charges.

To see how it works, look at the single cell in Figure 6-10(b). Observe that *P*-type semiconductor islands are grown into the *N*-type substrate to form the source and drain of a MOSFET device. The two islands are separated by an area of *N*-type material called a ***depletion region.*** The semiconductor material is then covered with an insulating layer of silicon dioxide, and two metal electrodes are placed over each *P* island to allow for electrical contact.

Light photons striking the array free electrons in the depletion region. This generates positive ions, resulting in a net positive charge in the depletion region. The charge is directly proportional to the light intensity striking the array. As a result, the array acts like a series of MOS capacitors, each holding a charge that is proportional to the light intensity at that point in the array.

The individual charges are transferred, or coupled, from one CCD cell to the next by applying clock pulses to the source and drain that are 180° out of phase. Hence, any given row of cells in the array acts like a serial shift register. Control

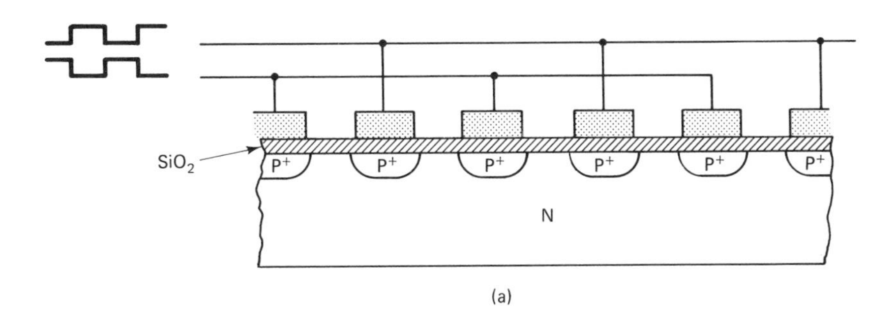

(a)

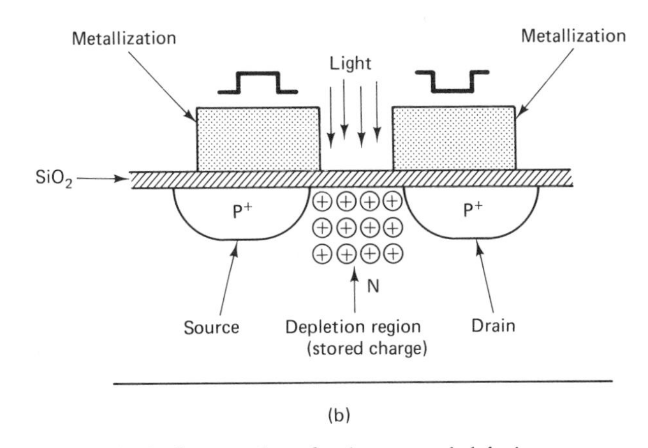

(b)

Figure 6-10 Construction of a charge-coupled device.

logic scans the array by shifting the charges in each row, one at a time, into an output buffer. The rows are scanned from top to bottom as shown in Figure 6-11 to produce a continuous analog output that forms a video signal.

The array scanning must be fast enough so that the stored charges do not dissipate. In fact, in most cases the video signal is generated much faster than it can be used. For this reason, a ***frame buffer*** is used to temporarily hold the output of the entire array until it can be stored by a computer.

Like photodiode imaging devices, CCD cameras are available in one- or two-dimensional arrays. The one-dimensional array is called a ***linear imaging device (LID)***, and the two-dimensional array is called an ***area imaging device (AID).*** Linear imaging devices contain from 256 to 2048 CCD cells. Two common sizes for area imaging devices are 190 by 244 cells and 380 by 488 cells.

Charge-injection Device Arrays

Charge-injection devices (CIDs) operate on a charge storage principle similar to CCDs. As light strikes a solid-state CID cell, a charge is *injected* into the cell that is proportional to the amount of light intensity at that point. The CID cells are

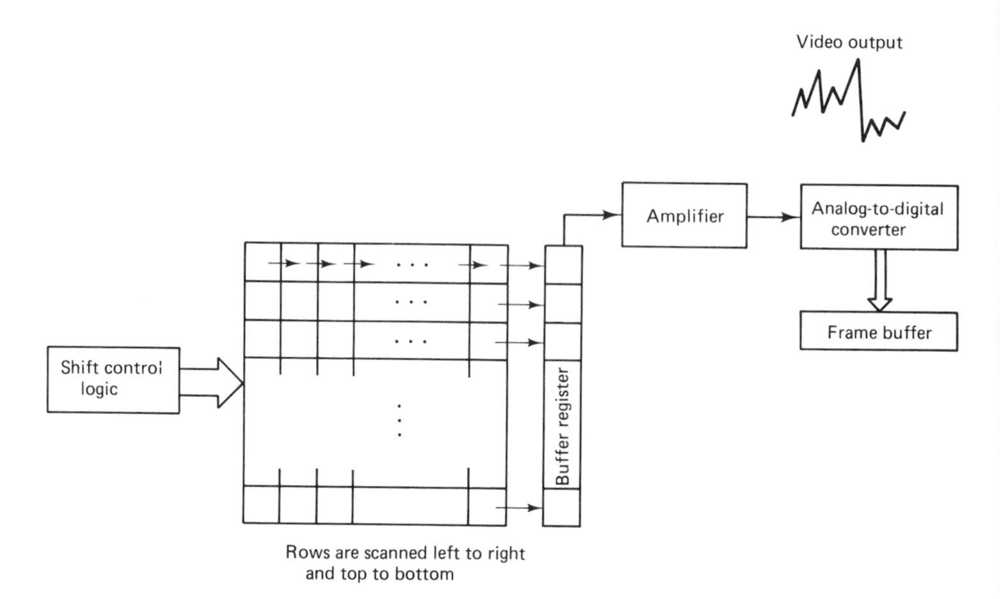

Figure 6-11 Control logic is used to shift the charges out of each row, one at a time, and generate a video signal.

arranged in a row/column array fashion and read using a row/column addressing technique, similar to the way that the bits of a computer memory are read. As a result, the stored charges do not have to be shifted as in the CCD array. They remain stationary and are read by the control logic. For this reason, you could say that the CID is a parallel access device, whereas the CCD array is a serial access device.

Both CCD and CID cameras have advantages of small size, light weight, low power consumption, high sensitivity, and wide spectral response. In addition, they are very cost competitive with other types of imaging devices. For these reasons, CCD and CID cameras are becoming increasingly popular for use in computer vision systems.

Digitizing the Image Signal

The output of the imaging devices discussed is a continuous analog signal that is proportional to the amount of light reflected from an image. To analyze the image with a computer, the analog image must be converted and stored in digital form. To do this, a rectangular image array is divided into small regions called ***picture elements***, or ***pixels***. Figure 6-12 illustrates this idea. With photodiode and CCD arrays, the number of pixels equals the number of photodiode or CCD devices.

The pixel arrangement provides a sampling grid for an A/D converter. At each pixel, the analog signal is sampled and converted to a digital value. Of course, control circuitry must be used to synchronize the pixel sampling with the image scanning.

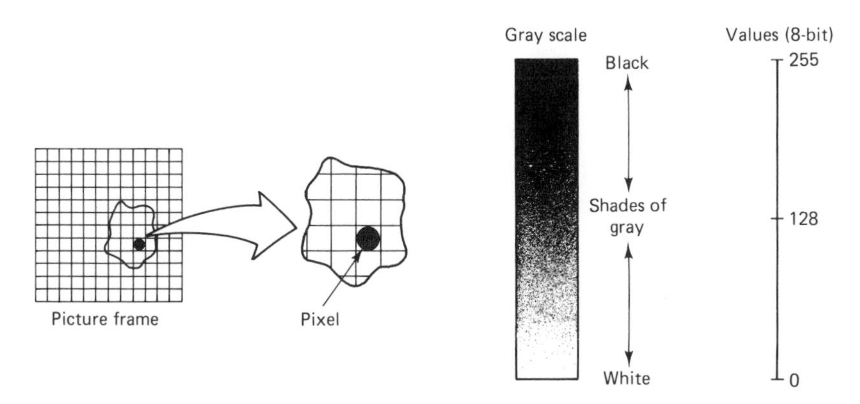

Figure 6-12 A picture frame is divided into picture elements, called pixels, for conversion to a gray-scale value.

With an 8-bit A/D converter, the converted pixel value will range from 0 for white to 255 for black. Different shades of gray are represented by values between these two extremes. This is why the term ***gray level*** is often used in conjunction with the converted values.

As the pixels are being converted, the respective gray level values are stored in a memory matrix. As a result, a given image is represented by a matrix of gray-level values. The resulting matrix is called a ***picture matrix***, for obvious reasons. Each gray-level value within the picture matrix has an associated row/column address so that it can be retrieved for image analysis.

With a TV camera system, up to 500 conversions are made for each horizontal scan line. To convert the image in real time, the sampling must follow the horizontal scan rate. A TV camera scans a horizontal line in about 63.5 microseconds (μs). Thus, up to 500 samples are converted to an 8-bit value every 63.5 μs. This translates to a data rate of approximately 63 megabits per second. Since a standard TV image frame has 525 lines, a single TV frame requires approximately 256k bytes of memory.

Example 6-1

Construct a 12 $\times$ 12 picture matrix for the digit 2 using an 8-bit gray-scale code.

Solution

The first thing you must do is divide the image into a 12 $\times$ 12 pixel grid as shown in Figure 6-13(a). We will assume that the digit is black on a white background. Thus, a black pixel is assigned a value of 255 and a white pixel is assigned a value of 0.

But what about those pixels along the *edge* of the digit? Think what the respective photodetector would "see." Wouldn't it detect the average brightness level over the pixel area? For instance, a pixel region that is half-black and half-white due to an edge passing through the region would be assigned a value of 256/2, or 128. Using this idea, try to estimate how much of the pixel is black and

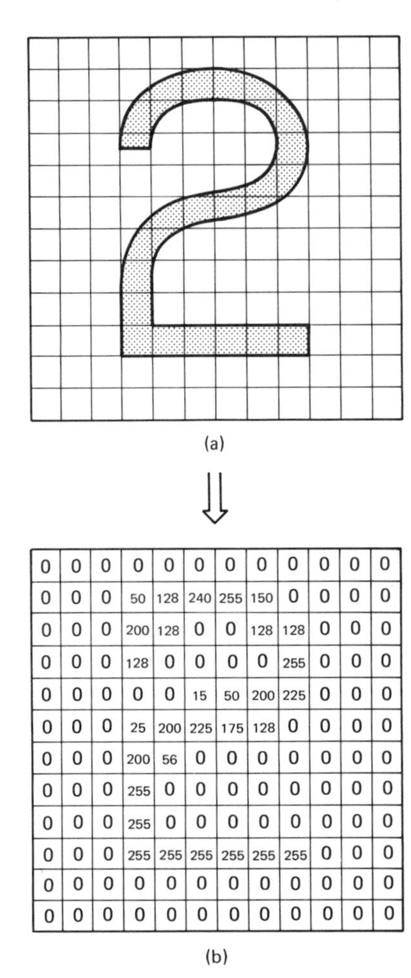

0	0	0	0	0	0	0	0	0	0	0	0
0	0	0	50	128	240	255	150	0	0	0	0
0	0	0	200	128	0	0	128	128	0	0	0
0	0	0	128	0	0	0	0	255	0	0	0
0	0	0	0	0	15	50	200	225	0	0	0
0	0	0	25	200	225	175	128	0	0	0	0
0	0	0	200	56	0	0	0	0	0	0	0
0	0	0	255	0	0	0	0	0	0	0	0
0	0	0	255	0	0	0	0	0	0	0	0
0	0	0	255	255	255	255	255	255	0	0	0
0	0	0	0	0	0	0	0	0	0	0	0
0	0	0	0	0	0	0	0	0	0	0	0

Figure 6-13 (a) a 12 × 12 pixel grid and (b) matrix for the number 2 (Example 6-1).

how much is white to arrive at an average intensity, or gray level, value. The resulting matrix is shown in Figure 6-13(b).

How might you detect the edges of the digit using a computer program?

Now that you are familiar with several imaging devices and how analog image signals are represented in a digital computer, it is time to learn about image analysis.

6-2 IMAGE ANALYSIS

Image analysis is a science in itself. Several volumes of text could be written on this subject alone. However, this complicated technology can be broken down into several fundamental topics that are common to all image-analysis systems.

Think, for a moment, how you go about recognizing images in a visual scene. What characteristics, or features, of the image do you key on to interpret what you see? How about image features such as edges, regions, shapes, colors, textures, and so on? These are all basic features that define an image. Thus, for a computer to understand an image, it must also analyze a given scene in terms of these fundamental image characteristics. Let's see how a computer might be programmed to analyze an image via these basic image features.

Edge Detection and Line Finding

Close your eyes for a few seconds and then open them. What image features do you first see that help you to identify the objects in your field of view?

Now, looking straight ahead, try to locate an object in your periphery vision. Again, what image features suggest the identity of a peripheral object?

One of the first things that you should notice are the edges of the objects in a scene. The edges of an object define its shape, thereby providing you with the basic information needed to identify the object. Think of a typical newspaper cartoon. Aren't the characters identified by simple line drawings, or edges? Don't these simple drawings provide you with enough information to understand the scene? Of course, other image information, such as color, makes the scenes more identifiable.

In the same way, a computer needs to locate edges in order to construct line drawings of the objects within a scene. Line drawings provide a basis for image understanding, since they define the shapes of objects that make up a scene. Thus, the basic reason for edge detection is that edges lead to line drawings, which lead to shape, which leads to image understanding. This idea is illustrated by Figure 6-14.

The starting point for edge detection is the digital picture matrix stored in memory. Recall that this matrix is composed of gray-scale values obtained from the variations in light intensity across the scene. Ideally, the picture matrix is an exact gray-scale representation of the scene. However, in reality the picture matrix includes signal noise that must be eliminated prior to edge detection.

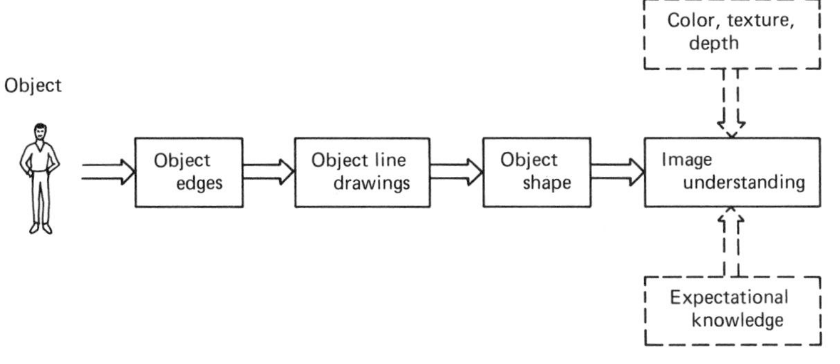

Figure 6-14 Edges lead to line drawings, which lead to shape, which leads to image understanding.

Noise Problem. Figure 6-15 illustrates the effect of noise on an edge signal. Notice that the noise distorts the precise location of the object edge. Ideally, the edge is represented by a distinct step between two intensity levels. Without noise, such a step would be easy to detect in the picture matrix, since there would be an abrupt change in gray-scale values across the edge. However, noise introduces intermediate gray-scale values that tend to make the precise detection of the edge difficult. In addition, the noise itself could be interpreted as a series of edges in a very sensitive system.

Picture matrix noise is usually the result of inaccuracies in the imaging device, illumination variations, or errors in the digitization process. Such noise results in small regions of the picture being much brighter or darker than they should be. There are basically two ways to eliminate noise: ***hardware filtering*** or ***software smoothing.***

Hardware filtering. As you can see from Figure 6-15, noise is usually seen as a high-frequency component riding on the gray-level signal. The obvious way to eliminate such noise is with a simple low-pass filter before the waveform is digitized. A low-pass filter will filter out these high-frequency noise components. However, most of the image detail, such as edges, is also represented by higher frequencies. Consequently, low-pass filtering might also filter out some of the edge detail, resulting in a blurred representation of the image.

If the noise is periodic and thus concentrated around a particular frequency, bandpass or bandstop filtering will do the job. This type of filtering can be used to eliminate noise frequency components without affecting the detail of the image.

Software smoothing. Smoothing is an image-enhancement technique that attempts to remove isolated bright and dark regions of a picture caused by signal noise. One way to smooth an image is to time expose the scene by generating multiple picture matrices of the same scene over a given time period. Each matrix will contain identical gray-scale picture information, but any random noise will not be represented the same in any two matrices. A composite matrix is then formed by

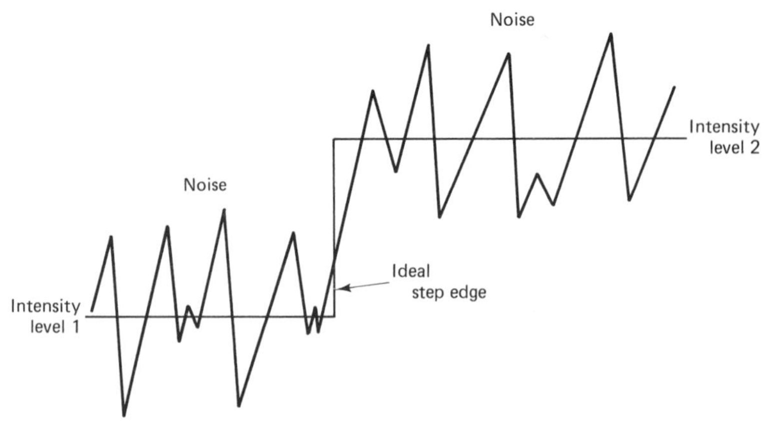

Figure 6-15 Effect of noise on an ideal edge.

averaging the corresponding gray-scale values of the multiple matrices. This will cancel out any ***salt-and-pepper noise*** due to random variations, without affecting the picture information. This technique is sometimes referred to as ***ensemble averaging.***

Another technique, called ***local averaging***, removes noise by replacing each gray-scale value in the matrix with the average of several values that occur in a surrounding pixel window. A ***pixel window*** is simply a small matrix of adjacent pixels that exists within the larger picture matrix. This technique will eliminate noise, but it also washes out and blurs the edges.

Example 6-2

Given a simple 4 X 4 picture matrix:

$$\begin{matrix} 9 & 9 & 9 & 3 \\ 9 & 9 & 3 & 3 \\ 9 & 3 & 3 & 3 \\ 3 & 3 & 3 & 3 \end{matrix}$$

Smooth this matrix using the local-averaging technique and a 3 X 3 pixel window.

Solution

There are four 3 X 3 pixel windows that can be identified in the matrix. If you replace the middle value in each window by the average of all the values within the window, you get

$$\begin{matrix} 9 & 9 & 9 & 3 \\ 9 & 7 & 5 & 3 \\ 9 & 5 & 4 & 3 \\ 3 & 3 & 3 & 3 \end{matrix}$$

The resulting matrix clearly shows the disadvantage of local averaging. There was an obvious, well-defined edge in the original picture matrix. However, local averaging produced a matrix that distorted, or blurred, the edge definition. This is the major problem with local averaging: noise is removed at the expense of blurring the image. You can decrease the window size to reduce the amount of blurring, but smaller windows remove less noise than larger windows.

The local-averaging smoothing technique operates on each element in the picture matrix, regardless of whether or not a given element is a noise element. As you observed in Example 6-2, a well-defined edge with no noise was blurred as a result of the smoothing process. To prevent blurring, salt-and-pepper noise can also be eliminated by finding those values that differ greatly from their surrounding values and replacing them with the average of the values in a surrounding window. This differs from local averaging in that only the extreme noise elements are replaced with an average value.

Finding the Edges. After smoothing, the next step is to find the edges within the picture matrix. The edges are found at the points that exhibit the greatest difference in gray-scale values within the matrix. Statistical techniques, such as

gray-scale transformation and ***histogram flattening***, can be applied at this point to enhance the contrast of the gray-scale matrix, thereby making the edges more detectable. Assuming the picture matrix has been smoothed and enhanced using one or more of the techniques described, how can you actually find the edges within the matrix? (*Hint:* What is the first derivative of a step function?)

Look at the ideal edge represented by the gray-scale intensity function in Figure 6-16(a). What is the slope of the function where the edge occurs? From calculus, you know that the slope of a step edge approaches infinity, as illustrated in Figure 6-16(b). Even if the edge were not ideal, the slope of the intensity function at an edge would be very high, since an edge is found where there is a large difference between adjacent gray-scale values. Using this idea, all you have to do to find edges is calculate the first derivative between adjacent gray-scale values. The first derivative of the gray scale intensity function is called the ***gradient.***

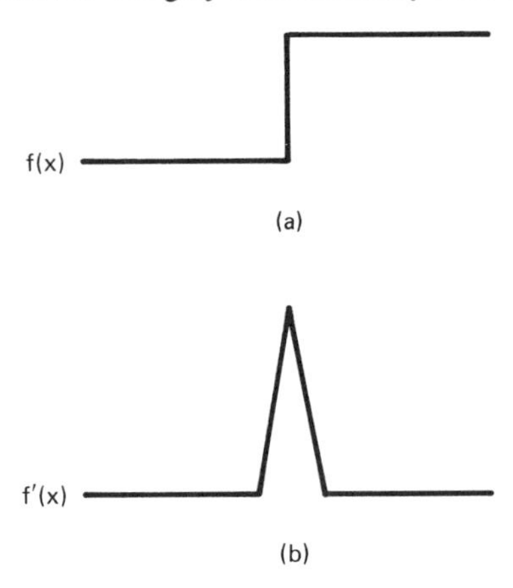

Figure 6-16 (a) An ideal step edge has a (b) first derivative approaching infinity.

Many successful image-analysis systems have employed this first derivative, or gradient, technique for edge detection. A block diagram of such an edge-detector system is shown in Figure 6-17. Observe that the edge detector consists of a ***pixel differentiator*** and a ***threshold detector.*** The function of the edge detector is to convert the gray-scale picture matrix into a ***binary picture matrix*** that can be used for finding lines that define the image.

Pixel differentiator. As shown in Figure 6-17, the pixel differentiator must operate on the digital gray-scale picture matrix stored in memory. The obvious

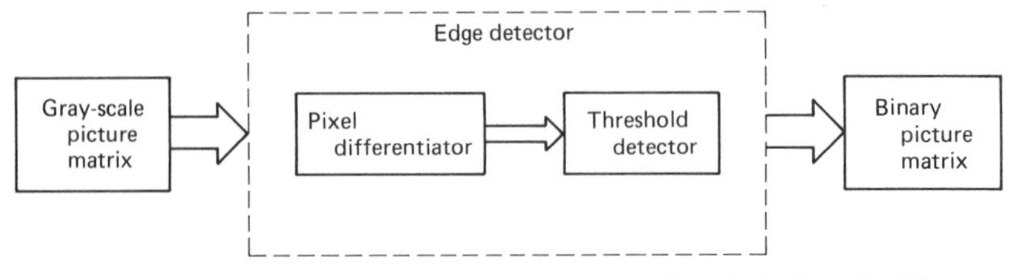

Figure 6-17 An edge-detector system based on the first-derivative principle.

question at this point is how to differentiate a digital image. In 1965, L. G. Roberts developed the ***Roberts cross operator***, which approximates the first derivative, or gradient, of a digital image. The Roberts cross operator is defined as follows:

$$R(i, j) = \sqrt{[i(m + 1, n + 1) - i(m, n)]^2 + [i(m, n + 1) - i(m + 1, n)]^2}$$

where $i(m, n)$ is the image intensity of pixel (m, n).

The operator computes the sum of the squares of the differences between matrix, or pixel, elements. Sound familiar? A similar technique is used for matrix comparison in speech recognition. In speech recognition the operator is applied to corresponding elements from two separate matrices. In image analysis, the operator is applied to diagonal pixels within a 2 X 2 window of a single picture matrix. The general 2 X 2 pixel window used by the Roberts operator is shown in Figure 6-18.

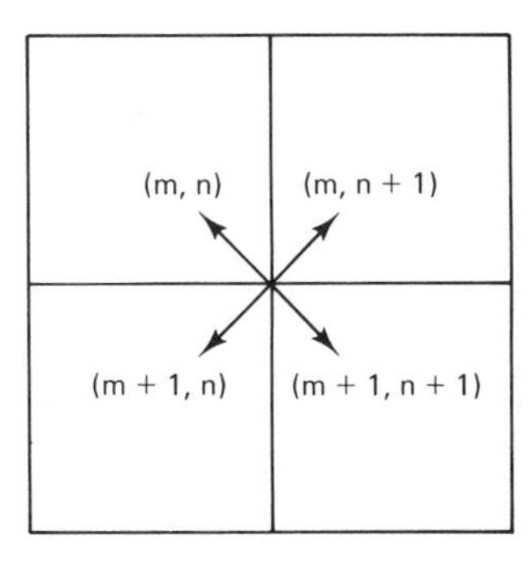

Figure 6-18 The 2 × 2 window used for the Roberts cross operator.

Example 6-3

Replace the gray-scale values of the following picture matrix with values obtained using the Roberts operator. If a Roberts operator value cannot be obtained for a given pixel, replace that pixel with an X.

```
9 9 9 3
9 7 5 3
9 5 4 3
3 3 3 3
```

Solution

The matrix is the one that you obtained during the smoothing operation in Example 6-2. Applying the Roberts operator formula to each 2 X 2 window in this picture matrix, you get

```
2.0  4.5  6.3  X
4.5  3.0  2.2  X
6.3  2.2  1.0  X
 X    X    X   X
```

For instance, the value 7 in the original matrix is replaced with a value of 3.0, since

$$\begin{aligned} R(i, j) &= \sqrt{i(m+1, n+1) - i(m, n)]^2 + [i(m, n+1) - i(m+1, n)]^2} \\ &= \sqrt{(4-7)^2 + (5-5)^2} \\ &= \sqrt{-3^2 + 0^2} \\ &= \sqrt{9} \\ &= 3 \end{aligned}$$

X's are placed where the required 2 X 2 window extends beyond the matrix.

Threshold detector. The function of the ***threshold detector*** is to decide which elements of the differentiated picture matrix should be considered as edge candidates. Edges are found by applying the Roberts operator to each intensity value in the picture matrix and comparing the resulting gradient approximation, $R(i, j)$, to a threshold level, T. An edge is present if $R(i, j) > T$. This technique is called ***thresholding.***

During the thresholding operation, the differentiated gray-scale picture matrix is converted to a binary picture matrix as follows:

- If the Roberts operator value exceeds the threshold level for a given pixel, that matrix element is set to a value of 1.
- If the Roberts operator value is less than or equal to the threshold level, the corresponding matrix element is cleared to 0.

Example 6-4

Using the Roberts matrix obtained in Example 6-3, construct a binary matrix using a threshold level of 4. Then construct a binary matrix using a threshold level of 6.

Solution

To construct a binary matrix, you must set a pixel value to 1 if the Roberts operator exceeds the threshold; otherwise it is cleared to 0. Using a threshold level of 4 on the differentiated matrix in Example 6-3, you get the following binary matrix:

$$\begin{matrix} 0 & 1 & 1 & X \\ 1 & 0 & 0 & X \\ 1 & 0 & 0 & X \\ X & X & X & X \end{matrix}$$

Here each matrix element over the threshold value of 4 is set to a 1, while those less than 4 are cleared to a 0.

For a threshold level of 6, the binary matrix is

$$\begin{matrix} 0 & 0 & 1 & X \\ 0 & 0 & 0 & X \\ 1 & 0 & 0 & X \\ X & X & X & X \end{matrix}$$

In both of the preceding binary matrices, the logic 1 elements are edge-point candidates. Observe that, if you connect the 1's in both of these matrices, you will see the edge. With a threshold level of 4, the edge is a bit distorted due to the blurring caused by the smoothing operation in Example 6-2. If the threshold level is increased to 6 to compensate for the blurring, gaps appear between the edge points.

The threshold level, T, is extremely critical. If images had perfect contrast and were completely noise free, T could be 0. Since this is not the case, a threshold level must be chosen small enough to detect all the actual image edges and large enough to eliminate most of the noise edges. A threshold level is normally chosen based on the analysis of typical images. In addition, statistical techniques can also be used to determine the threshold level for a given picture matrix, or window within the matrix, during the thresholding process.

As you can see from Example 6-4, once the edge points are detected, you must connect these points to find the lines that define the image.

Finding the Lines. Finding lines from possible edge points in a picture matrix is difficult. In fact, this is a major area for research in vision systems, since many of the image-understanding techniques that you will study in the next section are based on proper line extraction.

I will discuss three popular techniques for finding lines from an edge-point matrix: ***tracking, model matching,*** and ***template matching.*** Many successful line-finding systems have used one or more of these techniques.

Tracking. Tracking, or tracing, is simply a method of connecting the dots. Most current tracking techniques assume an ideal picture matrix. That is, light intensity is constant over an image until an edge is encountered, where the light intensity changes abruptly. This assumption is fine if the objects in the scene are smooth and nontransparent, the picture matrix is completely free from noise, and there are no shadows to contend with. As you are aware, this is hardly the case in the real world. Nevertheless, some vision applications can be controlled to generate these ideal conditions. Let's look at a simple tracking algorithm that operates under these ideal conditions as a starting point for further study.

Suppose that the edge points of a gray-scale picture matrix have been detected via thresholding, resulting in a binary picture matrix. The following three steps give a simple tracking algorithm that will connect the edge points to find the lines in a binary matrix.

1. Scan the binary picture matrix from left to right and top to bottom.
2. When an edge point, or binary 1, is encountered, connect it to the previous edge point and examine the matrix elements in a 3×3 window around the element for a binary 1 neighbor.
3. (a) If a single binary 1 neighbor is found, repeat step 2.
   (b) If two or more binary 1 neighbors are found, repeat step 2 for one of them. Remember the remaining neighbor locations within the window for future examination.
   (c) If no binary 1 neighbors are found, the present element is a line termination. Go and examine any previously found edge elements that have not been examined per step 2 or continue the scan per step 1.

This algorithm is simple and sounds like a reasonable way to connect edge points. Its major drawback is that it assumes that the lines are just one pixel wide. In addition, the algorithm tries to connect false edges due to noise elements and does not bridge large gaps between real line elements. For these reasons, the tracking algorithm must be much more sophisticated. In addition to using statistical analysis, such an algorithm might incorporate heuristic knowledge about what to expect in the scene. In this way, the tracking algorithm can verify whether or not a given line should exist through heuristic models. Moreover, heuristic knowledge can be used to bridge the gaps between line segments.

Another form of tracking used for line finding is the ***iterative end-point fit*** method. This method finds the two most extreme edge points in a matrix window and approximates a line connected between these end points, as shown in Figure 6-19(a). It then looks to see if edge points (binary 1's) fall on the line. If not, it chooses the most distant point from the line and replaces the single line with two

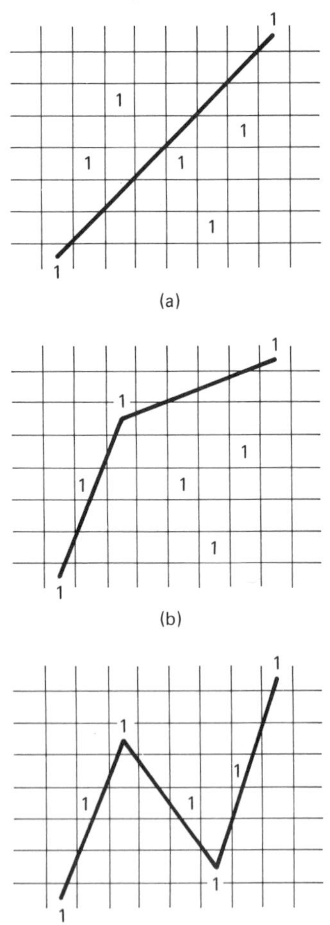

Figure 6-19 Iterative end-point method for connecting edge points.

lines, as shown in Figure 6-19(b). The process is continued as shown in Figure 6-19(c) until a series of line segments is found to match the edge-point pattern.

Model, or pattern, matching. Model matching is sometimes referred to as ***edge fitting,*** since the idea is to fit a general model of an edge to the picture matrix. The model, or pattern, is generally an intensity profile of an ideal edge.

A typical general-purpose edge model, called the ***Hueckel model***, is shown in Figure 6-20. Here the ideal edge is within a circular neighborhood. The edge is between two intensity levels, I and $I + h$. Its location within the neighborhood is defined by the vector $r\angle\theta$. The idea is to pass this general model over all areas of the picture matrix. At each point, the values of $I, I + h, r,$ and θ are varied in the model to achieve the best fit, or match, to the neighborhood under examination. A function, called the ***Hueckel operator***, is used to compute the difference between the ideal model and the area under examination. As the model variables are changed, the Hueckel operator generates a difference value. An edge exists where this difference value is minimum. The values of the model variables at that point define the edge location, as well as the intensity levels on either side of the edge.

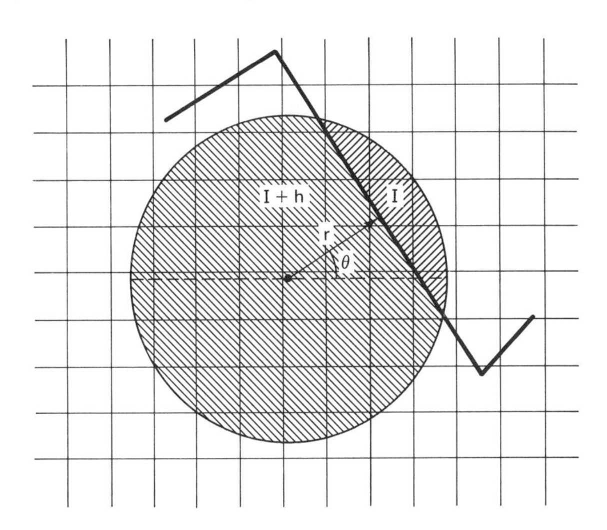

Figure 6-20 Hueckel ideal edge model.

Template matching. It is obviously impossible to store complete image templates for every possible image variation. However, edges can be found using edge masks like the ones shown in Figure 6-21. Such edge templates can be used to find horizontal, vertical, and diagonal edges using a template-matching technique.

As you can see, the edge masks form windows that are compared with every part of the binary picture matrix. A match is made when two or more 1's in the template match two or more 1's in the binary picture matrix. Template matching using edge masks is sufficient to locate the edges and determine the outline of simple block-type images.

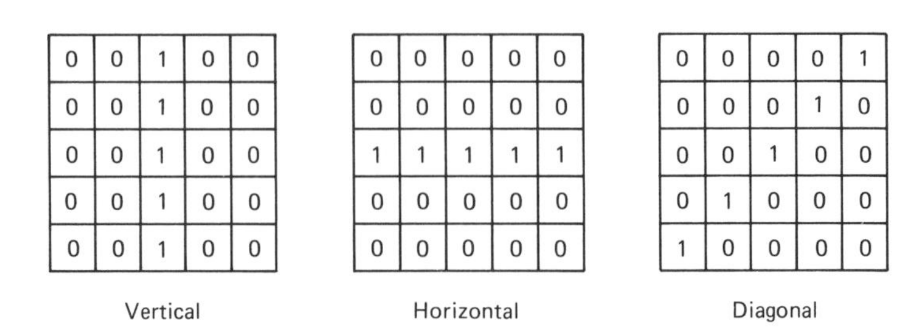

Figure 6-21 Templates used for line finding.

Region Analysis

The regions of a scene form another fundamental image characteristic. Region analysis is an alternative to edge detection, since regions can also be used to define shapes in an image. However, region analysis is just the opposite of edge detection, because it looks for similarities between pixels, rather than abrupt changes between pixels. A big advantage that region analysis has over edge detection is that there are no gaps in the shapes defined by regions, as there usually are with edge-based analysis. This aids in the image-understanding task.

The idea behind region analysis is to look for groups of pixels that are the same, relative to some image feature. Three image features are useful for region analysis: ***gray-scale intensity, color,*** and ***texture.*** For example, regions can be segmented due to color or texture, with different color or texture regions defining different surfaces in the image. The two most common methods used to segment regions are ***region splitting*** and ***region growing.***

Region Splitting. Region splitting involves thresholding and is the most obvious and simple approach to region analysis. The basic idea is to split a scene into several large regions made up of pixels with the same feature value, such as the same intensity range. These regions are then treated separately and split again using tighter constraints, or different image features such as color. Splitting continues until no more distinct pixel groups can be found.

With this method, pixels are grouped according to whether or not they fall within a certain image feature range. In its simplest form, all the pixels above a certain intensity threshold level form one group, while all those below the level form another group. Regions are then split off by collecting pixels of the same intensity values that are neighbors of each other.

For example, consider the two rectangle surfaces in Figure 6-22(a). An associated ***pixel histogram*** of this scene is shown in Figure 6-22(b). The histogram shows the number of pixels at different intensity levels. The two rectangles are two different shades of gray and thus are represented by two different gray-level intensity values, I_1 and I_2. Notice how the pixels collect around these two levels to form a bimodal, or "camel-back," type of distribution. A convenient threshold level is halfway between the two distributions, in the valley between the peaks. The pixels with intensity I_1 define region 1, while those with intensity I_2 define region 2.

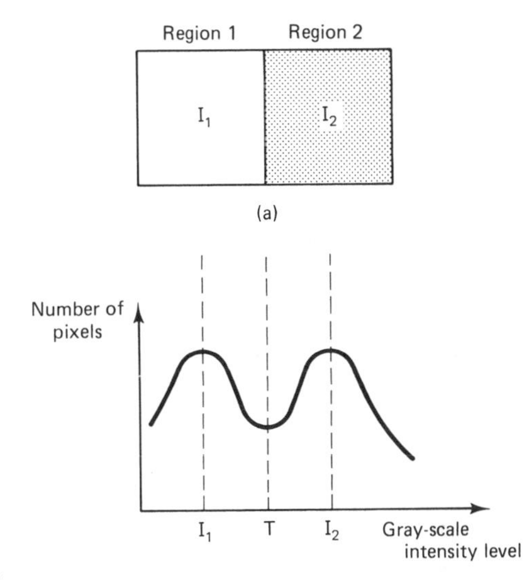

Figure 6-22 A pixel histogram (b) shows how intensity regions (a) can be split.

Now suppose the scene is a bit more complicated, like the one shown in Figure 6-23. Again, a pixel histogram is constructed. The peaks of the histogram indicate the individual regions, since each region will contain several pixels of the same intensity value. If there are several broad peaks in the histogram, each peak should be examined and split further, since a given region might contain several surfaces whose gray scale levels are similar. Additional histogram plots can be made

Figure 6-23 A complex scene candidate for region splitting.

according to other image features, such as color and texture. The peaks of these histograms will further split the scene into new regions. The pixels that make up a given peak should be split out and grouped by neighborhoods, since several regions might exist with the same feature. For example, there might be two red surfaces in the image that are not connected and thus belong to two different pixel neighborhoods.

As you can see, region splitting is a recursive process, since regions are split according to different features until no more splitting is possible. For this reason, region splitting is sometimes called ***recursive segmentation.***

Region Growing. Region growing is just the opposite of region splitting. With this method, neighboring pixels that fall within a given image feature range are grouped to form small regions called ***atomic regions.*** The predefined feature range, such as a narrow range of intensity levels, is used to control the amount of growth.

Neighboring atomic regions with similar features are then merged. The regions continue to grow until adjacent pixels are encountered whose features differ significantly from those of a given region.

The region-growing idea is illustrated by Figure 6-24. In general, the process begins by picking any atomic region within the scene. The atomic regions immediately adjacent to this one are then examined for possible merger based on a predefined feature range. If merger is possible, the respective adjacent regions are merged with the initial atomic region to form a larger region. Next, any atomic regions adjacent to this region are examined and merged if possible to form an even larger region. This growing process continues until no more adjacent regions meet the merger criteria. When this happens, a new atomic region is chosen for expansion, and so on, until all the atomic regions have been expanded or included as part of another region.

A tree structure can be constructed as shown in Figure 6-25 that represents

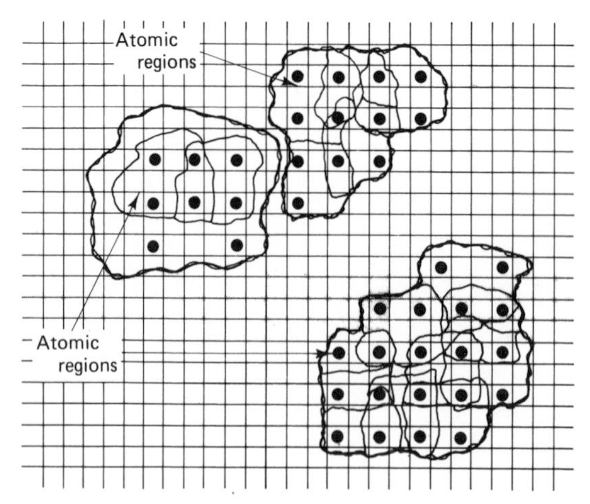

Figure 6-24 Region-growing process.

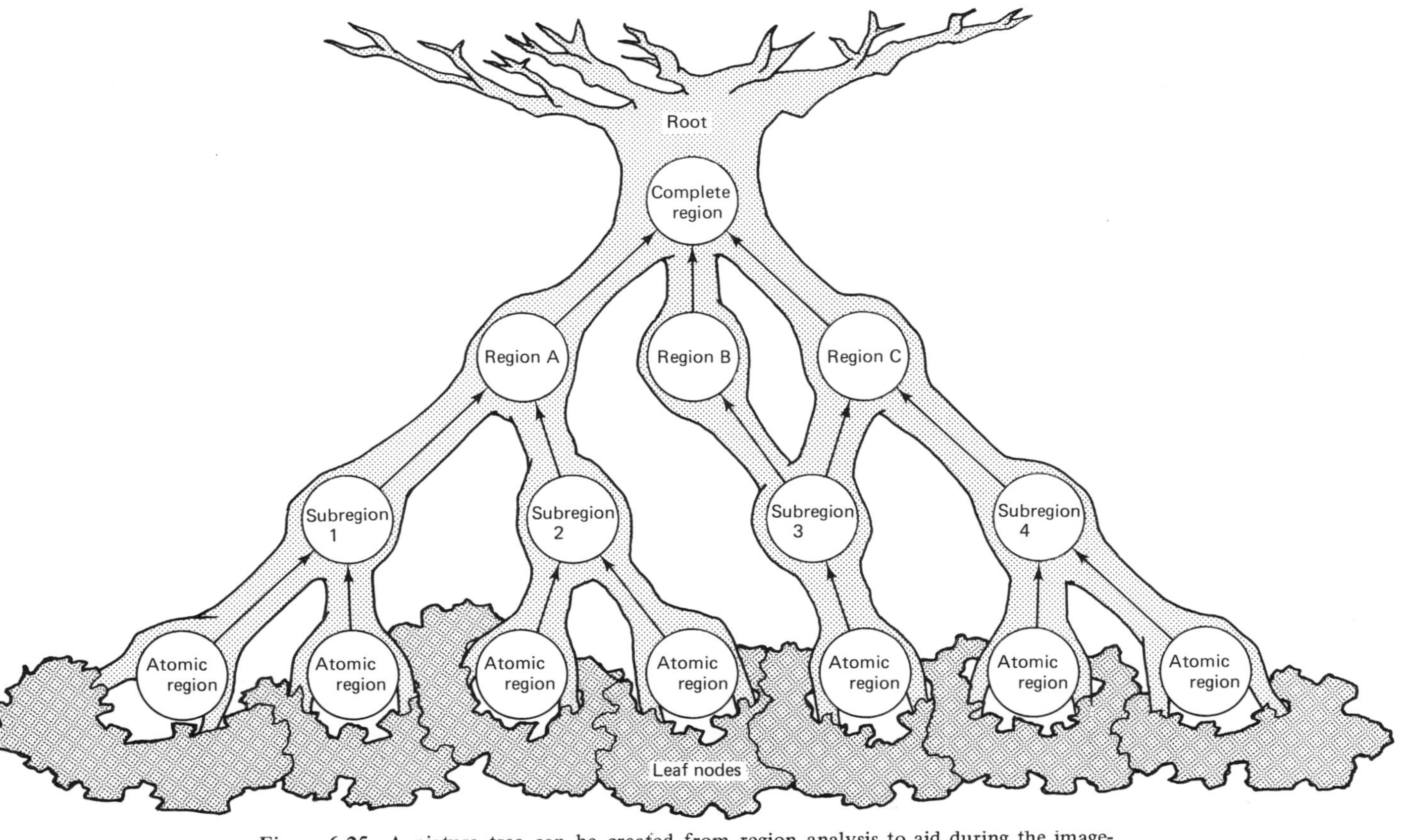

Figure 6-25 A picture tree can be created from region analysis to aid during the image-understanding task.

the output of the region-growing system. Here the leaf nodes represent the atomic regions, which are contained in several larger regions that make up the entire image, or root node, of the tree. Such a tree is called a ***picture tree***, for obvious reasons.

A picture tree is useful during the image-understanding process, especially if new information, such as heuristics, is introduced to suggest a different regional structure. If this happens, the image-understanding software can modify the picture tree according to any new information. Without the tree, the system is stuck with the interpretation provided during image analysis.

You have seen from our discussions that color and texture can be important during the image-analysis phase of a vision system. Thus, it is probably a good idea to take a brief look at these image features before leaving image analysis.

Color

Color information can be used in vision systems to detect both edges and regions, as well as to identify objects. People who are color-blind will quickly testify to the importance of color for visual perception. Likewise, a color-blind vision system is sometimes handicapped when attempting to detect edges, regions, and identify objects. In particular, large regions and surfaces can be identified more easily using general color information rather than detailed line drawings. Because of this, color is especially important in satellite vision systems.

Color information adds one more piece to the image-understanding puzzle. Unfortunately, most image-understanding research to date has been with black and white images. However, color is sure to aid in future research because of the additional image information it provides. Let's see how color can be converted into electrical signals and represented in a computer to aid in the image-understanding task.

Defining Color. Recall that the human retina is made up of rods and cones. The rods respond to brightness levels, while the cones react to color. It is thought that there are three types of cones that respond to the three primary colors of red, green, and blue. The graph in Figure 6-26(a) shows how the human eye responds to these three colors, while the color diagram in Figure 6-26(b) shows how any color can be made by various combinations of red, green, and blue. To interpret color, your brain averages, or integrates, the electrochemical signals generated by the retina color sensors.

A color is defined by three attributes: ***hue, saturation,*** and ***brightness.*** The color of an object is distinguished primarily by its hue, or ***tint.*** Different wavelengths of light produce different hues, which look like different colors. The ***shade*** of a color is determined by its saturation. The difference between red and pink is due to the saturation level of the primary hue, red. How would you make pink from red? Of course, by adding white. Thus, saturation is a measure of how little a primary color is diluted with white. Finally, the brightness of a color is a measure

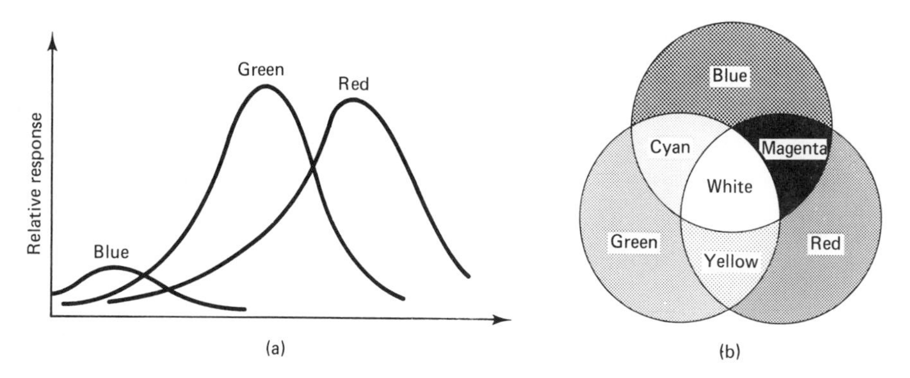

Figure 6-26 (a) Response of retina to red, green, and blue; (b) addition of the primary colors.

of the amount of light intensity present. You can see this when color images are converted to black and white, or gray-scale, images. Yellow objects look brighter than blue objects, since they have more light intensity.

The chart in Figure 6-27(b) is called a ***chromaticity diagram***. It shows how any color can be defined using a set of xy coordinates. Notice it forms a triangle with the three primary colors of red, green, and blue at the triangle vertices. Any color can then be presented by the coordinates (x, y), where

$$x = \frac{G}{R + G + B}$$

$$y = \frac{R}{R + G + B}$$

and R, G, and B are the amounts of red, green, and blue present, expressed as a fractional part of the mixture.

Example 6-5

Suppose a color is composed of equal amounts of red, green, and blue.
(a) What are the color coordinates on the chromaticity diagram?
(b) What is the color?

Solution

(a) With equal amounts of red, green, and blue, each color must contribute one-third to the mixture. Thus, $R = G = B = 1/3$, and

$$x = \frac{1/3}{1/3 + 1/3 + 1/3} = \frac{1/3}{1} = 1/3$$

$$y = \frac{1/3}{1/3 + 1/3 + 1/3} = \frac{1/3}{1} = 1/3$$

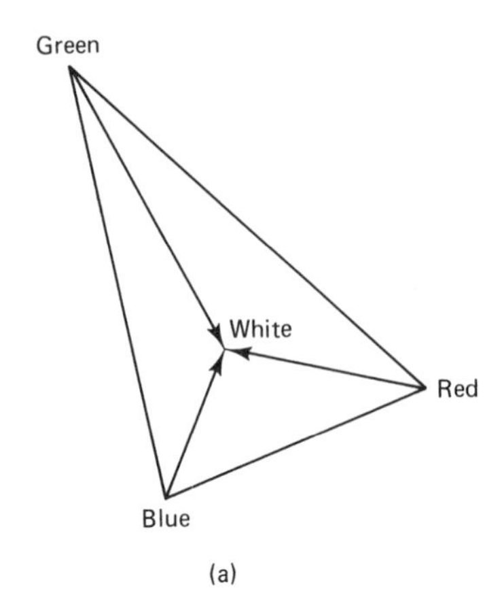

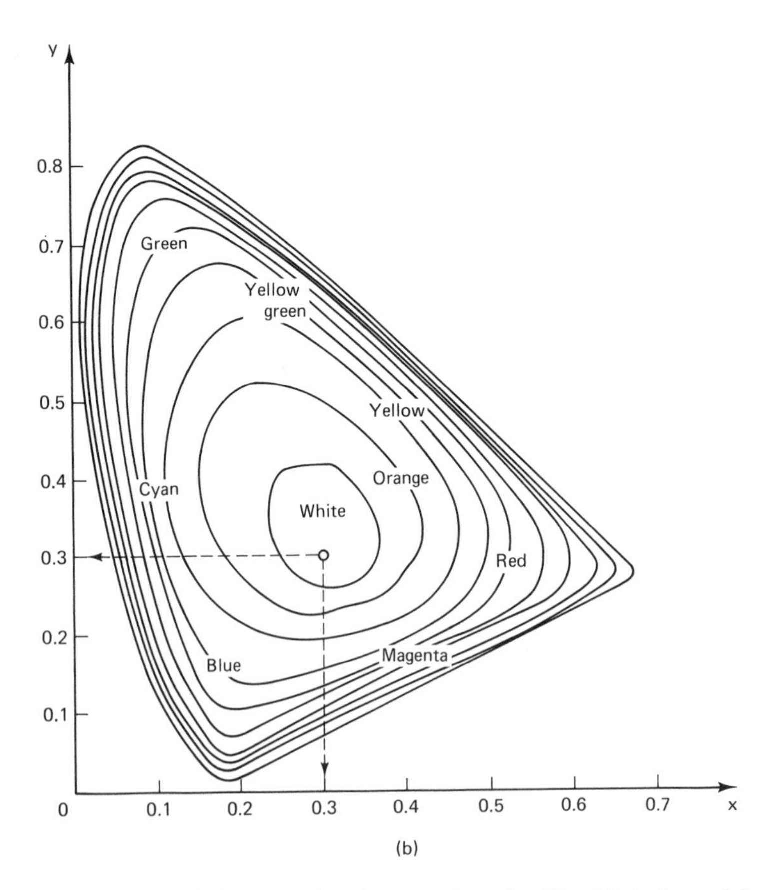

Figure 6-27 (a) Color saturation decreases toward white. (b) A chromaticity diagram is used to chart color.

(b) From the chromaticity diagram in Figure 6-27(b), you find that the color is white.

Now suppose that the axis in Figure 6-27(b) is transformed to make white the origin of the coordinate system. Also suppose that the chromaticity diagram is rotated to place red near the y-axis and blue near the x-axis. The resulting graph is called a ***color triangle*** and is shown in Figure 6-28(a). Notice that the y-axis has

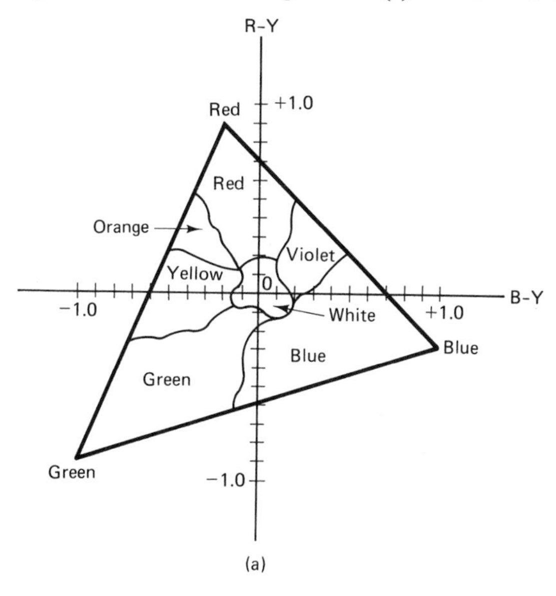

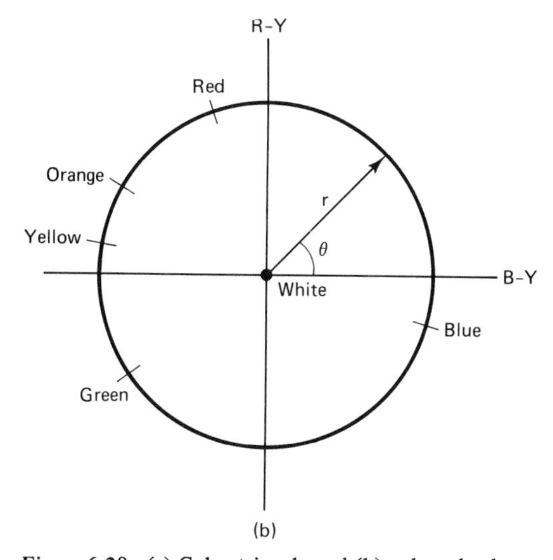

Figure 6-28 (a) Color triangle and (b) color wheel.

been relabeled *R-Y* and the *x*-axis relabeled *B-Y*. Next, if you cut out a circle in the triangle, you get the color wheel shown in Figure 6-28(b). Using the color wheel, any color can now be represented as a vector, $r\underline{/\theta}$, where the vector direction, θ, identifies the hue and the length of the vector, r, indicates the amount of saturation.

What about brightness? A chromaticity diagram assumes a constant brightness level. Consequently, brightness cannot be represented by a color triangle or circle. However, an approximation of the relative brightness, or light intensity, can be calculated as follows:

$$Y = 0.3R + 0.59G + 0.11B$$

where R, G, and B range from 0 to 1, depending on the saturation level of the respective color.

This equation says that color brightness ranges from 0 for black, since no colors exist, to 1 for white, since all three colors must be present. The value of Y is the gray-scale equivalent of the color brightness.

The reason for labeling the axis in Figure 6-28 as *R-Y* and *B-Y* is that you must subtract the brightness value, Y, from both the R and B levels to get the *R-Y* and *B-Y* coordinates required for the color triangle.

Example 6-6

Suppose that you have three TV cameras, one for each of the three primary colors of red, green, and blue. Each camera generates a signal between 0 and 1 volt (V), depending on the saturation level of the respective primary color it detects at a given point in its scan. Now assume that the output of the three cameras is as follows for a given point on an image:

$$\text{Red camera} = 0.8 \text{ V}$$
$$\text{Green camera} = 0.2 \text{ V}$$
$$\text{Blue camera} = 0.1 \text{ V}$$

(a) What is the gray-level intensity of the color?
(b) What is the color?

Solution

(a) Since each camera generates a signal from 0 to 1 V, depending on the respective color saturation, the given voltage outputs can be used directly in the brightness-level equation as follows:

$$\begin{aligned} Y &= 0.3R + 0.59G + 0.11B \\ &= (0.3)(0.8) + (0.59)(0.2) + (0.11)(0.1) \\ &= 0.4 \text{ V} \quad \text{(approx.)} \end{aligned}$$

(b) To determine the color, you must calculate the *R-Y* and *B-Y* coordinates and use the color triangle in Figure 6-28. Using the gray-level intensity value obtained in part (a), the *R-Y* and *B-Y* coordinates are

$$R\text{-}Y = 0.8 - 0.4 = 0.4$$
$$B\text{-}Y = 0.1 - 0.4 = -0.3$$

From the color triangle in Figure 6-28, you see that the color is orange.

To duplicate the action of the human retina, suppose that you take three vidicon tubes and place red, green, and blue color filters in front of each, as shown in Figure 6-29. The respective outputs will then be proportional to the amounts of red, green, and blue in a scene. This is precisely the way that color is detected for color television.

As the filtered vidicon tubes scan the image together, the output of each vidicon is proportional to the respective color detected at that point in the scene. The combination of the three color outputs is then used to determine the color at a particular point in the scene, as well as the brightness level at any point, just as you did in Example 6-6.

In a vision system, each vidicon output is digitized and stored separately as a red, green, and blue pixel matrix. The brightness, or gray-level intensity, at any point is calculated first using the brightness equation. Then the R-Y and B-Y values are calculated to determine the color hue and saturation. Consequently, the stored data contain both color and gray-scale intensity information, the combination of which can make the image understanding task a bit simpler.

Edges and regions can be detected based on the presence of individual red, green, or blue characteristics. Of course, the consistency/inconsistency of the combination of these three colors across an area can also be used to define edges and regions. However, the most promising use of color is to verify edges and regions defined from gray-scale analysis. For instance, color information might be used to fill in the gaps generated by a gray-scale edge detector. The obvious conclusion is that higher confidence results when color is used in combination with gray-scale intensity information to define both edges and regions.

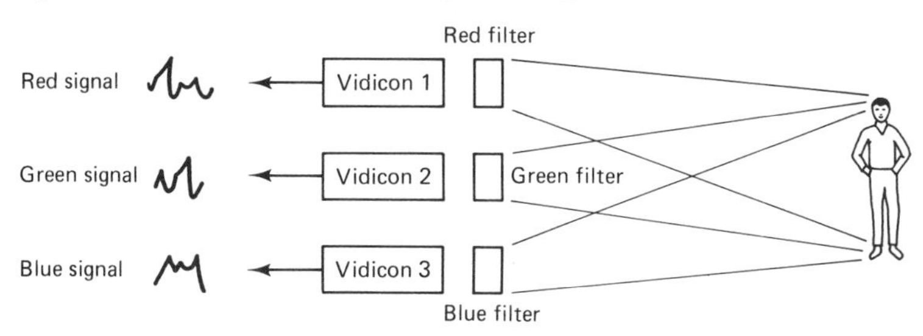

Figure 6-29 The idea behind a color TV camera is to place red, green, and blue filters in front of each of three vidicon camera tubes.

Texture

Like color, texture provides another clue to the identity of an image. For example, texture is extremely important in aerial surveys, since different terrains are seen as different textures. In real-word scenes, different objects and materials have different textures. Thus, texture analysis can aid in region segmentation.

The idea of texture is intuitive to all of us. But can you suggest a good working definition of texture? *Webster* defines texture as "the arrangement of the particles or constituent parts of any material . . . ." Many images can be broken down into arrangements of small elements called ***texture primitives.*** Examples of various texture-primitive arrangements are pictured in Figure 6-30. Can you pick out the primitives? The jelly beans and coins in Figures 6-30(a) and (c) are obvious texture primitives. But what about Figures 3-30(b) and (d)? In Figure 6-30(b), do you see a pattern of parallel lines, triangles, hexagons, or several stars of David? For the purpose of this introductory text, let's just say that texture relates to the pattern, or arrangement, of primitives within a region.

The next question is how textural features can be detected to aid in the image-understanding task. The problem is complicated by the fact that so many different textures are possible in real-world images. However, they all have one thing in common: the given texture primitives are repeated in some fashion throughout the pattern. Using this idea, statistics can be used to approximate the distribution of texture primitives. The distribution of primitive features, such as intensity or color, can be described by such statistical measures as mean, standard deviation, and

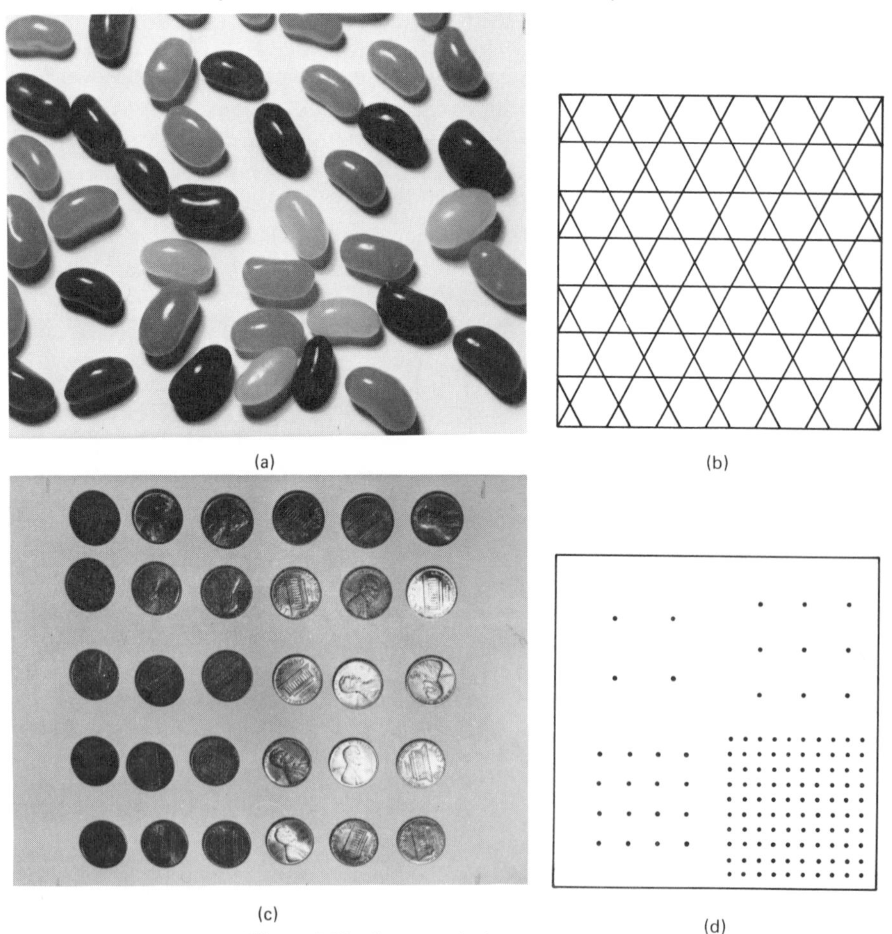

Figure 6-30 Some typical texture patterns.

variance. These measures then become the means by which regions within an image are segmented.

Another method is to use pixel histograms. Look at the image in Figure 6-31(a). Suppose that the pixels in this image are smoothed using local averaging; then a pixel histogram is constructed. Recall that local averaging replaces each pixel intensity value with the average of all the pixels in a given window. If a proper-sized window is chosen, local averaging will result in a pixel histogram like the one shown in Figure 6-31(b). Here it is obvious where to segment the regions. If thresholds are chosen between the pixel peaks, the image is segmented as shown in Figure 6-31(c).

The idea behind the local-averaging method is that pixels of a given texture

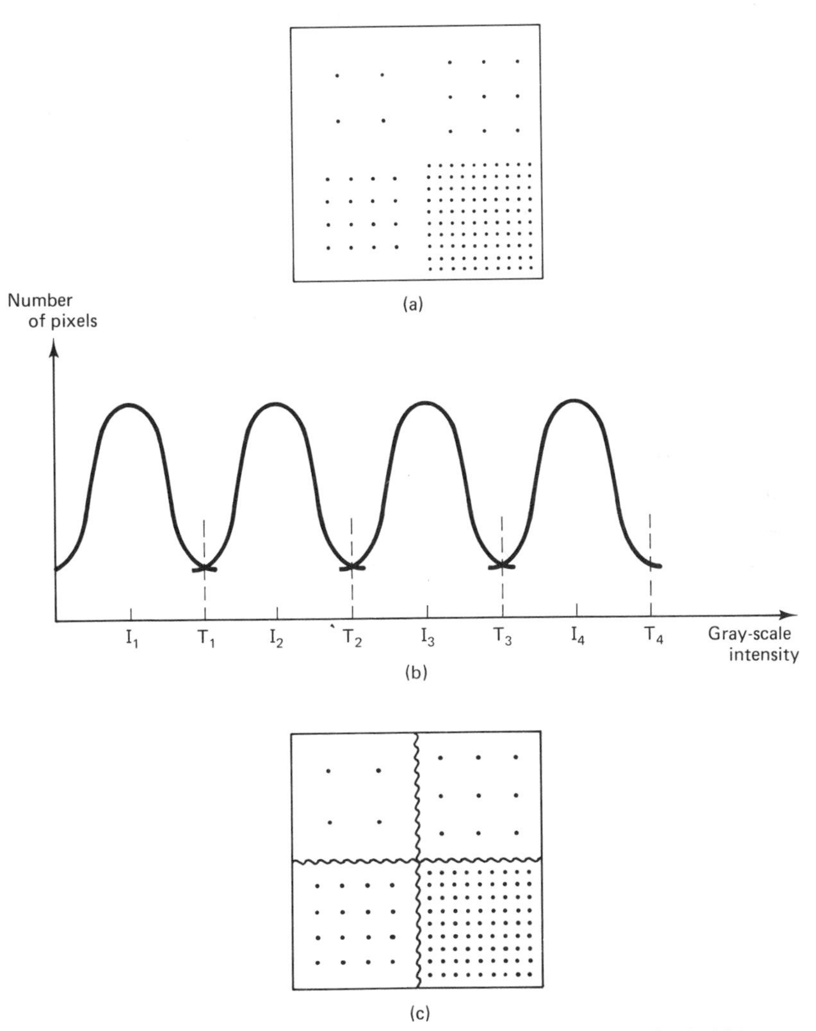

Figure 6-31 Texture analysis using local averaging: original scene (a), pixel histograms produced by local averaging (b), resulting segmentation (c).

will have similar intensity values after the averaging process. Of course, the trick is to pick the proper-sized pixel window. Too small a window results in too many local intensity variations, making it difficult to calculate thresholds. On the other hand, too large a window will blur the texture region boundaries.

It is now time to go on to the topic of image understanding using the image-analysis tools discussed in this section. Before going on, however, you might want to work the problems at the end of the chapter that are associated with image analysis, just to make sure you have a good understanding of this material.

6-3 IMAGE UNDERSTANDING

The final task of a computer vision system is to interpret the information obtained during the image-analysis phase. This is called ***image understanding***, or ***machine perception***, and is by far the most complicated of all machine vision tasks. The concepts used for image understanding and presented here are the direct result of research in the larger field of artificial intelligence.

Now that you have some image data to work with, how do you suppose this information can be used to interpret what the imaging device has "seen?" Let's begin by using the edge and line drawing information that was generated and stored during the analysis of the image. This information will be used to construct simple block images, which, together with color and texture information, might suggest the identity of an image. In addition, other sensing information and heuristic expectational knowledge about the visual environment can be injected along the way to guide the interpretation process.

Interpreting Line Drawings as Block Images

Much early image-understanding research was centered around the "blocks world." The blocks world is rather artificial, since it assumes that real-world images can be broken down and described by two-dimensional rectangular and triangular solids. Nevertheless, several AI programs have been developed that can reliably interpret real-world images using the blocks-world approach. In addition, block images coupled with color, texture, and heuristic knowledge often lead to an accurate interpretation of a complex real-world scene.

To begin with, you must assume that clearly defined lines have been extracted from the edge-point data during image analysis. Once the lines of a blocks-world object have been identified, they can be classified as either ***boundary lines*** or ***interior lines.*** The interior lines can be further classified as ***convex*** or ***concave*** lines.

This idea is illustrated by the line drawing in Figure 6-32. Here the lines defining the object have been labeled as either boundary, convex, or concave. The boundary lines are those that separate the object from the background or other objects. Boundary lines usually occur at the edges with the highest contrast on a gray-scale image, or a color edge on a color image. I will use an arrow (→) to mark these lines. The direction of the arrow will always be clockwise around the object boundary so that the object interior is to the right of the arrow direction.

Once the boundary lines are identified, the interior lines are labeled as either

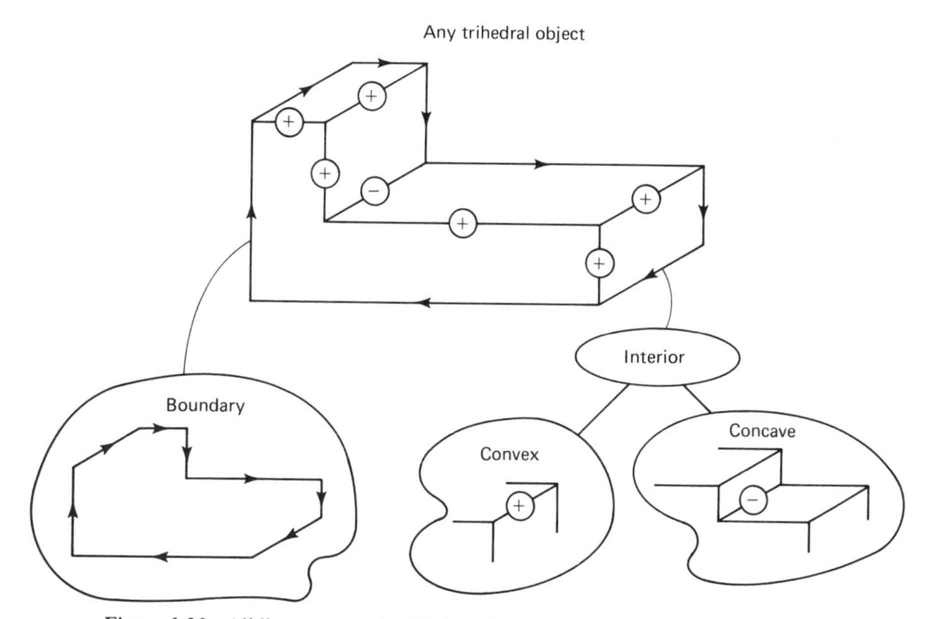

Figure 6-32 All lines can be classified as either boundary, convex, or concave.

convex or concave. A convex edge is where two object surfaces meet at an angle greater than 180°; a concave edge is where two surfaces meet at an angle of less than 180°. Convex edges will be labeled with a plus (+) sign and concave edges with a minus (−) sign.

Example 6-7

Using the line labels defined, mark the edges of the objects shown in Figure 6-33.

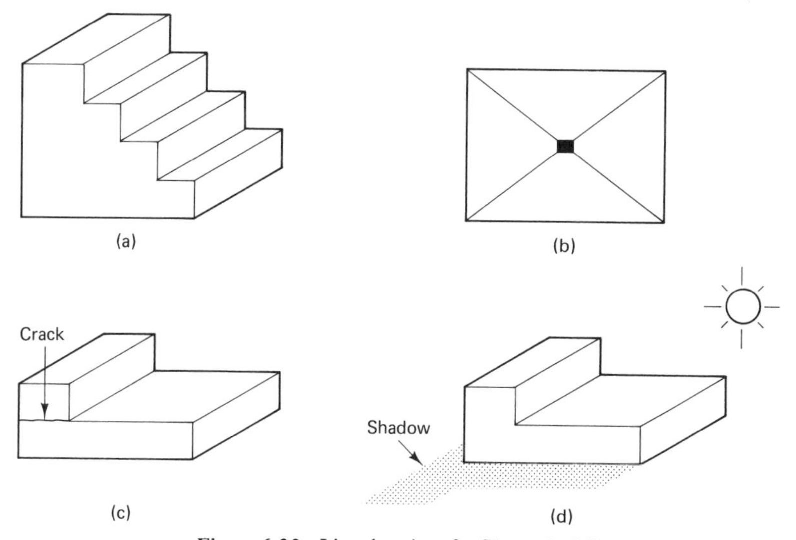

Figure 6-33 Line drawings for Example 6-7.

Solution

Labeling the staircase object in Figure 6-33(a) is rather straightforward, as shown in Figure 6-34(a). However, Figures 6-33(b) through 6-33(d) present problems.

In Figure 6-33(b), are you looking down a long dark hallway or at the point of a sharp object? Since, like a computer, you don't know, the interior lines cannot be accurately labeled.

The crack presents a problem in Figure 6-33(c). Is the crack a boundary common to two objects, or is it an interior line of a single object? If so, is it convex or concave?

How about the shadow in Figure 6-33(d)? Should it be labeled as part of the object boundary? Does it help in the interpretation of the scene, or does it simply create confusion for the image-understanding system?

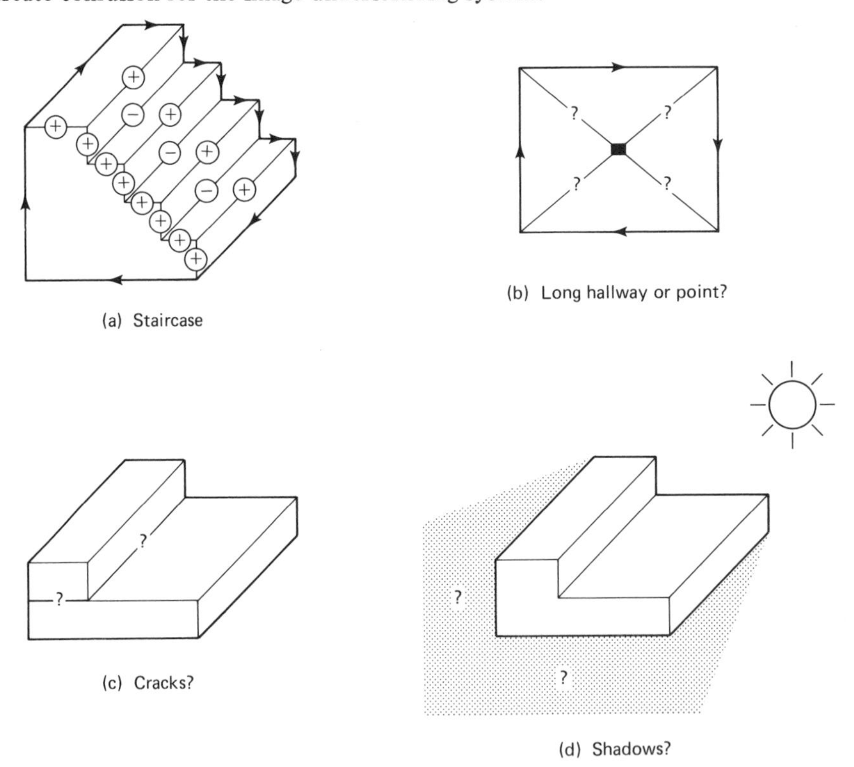

Figure 6-34 Line labeling for Figure 6-33 (Example 6-7).

The last three objects in Example 6-7 illustrate some of the difficulties associated with simple line labeling. To solve these problems, we must develop a series of ***constraints*** that will exclude certain labeling possibilities.

The Trihedral World. Simply stated, a vertex is a point, or junction, where two or more lines meet. Several types of vertices can be identified for blocks-world objects. The most significant are those shown in Figure 6-35. Let's start simply and

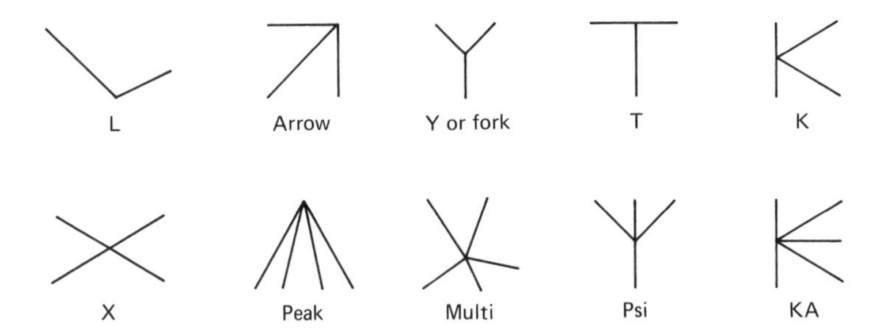

Figure 6-35 Types of vertices that exist in the real world.

limit, or constrain, the blocks world to ***trihedral*** objects. A trihedral object is one whose vertices are formed by the junction of at most three surfaces. In addition, we will not allow cracks or shadows on our trihedral objects. With these limitations, all the problems encountered in labeling the last three objects in Example 6-7 are eliminated. Of course, the price you pay is a more limited world model. However, the trihedral world is sufficient to describe many real-world objects. Once you see how objects can be interpreted in the trihedral world, we will lift some of the constraints to allow for more object possibilities.

The trihedral world limits the vertex possibilities in Figure 6-35 to four: the L, Y, T, and arrow. Let's see how many labeling possibilities can be generated using these four types of vertices. First, there are four ways to label any line: +, −, ←, or →. Since the L is where two lines meet, there are 4^2, or 16, labeling possibilities for the L vertex. Since there are three lines meeting at the Y, T, and arrow, there are 4^3, or 64, labeling possibilities for each of these vertices. This adds up to a total of 208 labeling possibilities for the four trihedral world vertices.

Look at the line labels in Figure 6-36. Do they make sense? Of course not! For instance, how can a convex edge meet a concave edge at an L vertex as in Figure 6-36(a)? If you explore all the 208 labeling possibilities in this manner and eliminate those like the ones in Figure 6-36 that cannot exist in the real world, you will reduce the original 208 labelings down to the 18 legal labelings shown in Figure 6-37.

Let's analyze these 18 legal trihedral labelings. First, the L labelings dictate that one leg of the L must always be a boundary. Thus, if one leg is convex or concave, the other leg must be a boundary.

Second, look at the legal Y labelings. Notice that if any of the edges are convex (+) all the edges must be convex. Also, if any of the edges are concave (−), there cannot be a convex edge.

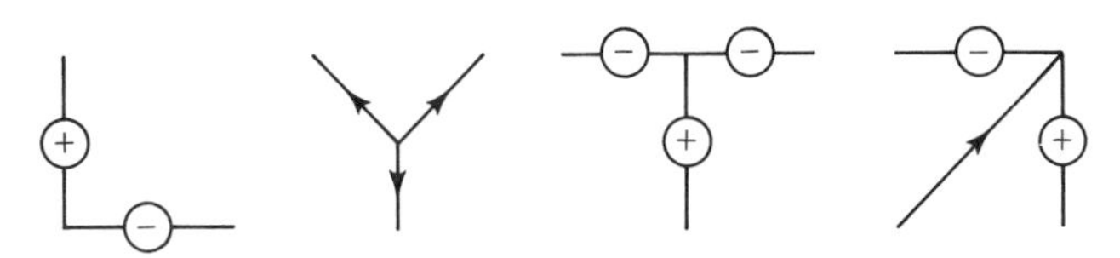

Figure 6-36 Illegal labelings in the trihedral world.

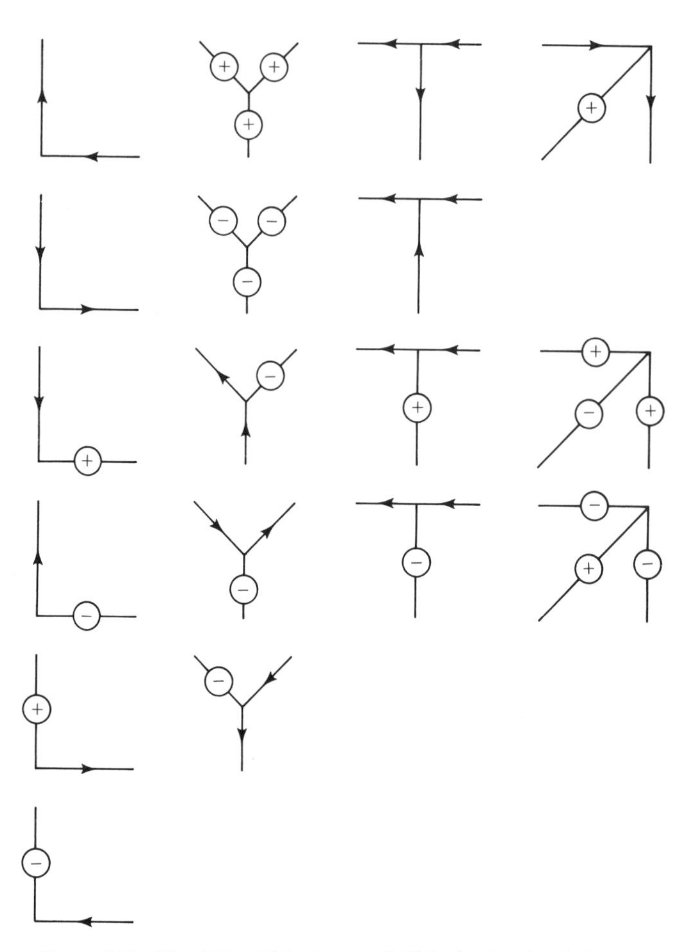

Figure 6-37 The 18 legal labeling possibilities in the trihedral world.

Next look at the T junction labelings. You notice that the top of any T junction must be a boundary, running from right to left.

Finally, with the arrow junction, a boundary on one fin forces the other fin to be a boundary edge and the stem to be a convex (+) edge. Also, a convex fin forces the other fin to be convex and the stem to be concave. Likewise, a concave fin forces the other fin to be concave and the stem to be convex. These observations are summarized in Figure 6-38.

Example 6-8

Using the implications developed in Figure 6-38, construct a short algorithm that can be used to label trihedral objects.

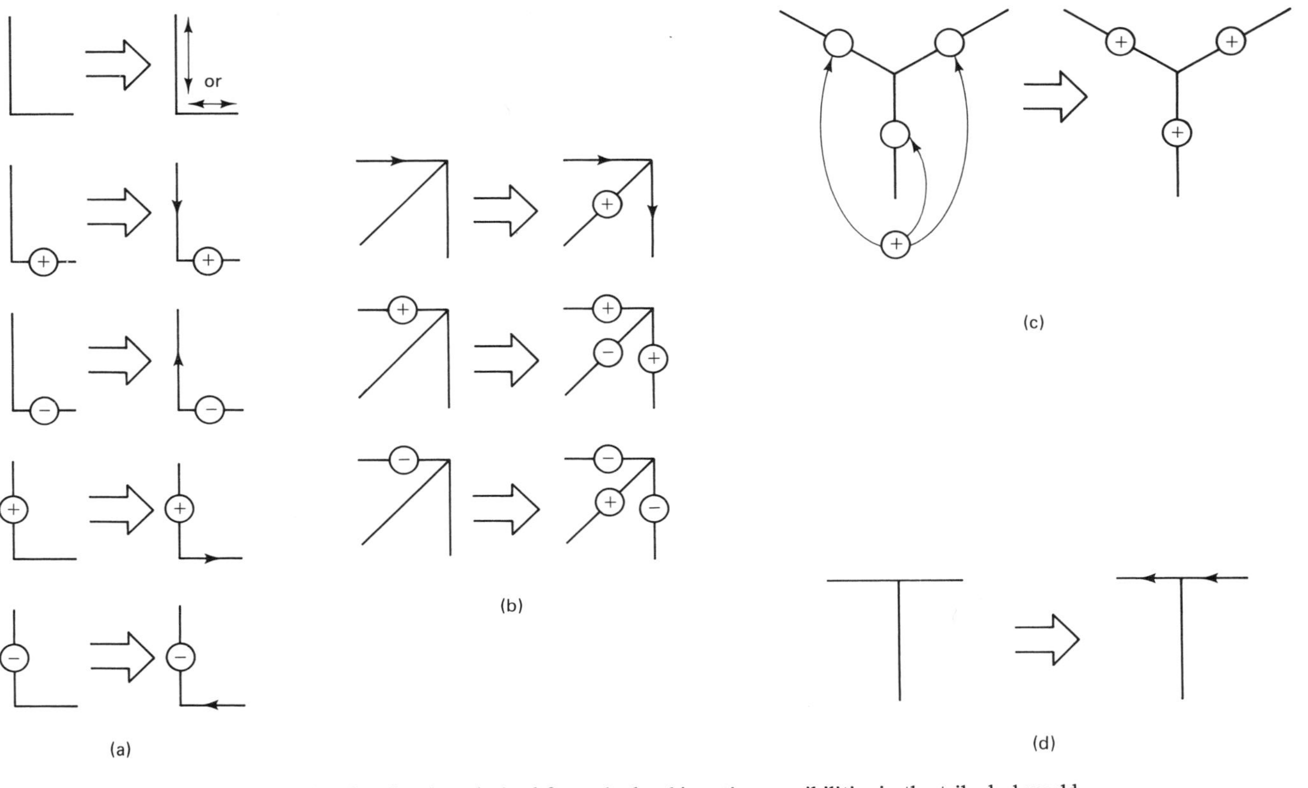

Figure 6-38 Implications derived from the legal junction possibilities in the trihedral world.

Solution

The obvious first step in such an algorithm is to label all the boundary lines with arrows in a clockwise direction. Then look at the interior junctions and apply each of the implications developed in Figure 6-38. The flowchart in Figure 6-39 summarizes such an algorithm.

Example 6-9

Using the algorithm developed in Example 6-8, label the edges of an ordinary cube.

Solution

The properly labeled cube is shown in Figure 6-40. Using the preceding algorithm, the boundaries of the cube are first labeled in a clockwise direction. Then the boundary is traced to find a junction.

Starting on the right boundary edge, the first junction encountered is an arrow junction at the bottom of the cube. Since the fins of the arrow are boundaries, the stem must be a convex edge. Therefore, the arrow stem is labeled with a plus (+) symbol.

Moving inward along the arrow stem, the next junction encountered is a Y junction. Since one edge of the Y junction is convex, the other two edges must also be convex. This completes the labeling process.

Another property of the implications developed in Figure 6-38 is that they exclude certain impossible objects. For example, look at the object in Figure 6-41. Observe that there are three arrow junctions on the object boundary whose stems come together to form an interior arrow junction. The stems of the boundary junctions must be convex, since their fins are all boundary lines. However, this requires all the edges of the interior arrow junction to be convex. This is an illegal labeling of an arrow junction and means that the object cannot possibly exist in the real world.

The obvious conclusion from this impossible object experience is that if the trihedral object is real then it can be labeled using legal junction labels. However, if the labeling process cannot be completed or results in illegal labelings, the object is impossible. When an image-understanding system concludes that an object is impossible, it could trigger another analysis of the image, since the current analysis has generated an impossible object. An impossible object could be the result of noise or inaccurate line extraction. Another more or less sensitive analysis might result in a legal object.

Example 6-10

Label the object in Figure 6-42 using the trihedral labeling algorithm developed in Example 6-8 (Figure 6-39).

Solution

The algorithm is followed as before, first labeling all the boundary edges. The boundary is then traced to find junctions. At each junction, the labeling implications are followed to label the respective edges.

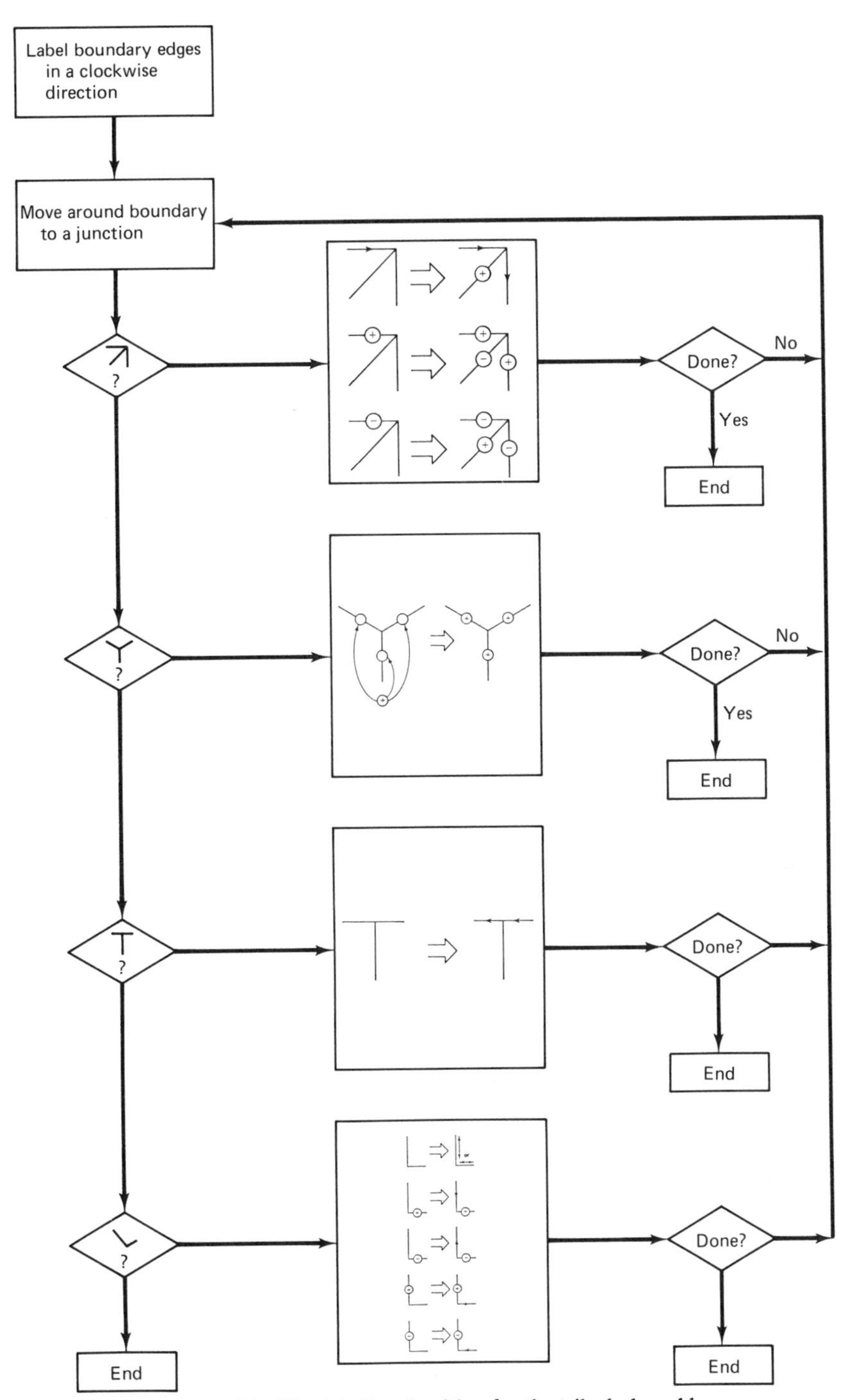

Figure 6-39 Line-labeling algorithm for the trihedral world.

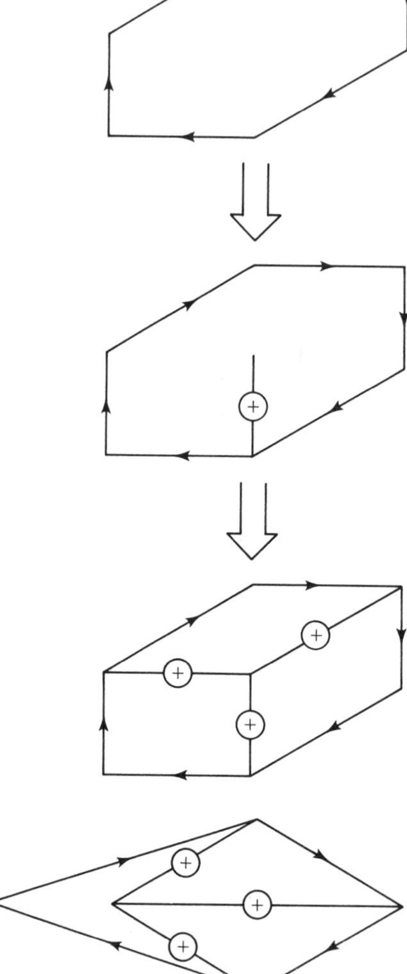

Figure 6-40 Labeling a cube using the labeling algorithm in Figure 6-39.

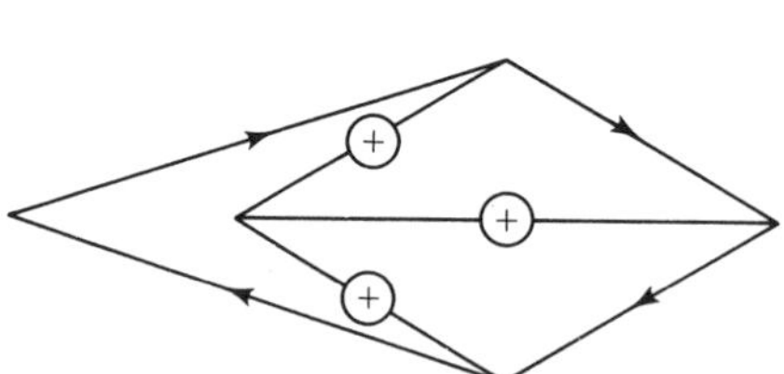

Figure 6-41 Line labeling detects impossible objects.

You will find no illegal labelings in the diagram in Figure 6-42. However, take a closer look at the object. You find that all the labelings are legal and conform to the rules developed previously. In fact, at first glance the object looks real! However, the two shaded regions are two separate planar surfaces. From geometry, you know that any two planes must meet along a common straight line. This is not the case in Figure 6-42, since the two surfaces meet along two different lines, A and B. In fact, the front planar surface would have to bend in order to meet the top planar surface. Consequently, the object must be mathematically impossible.

Example 6-10 proves that just because an object can be labeled does not necessarily mean that it is real. Without your knowledge of geometry, you might

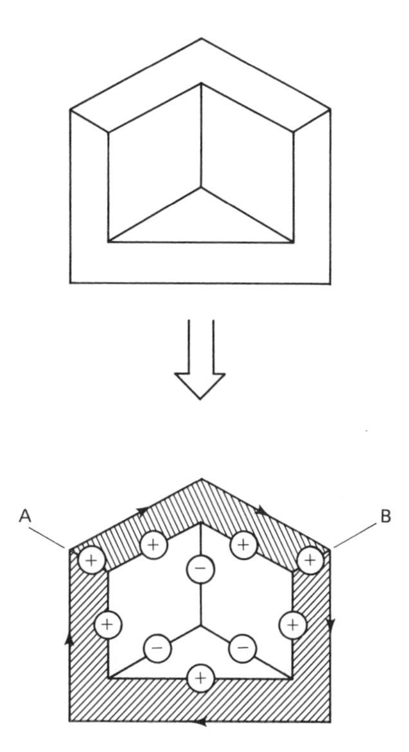

Figure 6-42 The legally labeled object in Example 6-10 turns out to be an impossible object.

assume that the object is real. The same is true of a computer line-labeling program. However, suppose the laws of geometry were programmed into the computer and used to guide the interpretation process. Such heuristic knowledge would result in more accurate image interpretations.

Using Shadows and Cracks. Up to this point, I have disallowed shadows and cracks. The idea was that they create confusion and hinder the interpretation process. However, rather than hinder the image-understanding task, both of these features can actually enhance the interpretation process.

Shadows. Consider the block image in Figure 6-43. Simple labeling of this object would indicate nothing about the fact that it is being supported by another object, such as a table. However, if the shadow of the object is detected and labeled as shown, it is clear that the object is being supported since the shadow would not exist if it were not. Even more basic to the image-understanding task is the fact that the presence of a shadow means that there ***must*** be another object in the scene.

Shadow information adds the following three important pieces to the image-understanding puzzle:

1. The existence of an object shadow indicates the existence of at least one other object in the scene.

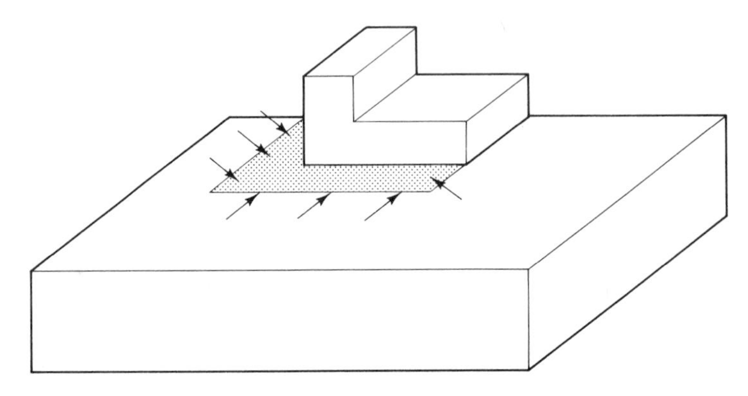

Figure 6-43 A shadow sometimes indicates (1) that an object is supported by another object and (2) its position relative to the supporting object.

2. Often the existence of an object shadow indicates that the object is being supported by another object.
3. An object shadow points to the location of other objects and indicates their position relative to the object casting the shadow.

Observe from Figure 6-43 that a shadow edge is labeled with arrows pointing into and perpendicular to the shadow.

Cracks. Cracks also add a piece to the puzzle, since they usually indicate the presence of several independent, but connected, objects. For example, the L-shaped object in Figure 6-44 might be constructed from one, two, or three separate blocks as shown. If the object is solid, there will be no crack lines and the object can be labeled as before. If the object is constructed from two separate blocks, there will be one crack line, as shown in Figure 6-44(b). This line is labeled with a *c* to indicate the crack. Two crack lines appear if the object is constructed from three blocks, as in Figure 6-44(c).

Now suppose the object is constructed from two blocks [Figure 6-44(b)]. Let's separate the two blocks as shown in Figure 6-45. Notice that, in each case, one of the blocks hides, or ***occludes***, part of a surface of the other block. The block that hides the other block is called the ***occluding*** block, while the block whose surfaces are partially hidden is called the ***occluded*** block. The separation clearly shows that the original concave edge and the crack are both boundary lines of the occluding object. To indicate this, a boundary symbol is placed on the crack line and the

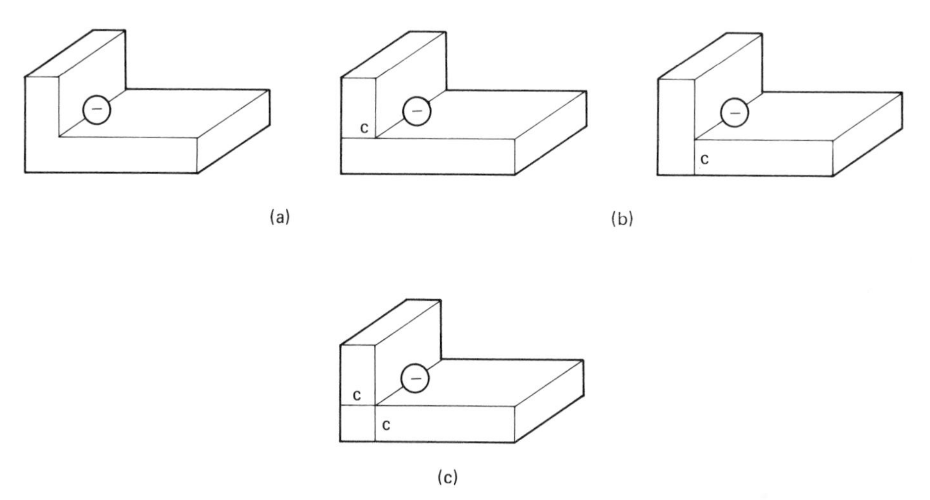

Figure 6-44 A single object might be constructed using separate blocks: (a) solid object; (b) two blocks; (c) three blocks.

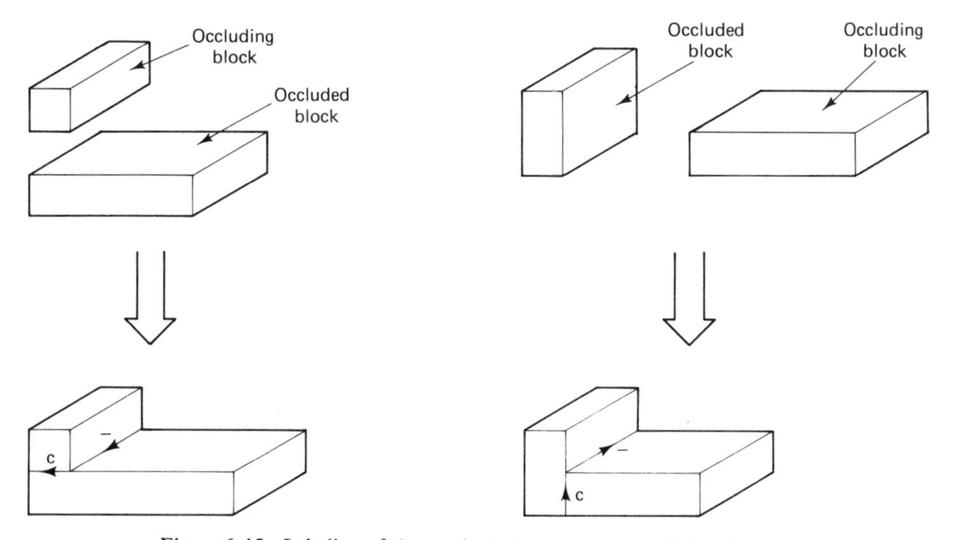

Figure 6-45 Labeling of the cracks between two separate blocks.

original concave edge, as shown. When the two blocks are placed back together, this labeling shows that there are two objects. Moreover, the labeling indicates which object is the occluding object and which object is being occluded.

The same object constructed from three blocks is shown separated in Figure 6-46. Again, the cracks and original concave edge are also boundary edges of the three component blocks. The boundary edges of the occluding blocks (1 and 3) are labeled and the object is placed back together. This results in the composite labeling shown. Notice that the original concave edge is created by the boundary lines of

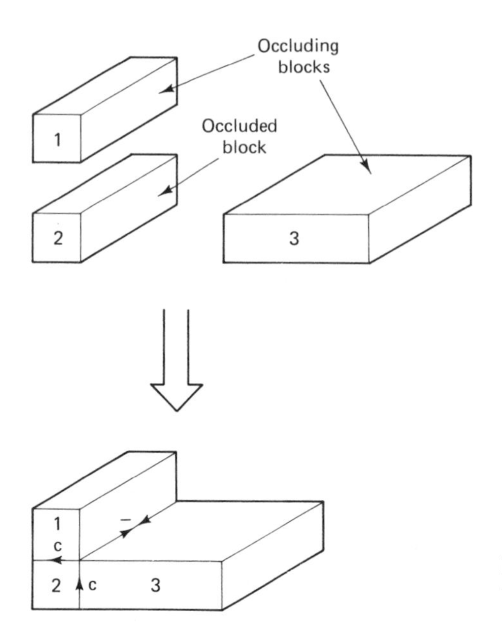

Figure 6-46 Labeling of cracks between three separate blocks.

two occluding blocks (1 and 3). The head-to-head arrows combined with the original concave symbol are used to indicate this construction.

We have now expanded our labeling possibilities with the addition of shadow and crack symbols. All the labels developed so far are summarized in Figure 6-47. Next suppose that we get away from the limited trihedral world and allow the remaining types of vertices found in the blocks world (see Figure 6-35). You might suspect that we have now expanded our labeling possibilities to astronomical proportions, and you are right. However, less than 3 percent of the total possible labelings are legal in the real world. Of course, this is still too many possibilities for labeling by hand, but it presents little problem for today's superfast computers. In fact, it has been shown that a computer can generate a single interpretation of a complex scene much more quickly using shadows and cracks and allowing nontrihedral vertices. In other words, more labeling possibilities for individual edges actually enhance, rather than hinder, the image-understanding process. This is because the labeling of an edge common to two or more junctions is limited to only that label that is legal for all the respective junctions. Said another way, the labeling of object edges is mutually constrained by the legal possibilities imposed by the real world.

Complex Images

Unfortunately, most real-world images do not fit conveniently in the blocks world. Nevertheless, the blocks world is a good place to begin learning how a computer can be programmed to recognize and understand simple images. It's like learning mathematics; you cannot begin to really learn and understand calculus until you have mastered algebra. Likewise, now that you have the basic image-understanding

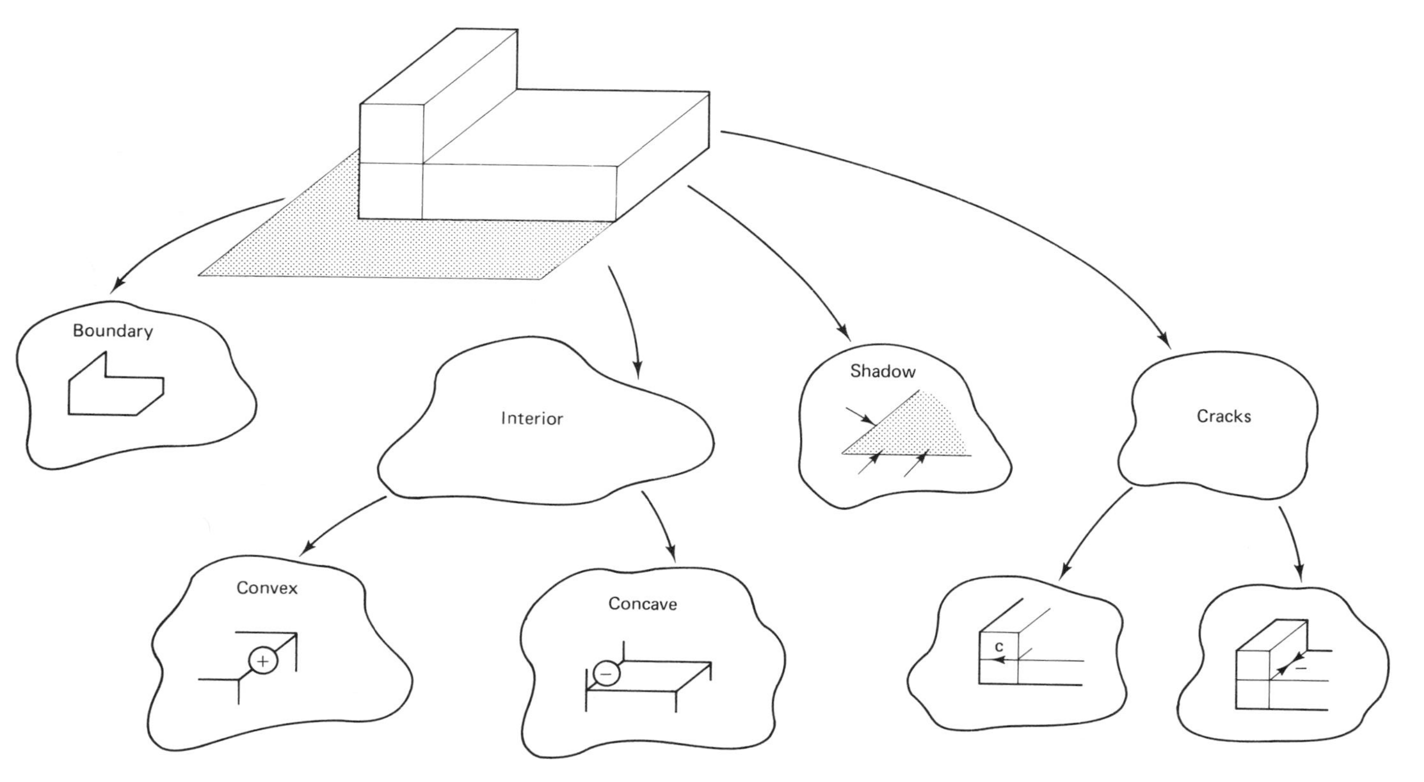

Figure 6-47 Summary of labeling possibilities.

tools developed in the blocks world, you are ready to expand on this knowledge and explore complex images.

Let's define a complex image as one that includes lines that are not straight. In other words, complex images include curved as well as straight lines. If all the lines are straight, it is a relatively easy task for a computer to interpret a scene using the blocks-world techniques discussed previously. However, curved lines make the image-understanding task much more difficult.

Recall that the key element to any intelligent task is knowledge. Suppose that we expand the blocks-world knowledge base of a computer to include image information other than gray-scale data. In addition, let's add heuristic ***expectational knowledge*** about the visual environment. The idea is that, for a computer to understand complex images, it must have gray-scale and other image information, as well as heuristic knowledge about what to expect in the scene. Such a knowledge base, coupled with an efficient knowledge organization and control strategy, is required for complex image understanding.

Using Additional Image Information. You are already aware of how color and texture information can enhance the image-understanding task. These are all things that you and I use for our own human visual perception. Can you think of any other cues that we use to interpret a scene? Or can you think of some things that might aid in the image-understanding task that we humans cannot detect, but that are easily detected and analyzed by machines?

Motion. Many times you cannot recognize an object unless it moves. For instance, a deer can be standing in the woods looking straight at you and you will not "see" it unless it moves. The same might be true of a frog, grasshopper, or walking stick. In fact, most animals rely heavily on motion to detect the presence of an object. Archery hunters will testify to the fact that a deer usually will not be frightened if you stand perfectly still, since it must see movement to confirm your presence. Once you move, the deer is able to *segment* you from the relatively stationary background. Thus, motion can be an important clue to image segmentation.

This same idea can be used by computers to segment images. For example, two overlapping objects in a single picture frame might be easily interpreted as a single object. However, movement of the objects or the imaging device will usually reveal two separate objects and lead to further segmentation of the image. In the extreme case, whole objects that are hidden in one picture frame can be detected in subsequent frames through movement of the objects and/or imaging device.

As another example, consider a vision system in a stationary weather satellite that is monitoring cloud movements. The first thing the system must do is to pick out the clouds from the background land masses and bodies of water. The relative cloud positions from one picture frame to the next allow for easy segmentation, since the background images remain stationary.

Another use for motion in vision systems is for depth perception. This comes from the idea that you and I can perceive three-dimensional images from objects moving in two dimensions. Think about watching your favorite TV program. As objects move across the two-dimensional screen, you see different perspective views of the objects that allow you to perceive three-dimensional images. Several attempts have been made to duplicate this human depth-perception process in computer vision systems, with limited success.

Detecting motion with a computer system is not an easy task. The basic idea is to find corresponding pixel points or regions in a series of picture frames. Thus, the system must determine which pixel points or regions on one picture frame correspond to the same points or regions on the next frame. This is referred to as the ***correspondence problem.*** The correspondence task is relatively simple if the image has already been segmented. However, this is usually not the case, since one of the main reasons for motion analysis is for image segmentation.

Another problem with using motion for image understanding is the amount of data that is generated and must be analyzed. Recall that a single TV picture frame requires about 256k bytes of memory to store. Multiply this times several frames and you begin to see the storage problem. In addition, in many applications, such as collision avoidance, the motion information is only useful if it can be analyzed in real time. Both the storage requirements and real-time analysis of large amounts of data stretch the limit of today's computers.

Nonhuman Image Information. Up to this point, we have only considered those fundamental image features that you and I as humans use for visual perception. Some of these things, like texture and motion, are easy for humans to perceive, but relatively hard for a computer to analyze. On the other hand, there are object features that we humans cannot detect that are relatively easy for a computer to detect and analyze.

Human vision provides a good working example of a vision system, but this does not mean that a machine vision system must duplicate all the aspects of the human system in order to "see" in a given application. With this in mind, can you think of any image features that humans cannot detect, but that are relatively easy for machines to detect and analyze, thereby enhancing the image-understanding task? (*Hint:* Superman was a super man because he had one of these sensing abilities.) You are right! Superman can see all kinds of things that you and I cannot by using his x-ray vision.

X-ray emissions can be used by machines to see parts of an object that we humans cannot. Other examples of nonhuman-type image data include infrared images, ultraviolet images, and thermographic images, just to mention a few. Even the magnetic properties of an image can be analyzed to enhance the image-understanding task. In fact, in some applications, gray-scale light-intensity images are not even necessary. A typical example is in the use of thermographic images to detect breast cancer.

Using Expectational Knowledge. Imagine going to your birthday party without any expectational knowledge whatsoever. You begin to open your presents, but being human you like to guess at their contents before opening them. You see a joystick poking through one of the presents, but without any expectational knowledge you cannot guess that it might be a new computer video game. On another present, you can barely see the name "Staugaard" through the wrapper, but cannot guess that it might be another exciting text by that famous technical author. What a bore your birthday would be without any expectational knowledge.

The obvious point I am trying to make is that expectational knowledge plays an important part in the human image-understanding process. For computers to truly see, and more importantly understand what they see, they must be programmed with knowledge about what they should expect to see.

Expectational knowledge in vision systems can come in many different forms, depending on the environment. This knowledge might include the expected characteristics of objects such as color, texture, size, and shape. In addition, the expected thermal, infrared, or magnetic properties of the objects might also enhance identification.

Positional knowledge is also an important type of expectational knowledge. When you see a toaster, you naturally expect it to be *on top of* the kitchen counter. Likewise, you usually expect to find a dishwasher *next to* the kitchen sink. Of course, the kitchen counter is usually *supported by* kitchen cabinets. These are all examples of knowledge about relative object position that you would expect to find in a given environment. Computers can also use this type of knowledge to identify objects. Table 6-1 lists some typical types of positional knowledge that might be important to the image-understanding task.

TABLE 6-1 Typical Positional Knowledge

In front of	On top of	Contained in
Behind	Below	Suspended from
Next to	Attached to	Inside of
Supported by	Part of	Near

Other types of expectational knowledge relate to common sense. Grass is green in the springtime, chandeliers hang from the ceiling, tables have four legs and flat tops, computer terminals have alphanumeric keys, cars have four wheels, and so on. These are all examples of commonsense knowledge that you and I take for granted, but that must be programmed into a computer to enhance image understanding. Imagine trying to identify a bicycle, for example, without having any idea of what you might expect to find. The task would be much easier if you expected to find two spoked wheels, a handle bar, two pedals, a chain, a small seat on top of an open frame, and so on. Does this commonsense type of knowledge suggest any particular knowledge-organization structure?

The preceding discussion is summarized by Figure 6-48. The idea is that image understanding must employ raw image data such as gray scale, color, and texture information, but it must also make use of nonhuman sensing information, as well as knowledge about what to expect in a given environment.

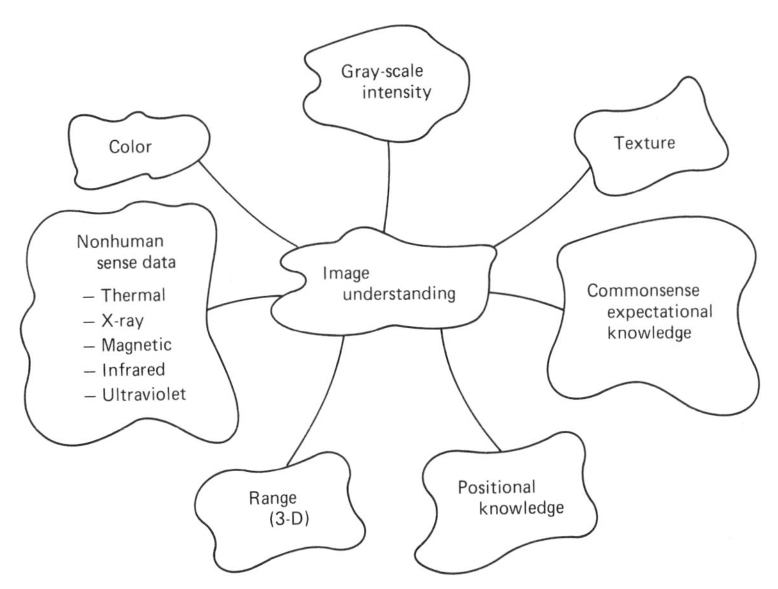

Figure 6-48 An image-understanding system must utilize several types of knowledge.

Knowledge Organization and Reasoning Strategies

Now that we have explored the various types of knowledge available for image understanding, it is time to see how this knowledge can be organized and controlled to construct an effective and efficient system. First, you can divide vision system knowledge into two broad categories: ***low-level knowledge*** and ***high-level knowledge.***

Low-level Knowledge Organization. Low-level knowledge in a vision system is generally associated with image and scene features generated by pixel data. As illustrated in Figure 6-49, this type of knowledge includes gray-scale pixel data, color data, texture data, and nonhuman sensory data. Given a pixel matrix, one

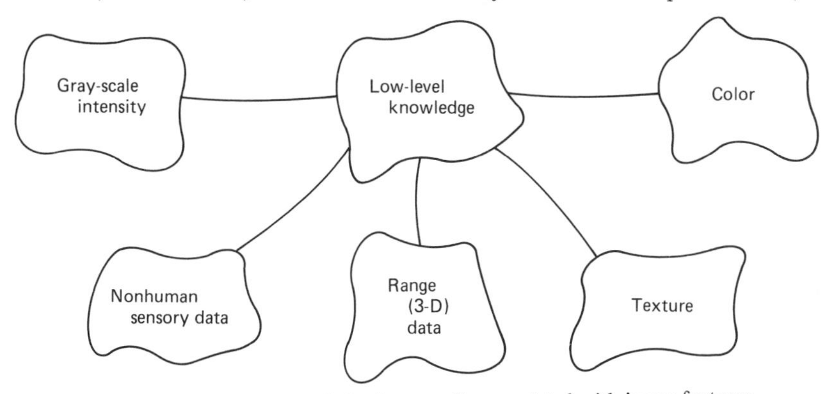

Figure 6-49 Low-level knowledge is generally associated with image features.

way to organize such knowledge is by using data structures called ***pyramids*** and ***quad trees.*** Both are used to organize low-level knowledge in a hierarchical manner.

Pyramids and Quad Trees. Look at the image in Figure 6-50. As you know, such an image can be represented in the form of a pixel matrix and stored in a computer's memory. Each cell within the matrix represents a pixel value.

Now suppose that the pixel matrix forms the base of the pyramid shown in Figure 6-51. To construct the next level of the pyramid, you must section the

Figure 6-50 Gray-scale picture matrix.

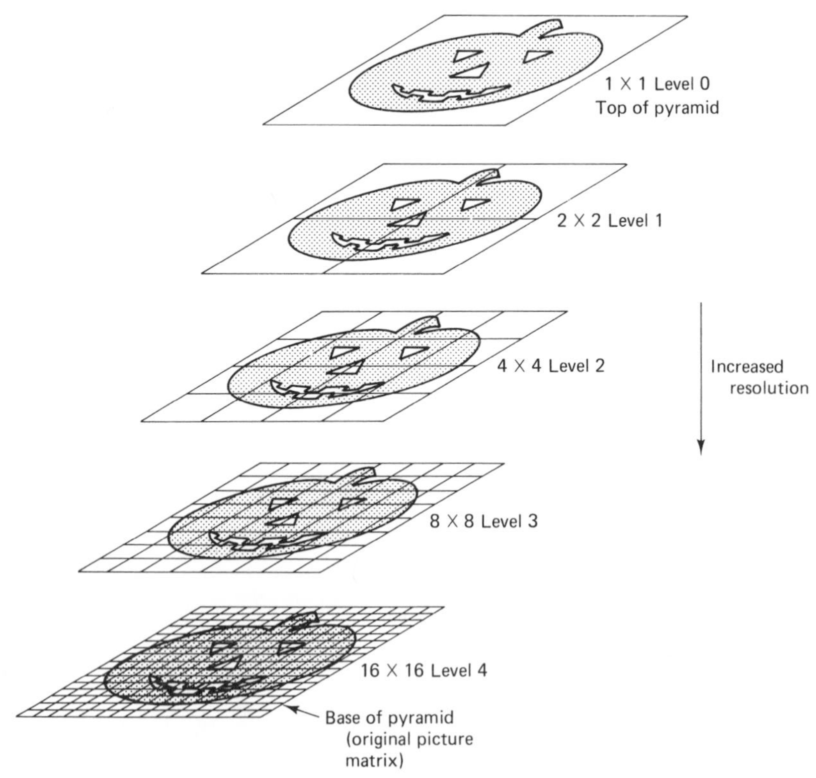

Figure 6-51 Gray-scale matrix pyramid.

original picture matrix into a new matrix whose dimensions are one-half that of the original matrix. For instance, if the original picture matrix is a 16k $\times$ 16k matrix, the next pyramid level will be an 8k $\times$ 8k matrix. Next, this level is sectioned into a new matrix whose dimensions are half again, and so on, until a 1 $\times$ 1 matrix results at the top of the pyramid.

For example purposes, I have assumed that the original image is represented by a 16 $\times$ 16 pixel matrix, as in Figure 6-51. Pyramiding this image results in 8 $\times$ 8, 4 $\times$ 4, 2 $\times$ 2, and 1 $\times$ 1 levels as shown. Of course, a pyramid made from an actual picture matrix will have many more levels.

Notice that the resolution of the pixel image is decreased as you ascend the pyramid. You might be wondering how this helps the image-understanding process. The idea is that you can begin searching for an expected object at the low-resolution levels at the top of the pyramid and then descend the pyramid until you reach the resolution required to identify the object. In most cases, this results in a more efficient search, since many times an object can be identified at the lower resolution levels.

A ***quad tree*** is a node structure that is generated from a pyramid. The nodes in a quad tree correspond directly to the cells of a pyramid. The root node corresponds to the top level of the pyramid. This node has four children corresponding to the four cells in the next level of the pyramid, and so on. Each node must have exactly four children, since each next-lowest pyramid level has four times as many cells as the preceding level.

A partial quad tree for the pyramid in Figure 6-51 is shown in Figure 6-52. Here's how it was constructed. The nodes have been labeled black, white, or gray. A black node is used to represent a given cell whenever all the corresponding cells in the next-lowest pyramid level are black. Likewise, a white node is used to represent a given cell if all the corresponding cells in the next-lowest pyramid level are white. A gray node is used to represent those cells whose corresponding cells in the next-lowest level are a mixture of black and white. The key in Figure 6-52 illustrates this relationship.

After a pyramid or quad tree is constructed, templates of expected images can be matched to the cells at each level in an attempt to identify the image. Here's how it works. The system first looks at the upper levels of the pyramid or quad tree to match overall boundaries or regions. High-level templates are used here. These templates describe the image in terms of overall features such an average intensity and/or color. Other sensory data and positional data can also be introduced at the upper levels to segment the image.

The upper-level analysis usually eliminates many image candidates and gives some clues to the real identity of the image. If the image cannot be conclusively identified at the upper levels, the image candidates are narrowed down further by examining the lower pyramid or quad tree levels until the image is identified. This can be done using more detailed, pixel-level templates. At each successive lower level, higher-resolution image data are introduced to verify and extend the conclusions made at the previous level.

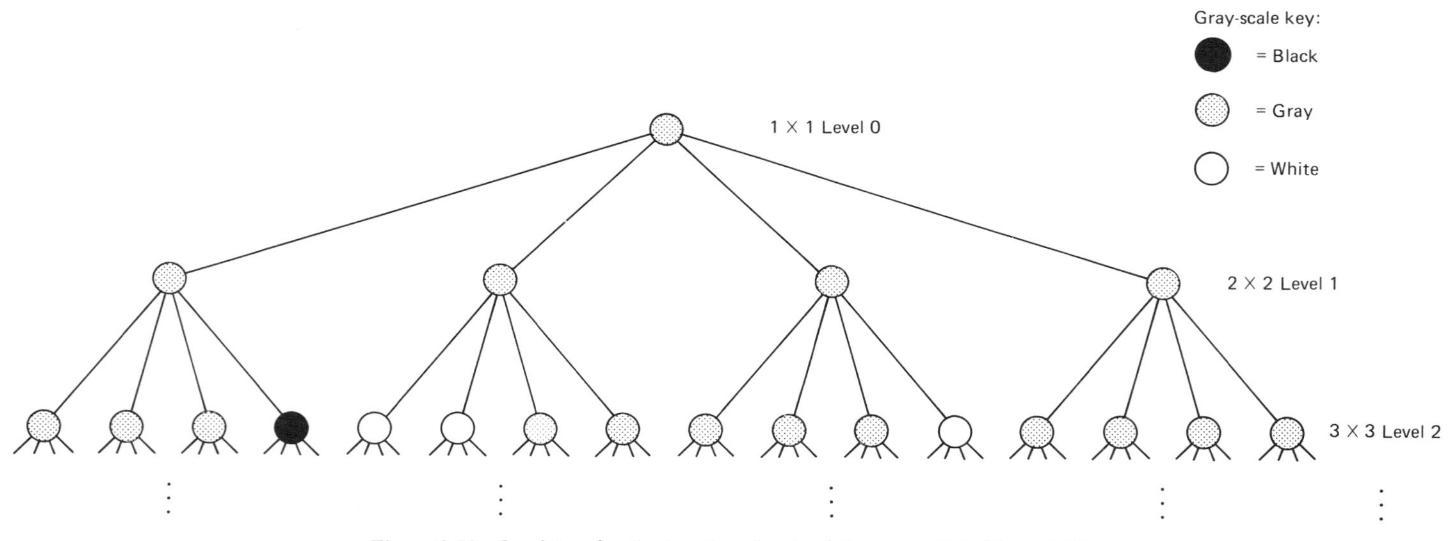

Figure 6-52 Quad tree for the top three levels of the pyramid in Figure 6-51.

High-level Knowledge Organization. High-level knowledge in a vision system is in the form of expectational knowledge. This includes positional as well as commonsense knowledge, as shown in Figure 6-53. High-level knowledge is used as a model of what the system should expect to see in a given environment. Thus, a typical vision system will use high-level knowledge to drive, or control, the interpretation process very much like you use past experience to control your visual perception. In AI jargon, such a system is said to be ***expectation***, or ***model, driven.***

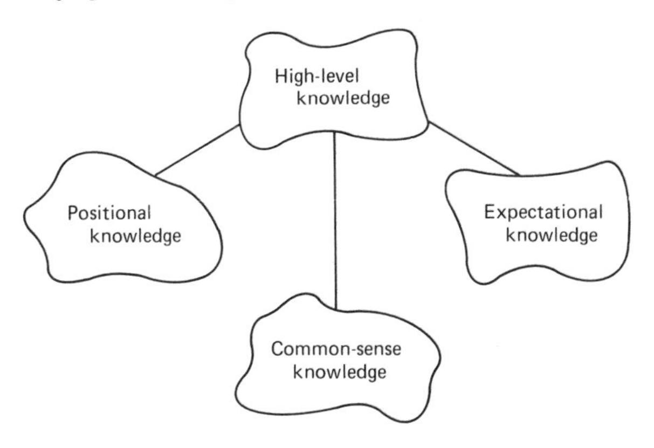

Figure 6-53 High-level knowledge sources.

Frame theory provides an ideal organization for expectational knowledge in vision systems. Recall from Chapter 3 that frames can be used to represent commonsense knowledge about what items you expect to find in a given environment. Remember the kitchen frame shown again in Figure 6-54. Here the kitchen frame provides knowledge about what to expect in a kitchen before the image data for a given kitchen is even analyzed. Such a frame network can be used to guide the reasoning process through expectation. In addition, frames facilitate learning through new experiences, since the frame network can be easily updated to adapt to a changing visual environment.

Reasoning Strategies. In the preceding discussions, I have purposely treated low- and high-level knowledge separately so that you could understand the difference between them. However, I think it is now obvious that the key to image understanding is the interaction between these two knowledge levels. Any intelligent reasoning strategy must employ both knowledge levels.

Think about what you do when you first see an unfamiliar image. Don't you first look at the overall outline, or shape, of the image, and then take a closer look at the image using past visual experiences to identify it? A computer can also use this idea to interpret an image. For example, the general outline of the pumpkin face back in Figure 6-50 might suggest a frame network of faces. This high-level knowledge network then directs a lower-level knowledge search to verify the location of the eyes, nose, and mouth. The system should not associate the face with

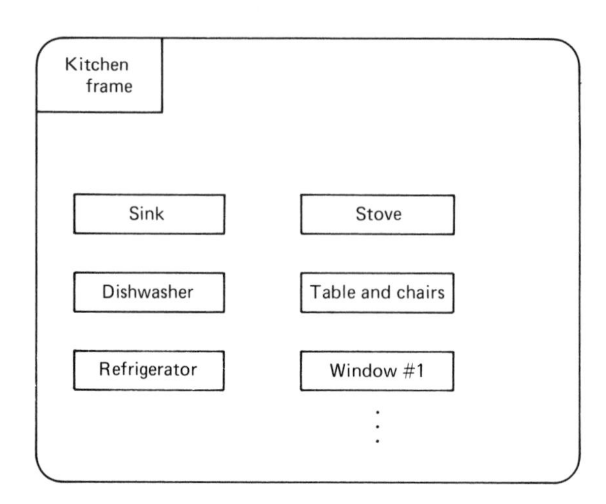

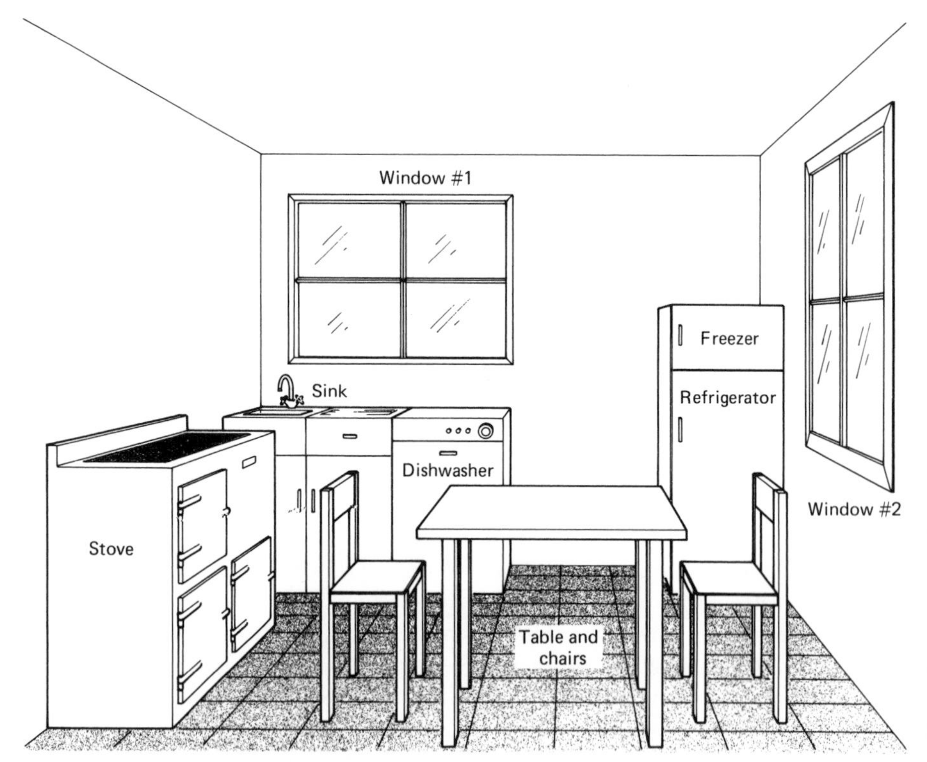

Figure 6-54 The kitchen frame.

any human, since it should identify the orange color and conclude that it is a Halloween pumpkin face.

This is a simple example of backward or expectation-driven reasoning. Of course, the low- and high-level knowledge sources also lend themselves to forward, or bottom-up, reasoning. The inherent advantages and disadvantages of both reasoning strategies as discussed in Chapter 2 also apply here. The main distinction between the two is that backward reasoning is more dependent on expectational knowledge.

Now go back and look at the diagram in Figure 6-48. Does this suggest any particular knowledge representation and control structure? You are right again—a blackboard structure! Like speech-understanding systems, image-understanding systems can integrate different sources of knowledge by means of a common structure called a blackboard.

The blackboard structure shown in Figure 6-55 shows how a typical image-understanding system might be organized. Here the various low- and high-level knowledge sources become system specialists for the blackboard model. The lower-level knowledge sources employ a pyramid/quad tree structure, while the higher-level knowledge is contained in frame networks. Constraining and expectational knowledge generated by the higher-level knowledge sources is used by the lower-level sources to streamline the reasoning process.

As new information is written on the blackboard, it is used as required by the independent knowledge sources. Consequently, the blackboard always represents the current description of the image. Each source processes the blackboard information independently and asynchronously. In this way, information is processed in parallel, resulting in improved system efficiency. Sound familiar? This is precisely the same knowledge organizational structure employed by several successful speech-

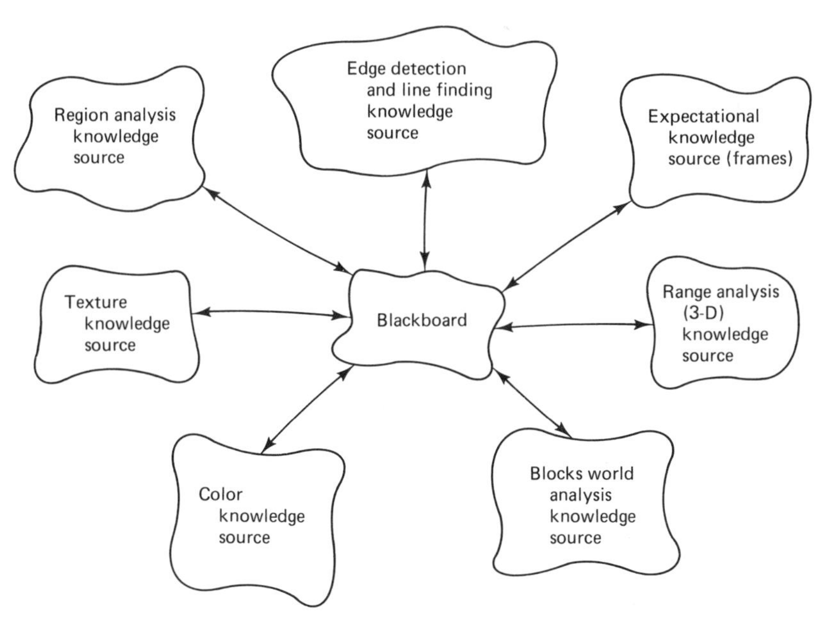

Figure 6-55 Blackboard knowledge organization for an image-understanding system.

understanding systems. It has also been applied successfully in image-understanding systems.

Before leaving computer vision, it's probably a good idea to take a brief look at some typical industrial vision systems in an attempt to bring things into perspective.

6-4 INDUSTRIAL VISION SYSTEMS

Before I even begin this discussion, I should caution you that existing industrial vision systems are nowhere near providing capabilities comparable to human vision, nor do they need to! Always remember that "the application dictates the system." Certainly, an intelligent vision system with humanlike abilities would be nice (and expensive), but is it really required to inspect for surface defects on printed circuit boards? Most existing industrial systems are not general-purpose vision systems. Rather, they are special-purpose systems that have been designed to satisfy a specific application.

Design Considerations

The primary considerations for an industrial vision system are ***cost, speed, accuracy***, and ***reliability***. The first consideration, cost, is probably the biggest limiting factor to using a vision system in industry. Over 75 percent of all machined parts produced are in batches of 100 parts or less. With today's technology, it is just too expensive to design a dedicated intelligent vision system for small production runs. This is why most of the existing industrial vision systems are found in large-volume production processes, such as automotive and electronic assembly. The bottom line is that, to pay for itself, an industrial vision system must be less costly than the human labor it replaces.

Speed, accuracy, and reliability are also important because an industrial vision system must outperform its human counterpart in all these categories if it is to be cost effective. In most cases, the vision system must perform a real-time analysis of the image to compete with humans. This requires that images be processed in less than 1 second. Of course, the simpler the system, the easier it is to do this. Finally, the system must also be accurate and make reliable judgments. To do this, all the system components, from the imaging device to the image-analysis software, must be carefully engineered and maintained.

Industrial Applications Areas

Present applications of computer vision in industry fall into two general categories: ***inspection*** and ***object identification.*** Product inspection is a natural applications area, since many inspection tasks require 100 percent visual inspection. This means that a production inspector must visually inspect every part for some defect or dimensional criteria. The obvious problem with human inspectors is that they are not always cost effective, accurate, or reliable. There are many reasons why, from fatigue to apathy. Regardless of the reasons, 100 percent inspection by humans usually results in an inspection rate somewhat less than 100 percent. On the other hand, computer vision inspection systems can offer a true 100 percent inspection and are fast, accurate, and reliable. These qualities make a vision system very cost effective where large-volume 100 percent inspection is the norm. Table 6-2 lists several areas where computer vision has been successfully applied to product inspection.

TABLE 6-2 Some Typical Inspection Applications of Computer Vision

Assembly and Processing	Surface Defects	Measurement
Major parts present	Pits	Diameters
Hole location	Scratches	Contours
Presence of holes, threads, nuts, washers, etc.	Voids	Overall part dimensions
	Cracks	
Proper conductor patterns	Discoloration	
Label verification		
Proper markings		
Foreign object detection		

The second major applications area for industrial vision systems is for object identification. In general, object-identification systems are more sophisticated than inspection systems, since they are required to perform more intelligent tasks. Most of the existing applications are in the manufacture and assembly of electronic components, such as transistors and integrated circuits. As vision technology improves, object-identification systems will be more cost effective for other applications, such as parts sorting and bin picking.

Another promising application for object-identification systems is in robot guidance. Such systems have been successful in guiding robots for spray painting, seam welding, and simple product assembly. Again, as machine vision technology improves, these systems will become more reliable and cost effective, allowing robots to be applied to many more production processes. Table 6-3 summarizes object-identification applications of vision systems.

TABLE 6-3 Some Typical Object-identification Applications of Computer Vision

Parts Handling	Robot Guidance
Component placement	Spray painting
Parts sorting	Seam welding
Bin picking	Product assembly

Levels of Design Difficulty

When engineering a vision system for an industrial application, you will always be confronted with a decision on how to control the position and appearance of the object to be inspected or identified. Your decisions about how to control the object's position and appearance have a direct impact on the vision system design and sophistication. The design of any industrial vision system falls into one of three levels of design difficulty:

1. The position and appearance of the object are both tightly controlled.
2. Either the position or appearance of the object is controlled, not both.
3. Neither the position nor appearance of the object is controlled.

First Level of Difficulty. The simplest vision system results when both the object position and appearance can be tightly controlled. The object's position must be controlled by fixturing and its appearance controlled through the use of structured lighting. In other words, the object must always be oriented and placed in the same position. Furthermore, it must have a predictable surface coloration and texture, and not cast any shadows. Applications that fit this category include parts measurement, verifying conductor patterns, foreign object detection, and label verification.

Most industrial vision systems that are designed around this level of difficulty use a template-matching technique called ***binary image subtraction.*** This technique is illustrated by Figure 6-56. Here a binary image of the object is subtracted, pixel by pixel, from a binary reference template.

As you can see, this binary image is different from those discussed earlier. Recall that, in Section 6-2, a binary image was formed by comparing the Roberts cross operator value to a threshold value. The process placed 1's only where the object edges occurred. On the other hand, binary image subtraction generates a binary image by placing 1's where the object exists and 0's for the background. The result is a binary 1 image silhouetted against a binary 0 background, as shown. Clearly, the lighting has to be well structured to create a high object/background contrast with no shadows.

Now, with a perfect match, no binary 1's would remain after subtraction. However, in reality, some binary 1 pixels remain due to part variations and slight differences in position and appearance. These remaining binary 1 pixels are counted, and the count is compared to a predetermined limit to inspect or identify the object. A match occurs when the subtraction results in a pixel count that is below the limit.

Sometimes, rather than storing reference templates in memory, two imaging devices are used as shown in Figure 6-57. Here, an image produced from a reference object is compared to the image on an unknown object. Again, the comparison is accomplished using binary image subtraction. Of course, the positions of both the reference and unknown object must be very tightly controlled, since the alignment of the two images is critical.

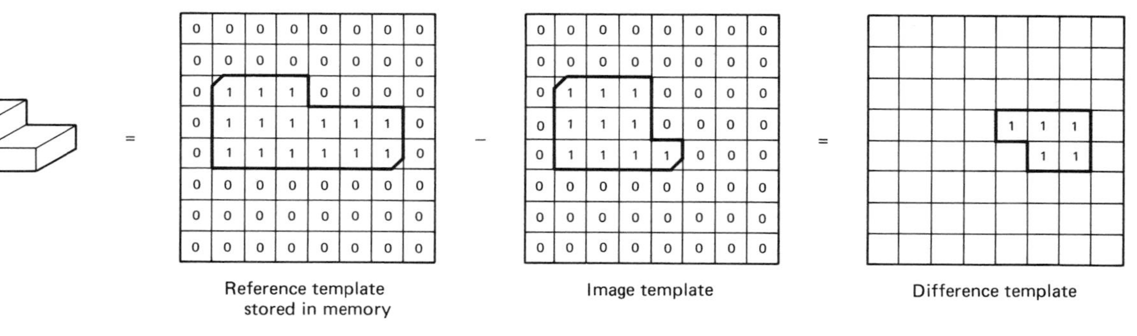

Figure 6-56 Binary image subtraction is used for level 1 industrial vision problems.

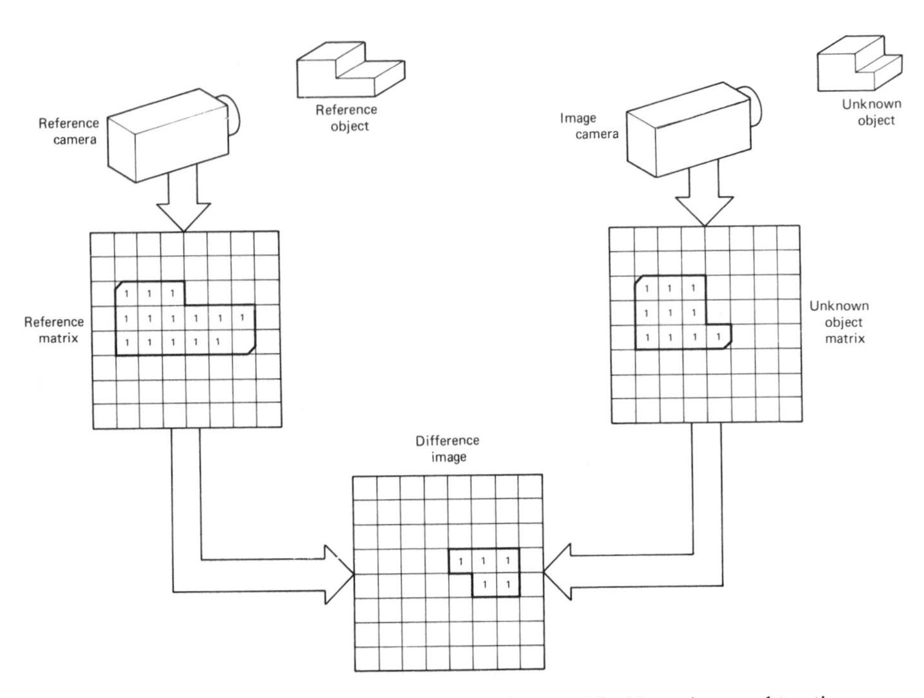

Figure 6-57 Two imaging devices are sometimes used for binary image subtraction.

Another technique, called ***window and pixel counting***, is also used for level 1 designs. With this technique, no comparisons are made to a reference template or object. Instead, a predefined optical window is passed over the binary image of the object or feature of interest. The binary 1's and 0's within the window are then counted and compared to a predetermined limit to inspect or identify the object. This technique is illustrated in Figure 6-58.

Second Level of Difficulty. The next level of difficulty occurs when either the object's position or appearance is controlled, but not both. Parts sorting and robot guidance fall into this level of difficulty. For instance, the appearance of parts to be spray painted can be controlled by proper lighting. However, the parts are usually dangling and swinging from a conveyor hook, making their position and orientation unpredictable. Here is where the challenge begins, since vision systems designed for this level of difficulty must utilize the edge and region analysis techniques described earlier in this chapter.

Third Level of Difficulty. Finally, the most difficult level is when neither the object's position nor appearance can be controlled. Such an application, like bin picking, calls for an extremely intelligent system. This is where all the image sensing, analysis, and understanding techniques discussed in this chapter must come together to produce a cost-effective, accurate, and reliable system. A typical application for such a system is robot navigation, which will be discussed in Chapter 7.

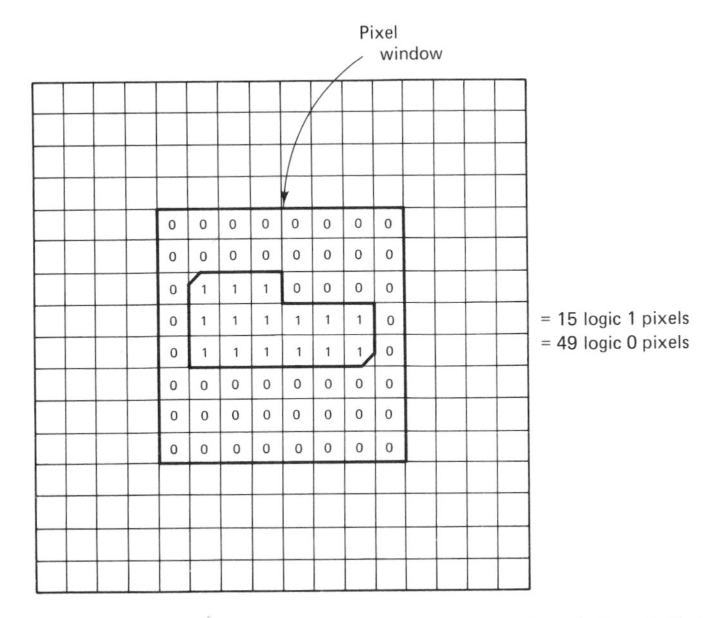

Figure 6-58 Window and pixel counting counts the number of 1's and 0's in a given pixel window. The pixel count is compared to a predetermined limit to identify the object.

Some Specific Industrial Applications. To conclude this topic, let's take a brief look at a few specific ways in which vision systems are being used in industry. A very simple system that is used to detect surface defects on cylindrical objects is shown in Figure 6-59. This system employs a light-line sensing method. A line of light is projected onto the rotating cylindrical object under inspection. The line of light is reflected by the object into a linear photodiode array. Any surface defects are detected as changes in the light intensity across the linear array.

A PC board inspection system is shown in Figure 6-60. This system uses two imaging cameras to compare the board under inspection to a reference board. The

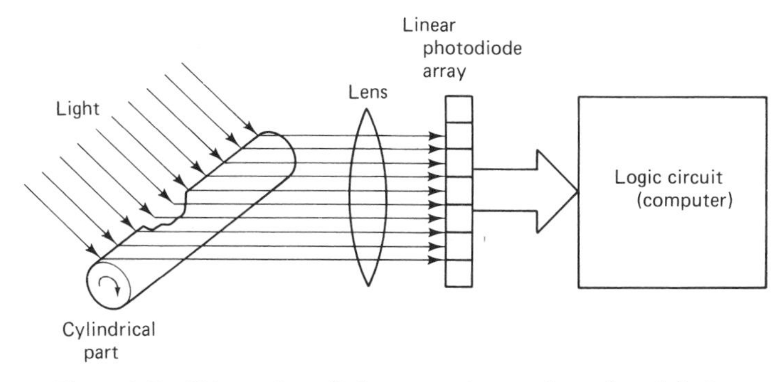

Figure 6-59 Using a photodiode array to inspect for surface defects.

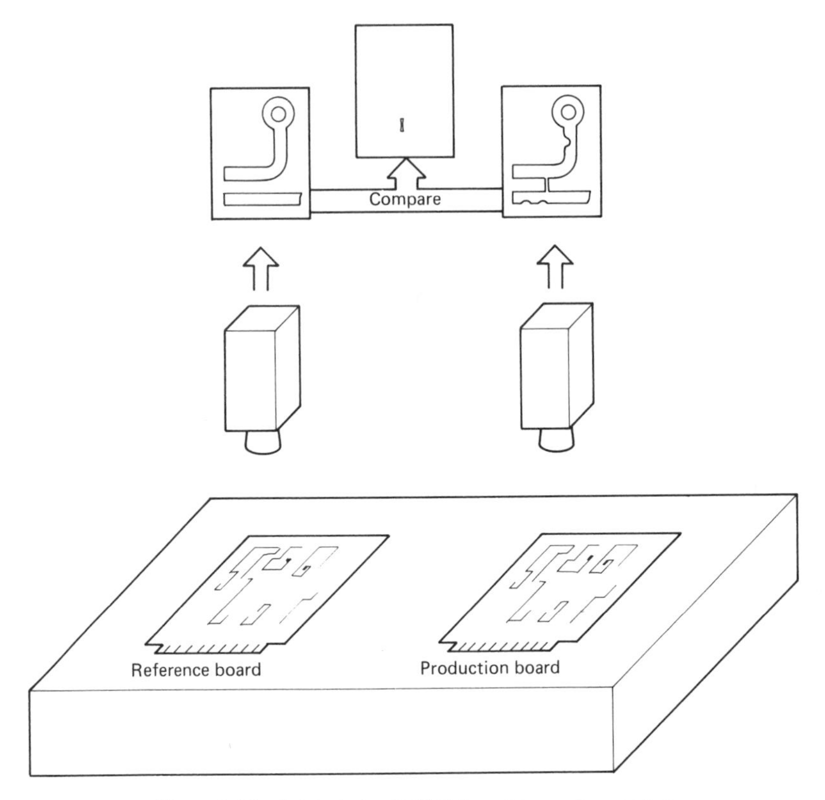

Figure 6-60 Printed-circuit (PC) board inspection system.

system can inspect for the proper conductor pattern, as well as breaks in a conductor or bridges between the conductors.

A system designed to sort parts on a conveyor belt is shown in Figure 6-61. The parts are randomly oriented and positioned on a high-speed conveyor belt. A special lighting system is used to direct light onto the part and create a high contrast between the part and the belt. Edge and region analysis are then used to identify the different parts and direct the sorting operation. As the parts are identified, a sorting mechanism directs a series of solenoids to kick the parts off the belt and into their respective bins.

Finally, a robot paint-spraying operation is illustrated in Figure 6-62. The addition of a robot guidance vision system allows the manufacturer to mix part types and colors on the same paint spray line. The vision system allows small batches of parts to be processed without stopping the line and reprogramming the system each time a different type of part comes along.

As you can see, the parts are suspended from conveyor hooks and passed slowly in front of the robot. Backlighting is used to control the part appearance by providing the high part contrast necessary for image segmentation. In most cases, the vision system is trained with each new part so that image templates can be

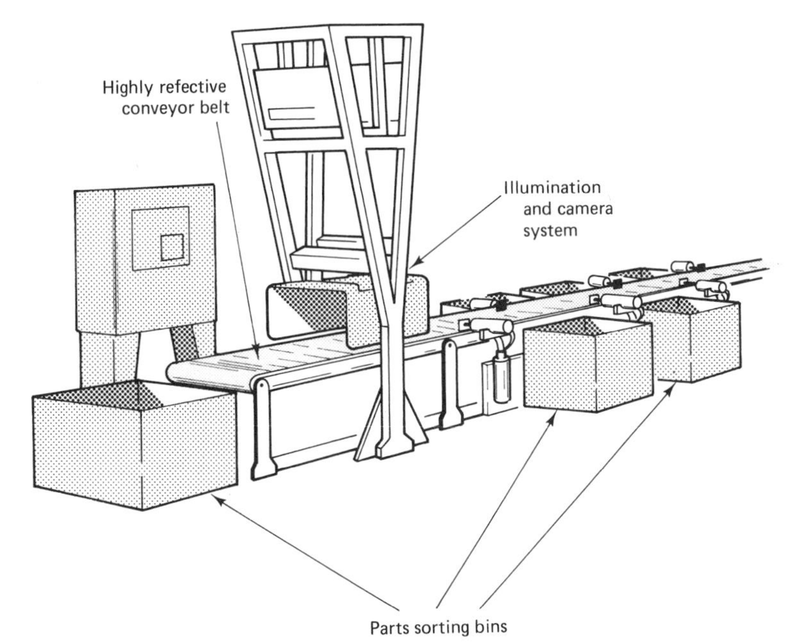

Figure 6-61 Parts sorting vision system.

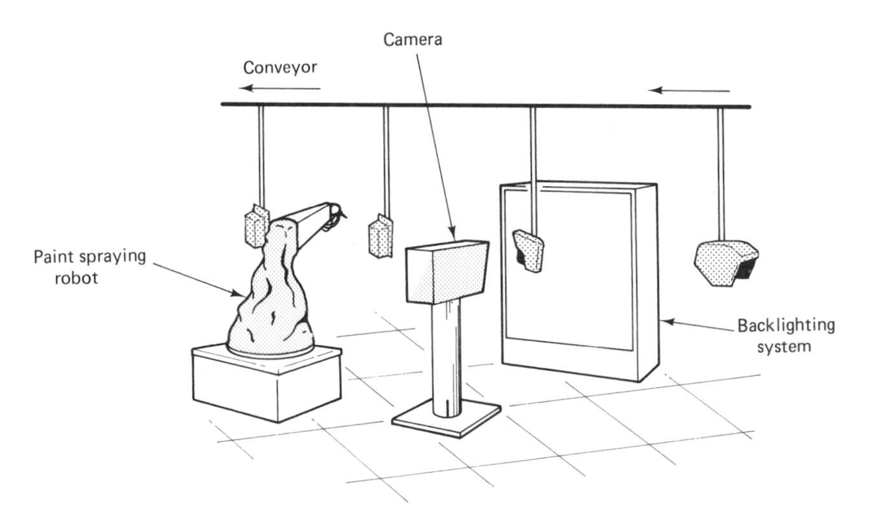

Figure 6-62 Robot paint-spraying operation uses a vision system for robot guidance.

stored and used during the identification process. Once the system is trained, it uses a combination of edge and region analysis to identify a given part and direct the painting operation. Color labels are sometimes attached to the parts and detected by the system to allow a given type of part to be painted with different colors.

SUMMARY

The three main areas of computer vision technology are image transformation, image analysis, and image understanding. Image transformation involves the conversion of image features, such as brightness, color, and texture, into digital signals that can be stored and analyzed by a computer. Imaging devices include vidicon TV cameras, linear photodiode arrays, diode-matrix cameras, and CCD or CID cameras. A computer stores image data in picture matrices made up of elements called pixels. For brightness data, the pixels represent the relative brightness levels at the respective points in the scene.

Image analysis must provide enough information to the subsequent image-understanding process so that it does not have to access any of the original image data. The first step in image analysis is to remove noise using hardware filtering and software smoothing techniques. Gray scale, color, and texture data are then analyzed to find edge points, lines, and regions. Edge detection and region analysis are the primary image-analysis techniques, since both edges and regions can be used to define the shape of objects.

Edge detection is accomplished by differentiating gray-scale picture data and using a subsequent thresholding process to generate binary images. Binary images are then used to find lines using tracking, model-matching, or template-matching techniques.

Region analysis is the opposite of edge detection, since it involves the analysis of pixel similarities rather than abrupt changes. The two general types of region analysis are region splitting and region growing. Region splitting requires the scene to be divided up into regions of pixels with similar image feature values. Region growing forms larger regions from smaller ones by merging atomic regions that have similar feature values.

The final task of a computer vision system is to interpret the information obtained during the image-analysis phase. This is called image understanding, or machine perception. Since the key element to any intelligent task is knowledge, image understanding involves the interaction of several types of knowledge, including gray scale, color, texture, nonhuman sensory data, motion, depth, and expectational knowledge. This knowledge is divided into low- and high-level knowledge sources. The high-level knowledge sources, like expectational knowledge, are used to guide the image-understanding process. A knowledge organization and control system that allows independent interaction of the low- and high-level knowledge sources is the blackboard model.

Most industrial vision systems are applied to product inspection and object identification. The primary design considerations for an industrial vision system are cost, speed, accuracy, and reliability. These design considerations are linked to three levels of design difficulty that control the position and appearance of the object. Simple industrial systems employ binary image subtraction, or window and pixel counting, to inspect or identify objects. More sophisticated systems must employ edge and region analysis, along with intelligent image-understanding techniques for object identification. Robot guidance falls into this category.

PROBLEMS

6-1 Name the three main tasks that must be performed by an intelligent vision system.

6-2 Describe the operation of a TV camera vidicon.

6-3 When operating as an imaging device, a photodiode must be ______________ biased.

6-4 Imaging devices that employ a rectangular photodiode array are called ____________________.

6-5 Explain the operating principle of a CCD camera.

6-6 Construct an 8 X 8 picture matrix for the letter A using a 4-bit gray-scale code.

6-7 Explain how ensemble averaging differs from local averaging.

6-8 Smooth the matrix you constructed in Problem 6-6 using the local-averaging technique and a 3 X 3 pixel window.

6-9 What is the purpose of the Roberts cross operator?

6-10 Generate a binary matrix from your smoothed gray-scale matrix in Problem 6-8. Use the Roberts operator thresholding technique with a threshold value of 4.

6-11 List three techniques that are used for finding lines from a binary matrix.

6-12 Explain how region splitting differs from region growing.

6-13 What is a picture tree and why is it useful?

6-14 How can color information be converted to gray-scale intensity data?

6-15 A color TV camera generates the following color signals: red = 0.5 V, green = 0.5 V, blue = 0.2 V.

(a) What is the amplitude of the gray-level signal?

(b) What is the color?

6-16 What do all textures have in common?

6-17 Label the object in Figure 6-63(a) using the line labels developed in this chapter.

6-18 Label the object in Figure 6-63(b) using the line labels developed in this chapter.

6-19 Explain how shadows and cracks can enhance the image-understanding process.

6-20 What clue does motion provide to the image-understanding puzzle?

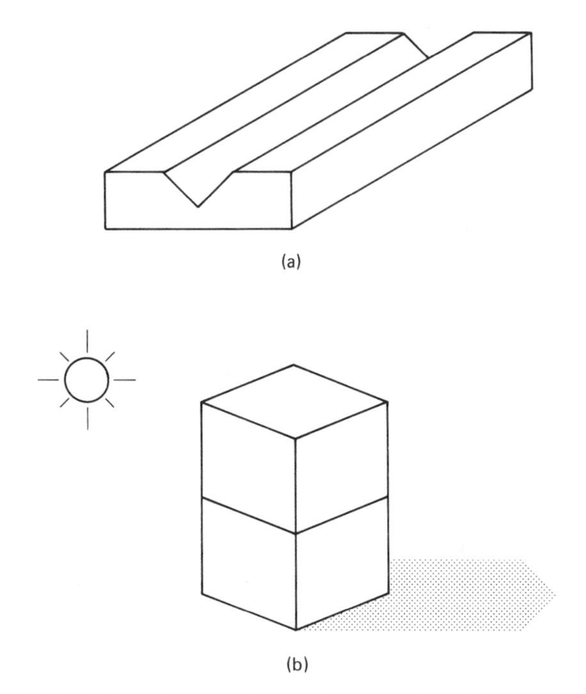

Figure 6-63 Objects to be labeled: (a) Problem 17, and (b) Problem 18.

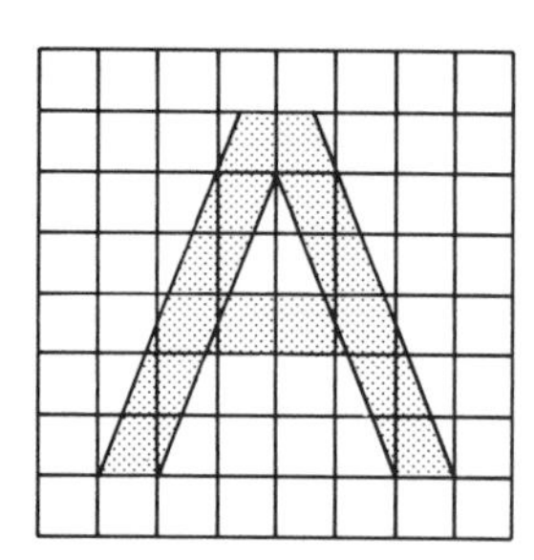

Figure 6-64 A simple 8 × 8 gray-scale picture matrix for the letter A (Problem 24).

6-21 List at least three nonhuman types of sensory data that can be applied in computer vision.

6-22 List at least two types of expectational knowledge that can be applied to computer vision.

6-23 Explain why expectational knowledge is important in computer vision.

6-24 Construct a pyramid for the letter A matrix shown in Figure 6-64.

6-25 Construct a quad tree from the pyramid you developed in Problem 6-24.

6-26 What type of knowledge organization and control system allows independent interaction of both low- and high-level knowledge?

6-27 Name the primary design considerations for an industrial vision system.

6-28 What are the two general applications categories that most industrial vision systems fall into?

6-29 Describe the three levels of design difficulty that apply to industrial vision systems.

6-30 Construct a binary image template that would be used to represent the letter A in an industrial binary image subtraction system. Use an 8 $\times$ 8 pixel matrix.

7 Range Finding and Navigation

Range finding and navigation are particularly important to robotics. For example, consider an industrial bin picking operation where a seeing robot must locate objects in a parts bin, not knowing exactly where they are. The sequence of operations for such a robot might go something like this:

1. Scan the parts bin and locate the object of interest in three-dimensional space.
2. Determine the relative position and orientation of the object.
3. Move gripper and/or body to the object location.
4. Position and orient the gripper according to the object's position and orientation.
5. Pick up the object.
6. Place the object at the required location.

The first step in this sequence presents you with a real dilemma, since the distance, or range, to the object must be determined in three-dimensional, or 3-D, space. Up to this point, you have not had to concern yourself with the third dimension, because all the image analysis and understanding techniques discussed in Chapter 6 dealt with 2-D images. Many vision tasks can be performed with 2-D images, but real-world operations like bin picking must incorporate 3-D information.

As it turns out, many 2-D image-understanding problems are caused by the lack of range information. There are techniques that can be used to infer depth from 2-D images. The resulting image is called a ***2½-D image.*** However, 2½-D analysis does not provide the range information required for most real-world operations.

The third dimension involves determining the distance, or range, of all the points that define a scene, the result being a range image that complements the 2-D image. The beginning two sections of this chapter are devoted to range finding. Two fundamental techniques will be discussed: ***triangulation*** and ***time of flight.***

You might also think of the robot pick-and-place operation as a navigation problem. If the robot is mobile, it must navigate its body to the object location, and then navigate its gripper to grasp the object. Even if the robot is stationary, it must navigate its gripper to the object. When you think about it, this type of navigation actually reduces to position and proximity sensing. The robot must be capable of sensing the position of objects, as well as its proximity to those objects, in order to navigate its gripper and/or body accurately. Of course, a 3-D vision system can be applied to this task. However, simpler position and proximity sensing techniques can be used to replace or complement the vision system. You will explore these sensing techniques in Section 7-3.

7-1 TRIANGULATION

You are probably familiar with the idea of using triangulation to measure distance. Just for review, suppose two triangulating devices are a certain distance apart and both devices can get a "fix" on an object as shown in Figure 7-1. As you can see, the two devices and the object form a triangle whose one leg, d, and two angles, θ_1 and θ_2, are known. The third angle is easily found by subtracting the two known angles from 180°. The distance from each device to the object can then be found using the law of sines. In vision systems, the geometry gets a bit more complicated, but the idea is the same.

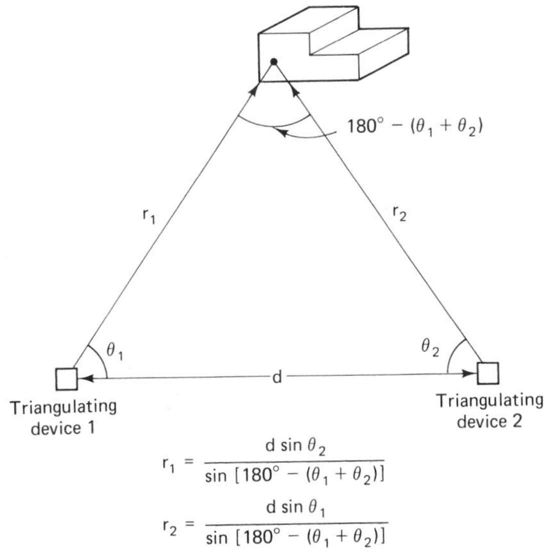

$$r_1 = \frac{d \sin \theta_2}{\sin [180° - (\theta_1 + \theta_2)]}$$

$$r_2 = \frac{d \sin \theta_1}{\sin [180° - (\theta_1 + \theta_2)]}$$

Figure 7-1 Simple triangulation.

A vision system can employ one of two different triangulation techniques for measuring the distance, or range, to an object. One technique, ***passive triangulation***, uses two imaging devices. The other technique is referred to as ***active triangulation*** and employs a single imaging device and a controlled, or structured, light source.

Passive Triangulation

Passive triangulation is also referred to as ***stereo*** or ***binocular vision.*** You could say that it is biologically motivated, because this is the way that humans and most animals perceive depth. Two imaging devices are placed a known distance apart, as shown in Figure 7-2. In a machine vision system, the two imaging devices might be two TV, diode-matrix, or CCD cameras. Of course, in a biological system they are two eyes.

Here's how it works. Two things in the system are known: the distance between the cameras, d, and the focal length of the cameras, f. The idea is to calculate the range, r, from the cameras to a given point, P, on the object. Both cameras scan the scene and generate a picture matrix. Given any point in the scene, such as point P, there will be two pixels representing that point. One pixel is in the left camera image and the other is in the right camera image.

Each pixel is located a given distance from the center of its image. Let x_1 be the distance that the left camera image pixel is located from the center of its image. Likewise, let x_2 be the distance that the right camera image pixel is located from the center of its image.

If you overlap the two camera images, the two image points, x_1 and x_2, will not coincide. Rather, there will be a certain distance between them. This distance is calculated by taking the absolute value of their difference. The resulting difference is called the ***disparity*** between the two image points.

The range, r, from the cameras to the object point is inversely proportional to the disparity between x_1 and x_2. For instance, as the disparity approaches 0, the range becomes infinite, since there is no distance between the two matching image pixels. Conversely, the range gets smaller as the disparity gets larger.

Given the stereo system in Figure 7-2, the range of any point on the object can be approximated by the following equation:

$$\text{Range} = r = \frac{d\sqrt{f^2 + x_1^2 + x_2^2}}{|x_1 - x_2|}$$

where r = range from the left camera lens if the object point is in the right side of the scene
= range from the right camera lens if the object point is in the left side of the scene
= range from either camera if the object point is directly in the middle of the scene

d = distance between camera lens centers

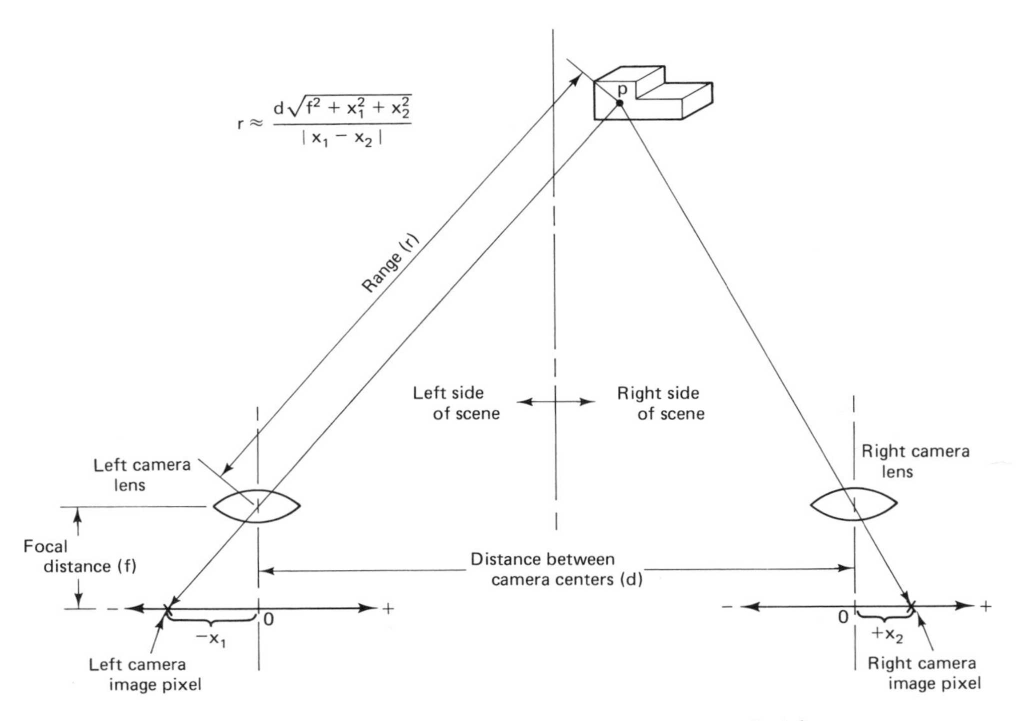

Figure 7-2 Stereo vision system using the triangulation principle.

f = focal length of cameras
x_1 = distance of the image pixel from the center of the left camera lens
x_2 = distance of the image pixel from the center of the right camera lens

Note from the definition of range in this equation that the range value, r, can be the distance from the left, right, or either camera, depending on where the object point is located in the scene. For instance, if the object point is located in the right half of the scene, r is defined as the range from the left camera lens. The left and right halves of the scene are divided by an imaginary line located exactly halfway between the two cameras.

Also, the individual values of x_1 and x_2 can be positive or negative, depending on the location of a given image pixel relative to the center of its respective image. For example, if the image is between the two cameras, x_1 would be negative and x_2 positive. Note, however, that the disparity is always the absolute value of the difference between the two image points and is used in the denominator of the range equation.

Example 7-1

Two cameras having a focal length of 5 centimeters (cm) are located 30 cm apart. An object is detected in the right half of the scene, between the two cameras. A pixel point representing the object is located at a distance of 0.4 cm from the center of the left camera lens. The corresponding right camera pixel point is located 0.2 cm from the center of its camera lens. Calculate the distance, or range, of the object point.

Solution

The following information is given directly in the problem statement:

$$d = 30 \text{ cm}$$
$$f = 5 \text{ cm}$$

Since the object is located between the two cameras, x_1 must be negative and x_2 positive. Recall that x_1 applies to the left camera and x_2 applies to the right camera. Thus the disparity value is

$$|x_1 - x_2| = |-0.4 \text{ cm} - (0.2 \text{ cm})| = 0.6 \text{ cm}$$

Using these values in the range equation, you get

$$r = \frac{30 \text{ cm} \sqrt{(5 \text{ cm})^2 + (0.4 \text{ cm})^2 + (0.2 \text{ cm})^2}}{0.6 \text{ cm}}$$

$$= \frac{(30 \text{ cm})(5.02 \text{ cm})}{0.6 \text{ cm}}$$

$$= 251 \text{ cm} \quad \text{(approx.)}$$

Since the object point is located in the right half of the scene, this range value is the distance from the center of the left camera lens.

The closer the two cameras are placed together, the less important it is to distinguish between the left and right halves of the scene. However, decreasing the distance between the cameras also decreases the resolution of the range measurement. Of course, decreasing the distance too much results in no stereo at all.

Correspondence Problem. From the preceding discussion, you can see that once a stereo system is in place its accuracy is totally dependent on the relative positions of the two matching image points. Hence, the position of these two points must be precisely determined. As it turns out, this is the hardest part of the stereo range-finding process. Matching corresponding image points of an object from two different images is referred to as the ***correspondence problem.***

Ideally, it would be nice to find individual pixels in one camera image that matched those of a second camera image. However, in reality, you cannot guarantee that two pixels with the same gray scale or color values were produced by the same object point. For this reason, stereo vision systems often search for similar edge or region features between two images to locate corresponding pixels.

Edge-based stereo attempts to match stereo images by detecting intensity or color edges. The edge-detection techniques discussed in Chapter 6 are employed. As you know, edges define lines that define regions. Most likely, two regions separated by an edge are at two different depths. Thus, an edge in one stereo image that matches an edge in a second image is a good place to begin calculating range values.

Another matching technique is to take a pixel window, or ***patch***, from one image and pass it over the same general region of the second image until the best match is found. A displacement, or disparity, value can then be determined based on how much the window must be displaced from the first image to match the second image. This value is then used to calculate range.

Finding matching points, edges, or regions between two stereo images is hard enough if the objects look the same in both images. However, one of the biggest problems with stereo vision is the fact that objects don't look the same from two different views. At worst, objects in one image might be obscured and not even appear in a second image. Of course, you say that you can decrease the distance between the cameras to reduce the correspondence problem, but remember that this reduces the accuracy of the system.

The correspondence problem with stereo vision can be reduced significantly by eliminating one of the imaging devices. What? You say that this results in no stereo, and you are right! However, triangulation is still possible by replacing the second camera with a moving, or active, light source. The resulting ranging technique is called ***active triangulation*** and is the topic of my next discussion.

Active Triangulation

Active triangulation involves the movement of the imaging device or the light source, not both. Either the imaging device is stationary and the scene is scanned with a projected beam of light, or the illumination is constant and the camera is moved to generate different perspectives of the scene. Let's first consider the case where a single imaging device is stationary and a projected light source is used to scan the scene. This type of system can employ one of two techniques for depth measurement: ***spot sensing*** or ***light-stripe sensing.*** These are sometimes referred to as a ***structured lighting*** techniques.

Spot Sensing. Suppose that you project a single beam of light onto an object in a darkened room, as illustrated in Figure 7-3. The projected light beam creates a light spot on the object that is reflected into a camera positioned a known distance, d, from the spot projector. This creates a triangle between the projector, object, and camera. Using this triangle, you can calculate the range, r, of the object spot from the camera.

Here's how: the reflected light spot produces an image point, x, in the camera image. This image point is easily detected, since it will be the brightest "spot" in the image. The distance of the image point from the center of the camera image can be determined. Furthermore, the camera focal length, f, is fixed. Since the focal length, f, and the image point distance, x, form the sides of a right triangle, the angle θ_2 can be calculated as

$$\theta_2 = \tan^{-1}\frac{f}{x}$$

where f = focal distance of the camera
x = distance of the image point from the center of the image

Next, the distance between the projector and the image point can be calculated as follows:

$$D = d + x$$

where d = fixed distance between the projector and camera
x = distance of the image point from the center of the image

In both of these relationships, x can be positive or negative, depending on whether the image point is to the right (+) or left (−) of the center of the camera lens.

Now, since the precise position of the light beam is controlled by the projector, the beam angle θ_1 is known. Consequently, the range, r, can be calculated using the law of sines as follows:

$$\frac{r}{\sin\theta_1} = \frac{D}{\sin[180^\circ - (\theta_1 + \theta_2)]}$$

$$r = \frac{D\sin\theta_1}{\sin[180^\circ - (\theta_1 + \theta_2)]}$$

Example 7-2

A camera with a focal length of 5 cm is located 30 cm to the right of a spot-scanning projector. A beam of light is projected at an angle of 45° onto an object and reflected into the camera. Find the range of the object point if it produces an image point that is located 2 cm to the right of the image center.

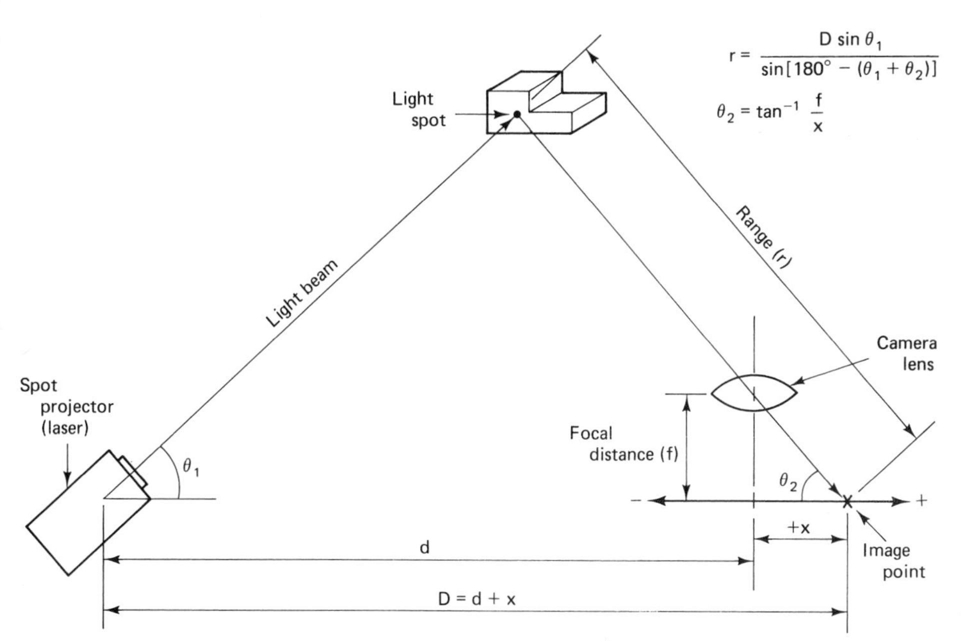

Figure 7-3 With spot sensing, the camera is stationary and the scene is scanned with a spotlight beam.

Solution

The following information is given directly in the problem statement:

$$f = 5 \text{ cm}$$
$$d = 30 \text{ cm}$$
$$x = 2 \text{ cm}$$
$$\theta_1 = 45°$$

From this information, you can construct the diagram in Figure 7-4. The first thing you must do is calculate θ_2, as follows:

$$\begin{aligned}\theta_2 &= \tan^{-1} \frac{f}{x} \\ &= \tan^{-1} \frac{5 \text{ cm}}{2 \text{ cm}} \\ &= \tan^{-1} 2.5 \\ &= 68.2°\end{aligned}$$

Next, the distance between the projector and image point is

$$\begin{aligned}D &= d + x \\ &= 30 \text{ cm} + 2 \text{ cm} \\ &= 32 \text{ cm}\end{aligned}$$

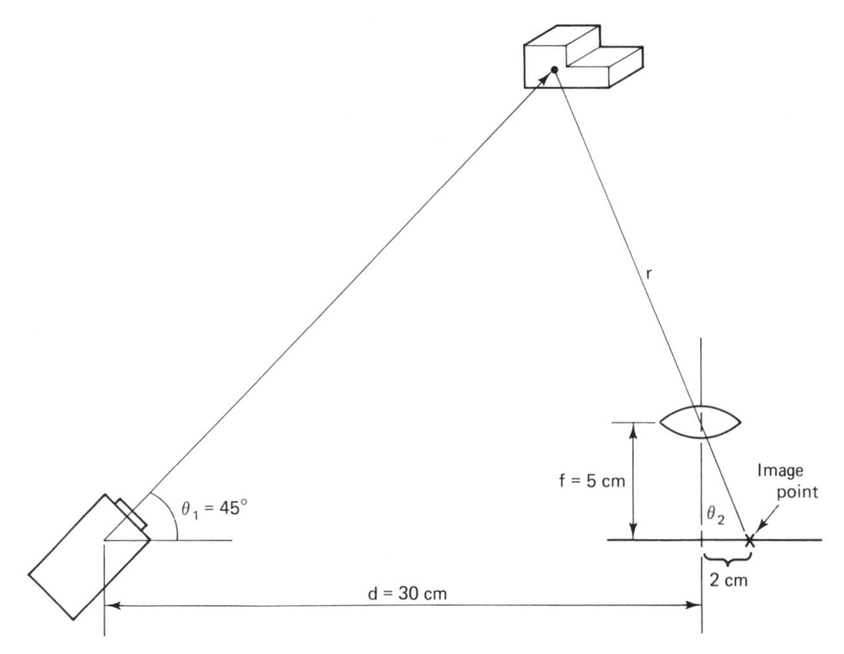

Figure 7-4 Problem definition for Example 7-2.

Substituting these values in the range equation, you get

$$r = \frac{32 \text{ cm} \sin 45°}{\sin[180 - (45 + 68.2)]°}$$

$$= 24.62 \text{ cm}$$

Notice that this range value is the distance between the image point and the object point. To get the range from the camera lens, you must subtract the distance between the lens and the image point. This distance can be found using the Pythagorean theorem.

The preceding discussion only applies to a single point on the object. To produce the 3-D, or range, image, the light spot must be scanned over the entire scene and the range calculation made at each point in the scan. In other words, the light spot must scan the scene from right to left and top to bottom. The simplest way to do this is by using two rotating mirrors. One mirror rotates in a direction required to produce the right-to-left scan, while the other mirror rotates in a direction perpendicular to the first to provide the top-to-bottom scan.

Different types of light, like infrared or laser light, could also be used in the preceding system. Neither of these would require a darkened room, but filters must be placed in front of the camera to filter out the room lighting. Laser light is particularly useful, since the laser beam produces a very small spot, thereby providing greater depth resolution. More about this later.

Light-stripe Sensing. Light-stripe sensing is just an extension of spot sensing. Rather than projecting a spot, a stripe of light is projected on the scene, as Figure 7-5 shows. As a result, the stationary camera sees a line, or stripe, of light. The stripe can then be divided into individual image points and the range calculated for each point along the stripe. The range calculation is identical to that for spot sensing.

The light stripe can be formed by passing ambient or infrared light through a slit on the projector. A laser stripe can be formed by scanning the laser beam rapidly in one plane or passing it through a cylindrical lens. In any event, the scene is scanned in a direction perpendicular to the stripe, resulting in a complete range mapping of the scene.

One problem encountered with light-stripe scanning is that poor depth resolution is obtained for object surfaces that are parallel to the light stripe. This problem

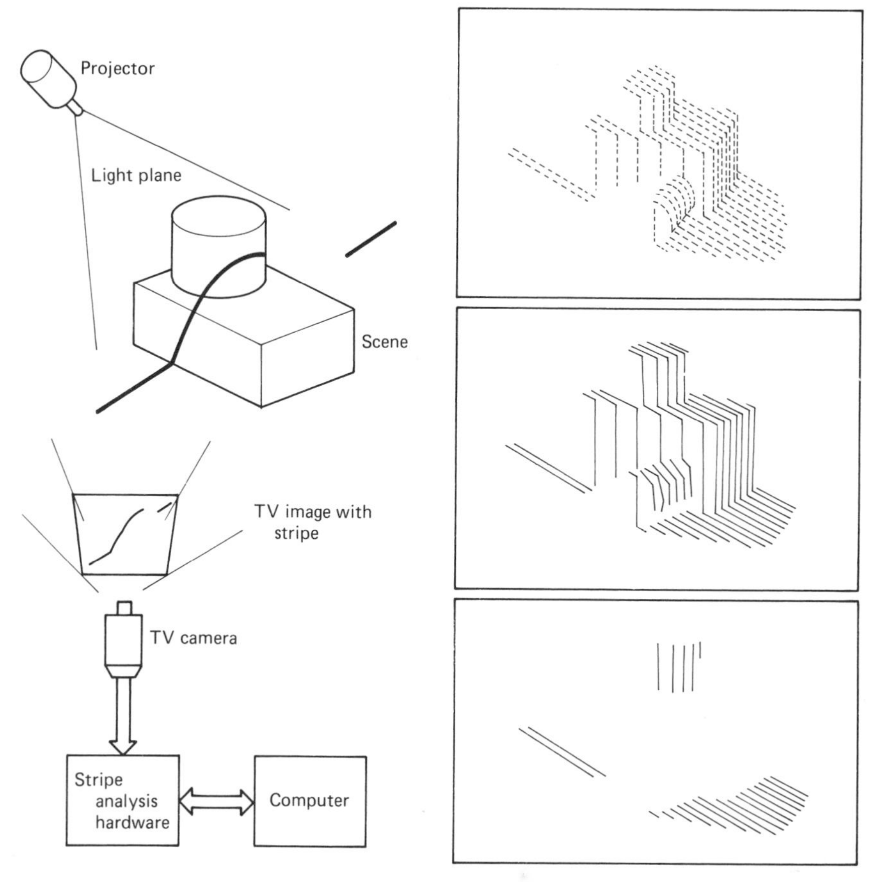

Figure 7-5 Light striping. (Reprinted with permission from *Computer Vision* by Dana H. Ballard and Christopher M. Brown, Prentice-Hall, Inc., Englewood Cliffs, N.J.)

can be overcome by scanning the image in two directions, one perpendicular to the other.

The advantage of light striping is that it is relatively simple and fast, as opposed to spot sensing. A bonus feature to light striping is that object boundaries and regions can be determined by connecting the end points of the light-stripe images. Thus, light striping aids in the image-segmentation process. This can be seen by examining the series of light-stripe images in Figure 7-5.

Camera Motion. This final form of active triangulation involves moving the camera as illustrated in Figure 7-6. Here a single camera is moved a given distance to produce two stereo images of the scene. In other words, a single moving camera takes the place of the two stationary cameras in a stereo system. Once the two images are obtained, the range calculations are made using the disparity between the two images as in stereo vision. Again, the major drawback to this type of system is the image correspondence problem.

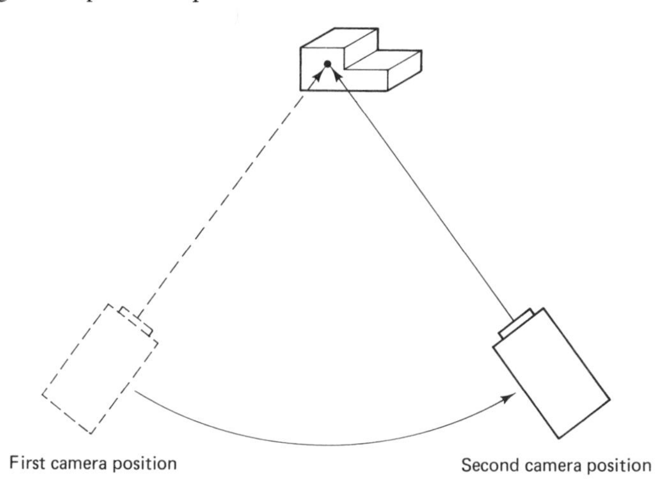

Figure 7-6 A moving camera system generates stereo images at two consecutive positions.

7-2 TIME-OF-FLIGHT RANGING

Time-of-flight (TOF) ranging involves calculating the time it takes for a signal to reach and return from an object. Since distance equals the product of velocity and time, the range of an object must be

$$\text{range} = \frac{v \times t}{2}$$

where r = range from the ranging device
v = velocity of the transmitted signal
t = time it takes the signal to reach and return from the object

Three types of signals are used for TOF ranging: ***electromagnetic, light,*** and ***sound.*** The determination of range using the preceding relationship is the same for each type of signal. However, each type of signal has its own characteristics that affect the accuracy of the range data.

TOF Signal Characteristics

Two important signal characteristics must be kept in mind when employing TOF ranging. First, the width of the signal beam determines the amount of detail that can be recovered during the ranging process. A wide signal beam will not produce accurate range data for small object details, since it covers a larger area as compared to a narrower beam. Narrower beam widths result in higher object resolution.

For instance, compare the wide beam in Figure 7-7(a) to the narrow beam in part (b) of the figure. The wider beam covers much of the object surface, including the hole in the middle of the object. The depth of the hole will not be detected, since the signal is being reflected by the surrounding object surface. In fact, the TOF range data might not even indicate the presence of a hole! However, if a narrower beam is scanned across the object, as shown in Figure 7-7(b), the time of flight will change when the beam enters the hole. As a result, the hole presence can be detected and its depth measured.

The second important signal characteristic is its speed. The faster the signal reaches and returns from an object, the harder it is to determine its range. For example, a sound signal travels at about 1100 ft/s, whereas a light signal travels

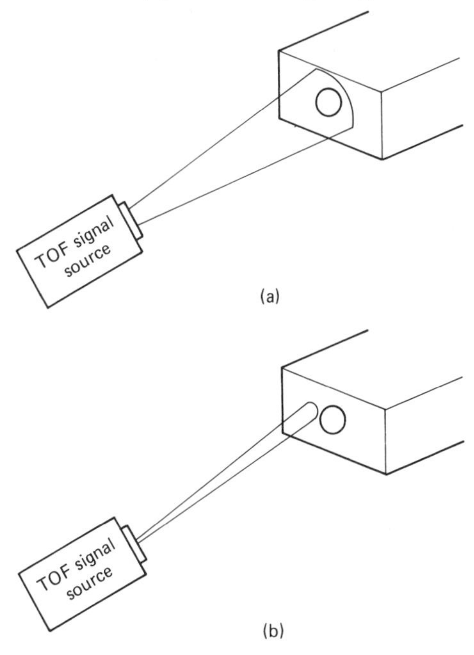

Figure 7-7 (a) A wide beam will not produce as detailed object range data as a (b) narrow beam.

about 982,080,000 ft/s. Using these figures and solving for time in the preceding equation, each foot of range equates to about 1.8 ms in a sound system, as compared to 2 billionths of a second in a light system. Thus, for the range data to be accurate to within 1 ft, a light system must be able to measure time within billionths of a second. Such a system might be fine for determining the depth of a moon crater, but it would be highly impractical in a short-distance robot vision system.

Now let's explore each type of range-finding system, keeping in mind the signal characteristics. In most cases, you must trade off one signal characteristic to get the other.

Electromagnetic. Electromagnetic range finding involves the use of radio signals and is commonly called ***radar.*** Radar has become rather commonplace for both military and civilian (police) applications. The idea is to transmit a radio signal into the atmosphere. If an object is present, the signal is reflected back and the distance, or range, to the object is determined using the time-of-flight relationship.

Radar works fine to measure the range of highly reflective metallic objects over relatively long distances, but it is impractical for measuring relatively short distances of nonmetallic objects. For one thing, the radio signal must be concentrated enough to produce a narrow beam width. Since radio beam width is inversely proportional to antenna size, it requires a relatively large antenna to transmit a narrow radio beam. Moreover, high-frequency radio signals travel at almost the speed of light. Thus, accurate depth measurement is difficult over short distances. Finally, nonmetallic objects will absorb most of the high-frequency radio signal. To get some measurable reflection, the radio signals must be relatively strong, requiring high-power transmitting equipment. For these reasons, radar is not commonly used in computer vision or robot navigation systems for range finding.

Light. Time-of-flight ranging using light is referred to as ***lidar*** (light detection and ranging). A LIDAR system generates a short burst of light and measures the time it takes the light to be reflected off an object and return. Laser systems are usually employed because they are capable of producing very accurate object resolution, due to the narrow beam width of laser light. In addition to measuring time of flight, the intensity of the reflected laser beam can also be measured and used to generate a gray-scale intensity image of the scene. The reflected beam intensity will change in proportion to the amount of light absorption of different objects. The range data coupled with the intensity data provide most of the information needed for complete 3-D image understanding.

The only problem with a LIDAR system is trying to measure the speed of light. Accuracies of less than 1 ft require time measurement in billionths of a second. One solution to this problem is to measure the phase shift of the reflected laser light, rather than the time of its flight.

In a phase-shift LIDAR system, the laser beam is amplitude modulated and transmitted, as shown in Figure 7-8. When the light strikes an object, the reflected signal is out of phase with the original signal. The amount of phase difference is proportional to the range of the object. Such a phase-shifting system has demon-

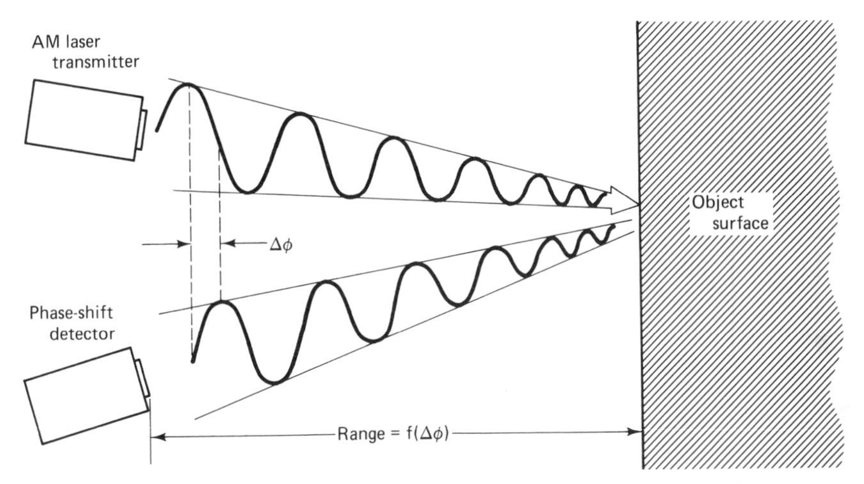

Figure 7-8 The phase shift of a reflected amplitude-modulated laser beam is proportional to range.

strated depth resolution within 1 cm over short distances. This, combined with its narrow beam width, makes laser range finding more practical for computer vision and robot navigation applications.

Sound. Sound, or ultrasonic, ranging permits very accurate depth resolution due to its relatively low velocity. However, it is difficult to concentrate the sound energy into the narrow beam required to produce high object resolution for 3-D vision. Nevertheless, ultrasonic ranging is useful in robot navigation to detect the presence and range of whole objects, even though it might not be capable of generating a detailed depth map of the objects.

Ultrasonic ranging is useful for measuring distances from 1 to 50 ft. In most cases, accuracies of 1 or 2 percent are possible. The idea is to transmit several cycles of high-frequency sound and use a simple hardware or software counter to measure the amount of time it takes for the sound to return. The counter is started when the sound is transmitted and stopped when the reflected sound is received. The resulting count will be proportional to the range of an object.

An ultrasonic system requires a sound transducer that can act as both the sound transmitter and receiver. Two types are common: the ***piezoelectric transducer*** shown in Figure 7-9(a) and the ***electrostatic transducer*** shown in Figure 7-9(b). In the transmit mode, the transducer acts like a speaker to generate a short burst (5 to 20 cycles) of ultrasound. The ultrasound should be above the audible range, usually between 30 and 50 kHz. The sound is produced by applying an ac excitation voltage to the transducer. With a piezoelectric transducer, the ac excitation causes the piezoelectric crystal to vibrate at its resonant frequency. The crystal vibration is then transferred to its metal plates or a diaphragm to produce the ultrasound waves. With an electrostatic transducer, metal foil is bonded close to a metallic backplate, as shown in Figure 7-9(b). The two metallic plates are insulated from each other

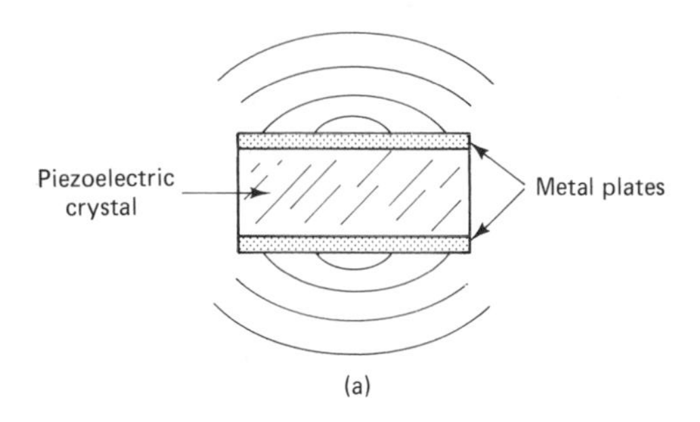

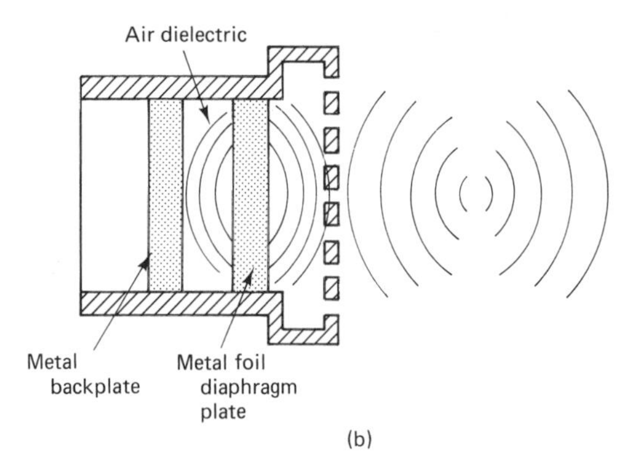

Figure 7-9 (a) A piezoelectric sound transducer and (b) an electrostatic transducer.

and, thus, form a capacitor. An ac excitation signal applied across the plates causes the foil to vibrate and generate the ultrasound waves.

The transducers reverse their roles and act like microphones during the receive operation. The returning sound waves cause the metal plates of either type of transducer to vibrate. In the piezoelectric transducer, the vibrations compress the crystal at a rate equal to the ultrasound frequency. This generates a small ac signal due to the piezoelectric effect, which is detected and used to stop the counter. In the electrostatic transducer, the plate vibrations change the capacitance of the transducer. This change in capacitance is detected and used to stop the counter.

Ultrasonic range finders are commercially available as complete modules. In most cases, a digital *start* pulse must be applied to the module to generate the sound burst. At the same time, the pulse enables a hardware or software counter, as shown in Figure 7-10. When the range finder receives the reflected sound, it generates an *echo* pulse that is used to disable a hardware counter or interrupt a software counter.

Example 7-3

A commercial ultrasonic range finder generates a sound burst that consists of 16 cycles at 50 kHz. At the same time, a 16-bit counter is enabled to count at a 1-MHz rate.

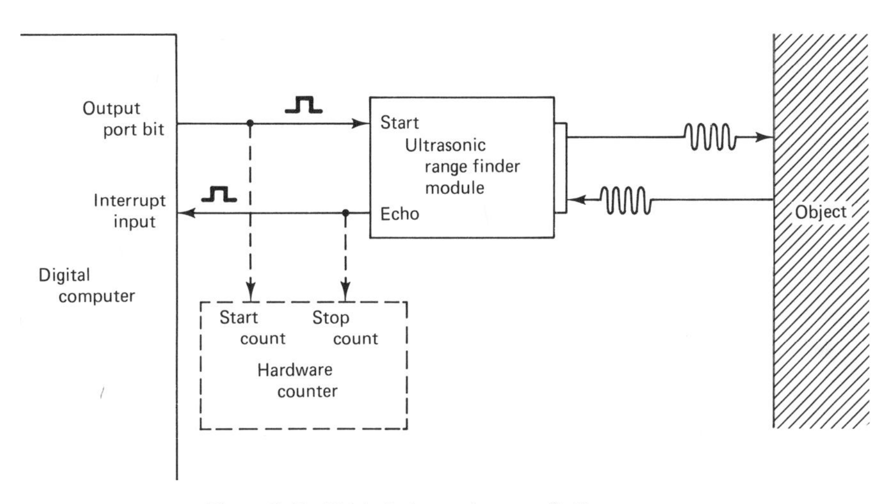

Figure 7-10 Typical ultrasonic range-finding system.

(a) What is the maximum range of this system?

(b) What is the maximum possible resolution of the system?

(c) What is the range of an object if the counter contains the value 0011 1101 0010 1111 when it is disabled by a returning signal?

Solution

(a) The maximum count value for a 16-bit counter is 1111 1111 1111 1111, or 65,535. With a counting frequency of 1 MHz, the maximum amount of time that the system can measure is 65,635 X 1 μs, or 65,535 μs. As stated earlier, sound travels at about 1100 ft/s. This equates to 1.8 ms for each 1 ft of range. Consequently, the maximum range of this system is

$$\frac{65{,}535\ \mu s}{1.8\ \text{ms/ft}} = 36.4\ \text{ft}$$

(b) The maximum possible resolution of the system is proportional to the contribution of the least significant bit (LSB) of the counter. Since the LSB contributes 1 μs to the timed interval, the maximum possible resolution of this system is

$$\frac{1\ \mu s}{1.8\ \text{ms/ft}} = 0.00055\ \text{ft}$$

This translates to a resolution of 0.055 percent. *Note:* It is highly unlikely that an ultrasonic system will produce such a high resolution, due to other variables in the system.

(c) A count value of 0011 1101 0010 1111 equates to a timed interval of 15,663 μs. Since each foot of range is equivalent to 1.8 ms, the range of the object is

$$\frac{15{,}663\ \mu s}{1.8\ \text{ms/ft}} = 8.7\ \text{ft}$$

Range Images

For image-understanding purposes in vision systems, the range data obtained using the preceding techniques are compiled into a range image, sometimes called a ***depth map.*** A range image is similar to an intensity image in that gray-scale levels are used to represent the range values of a given scene. Thus, a range image is constructed by converting range values to a gray scale and storing the corresponding gray-scale values in a pixel matrix. The stored matrix becomes the range image, or depth map, of a given scene.

An ideal gray-scale range image and its associated intensity image are shown

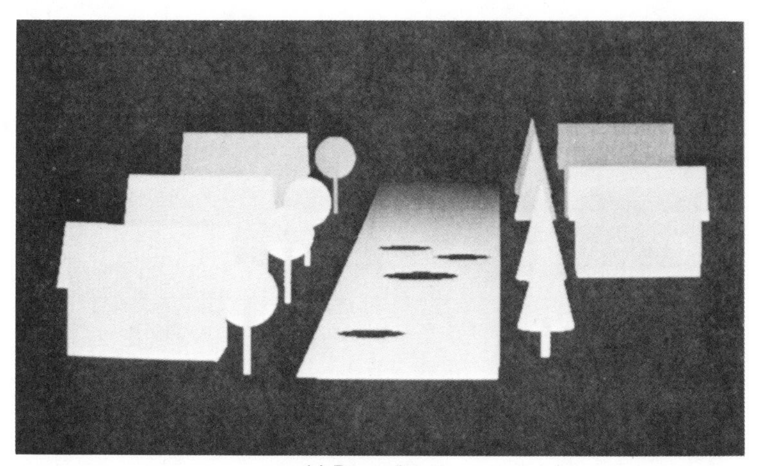

(a) Range Image

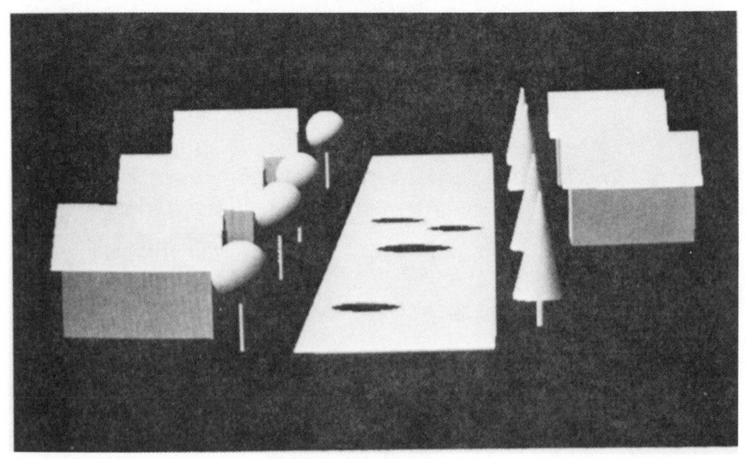

(b) Intensity Image

Figure 7-11 (a) A range image, and (b) corresponding intensity image. (Reprinted with permission from *Computer Vision* by Dana H. Ballard and Christopher M. Brown, Prentice-Hall, Inc., Englewood Cliffs, N.J.)

in Figure 7-11. Observe that the range image in Figure 7-11(a) uses different levels of gray to represent different distances. The objects get darker as their distance from the range-finding device increases. It is easy to see that the corresponding intensity image in Figure 7-11(b) segments the roofs and walls of the houses by showing them at different gray levels. However, the intensity alone cannot be used to segment one house from another. Using just the intensity image, the houses on the left side of the street appear as a single object, as do those on the right side of the street. On the other hand, the range image in Figure 7-11(a) clearly shows that the houses are separate, since the gray-scale range values differ from house to house. As a result, the range image segments one house from another, and the intensity image segments the different parts of each house.

The point is this: intensity data coupled with range data allow for more accurate image understanding of real-world scenes. As you can see from the preceding illustration, the addition of 3-D range information allows further segmentation of the scene. Range images can be segmented by applying the same image-analysis techniques that you learned earlier to segment intensity images.

7-3 ROBOT POSITION AND PROXIMITY SENSING

The vision and ranging techniques discussed so far can all be applied to robot navigation, whether the robot is stationary and navigating its gripper or mobile and navigating its body. Most of these vision-related techniques, however, are relatively sophisticated, except for possibly the ultrasonic technique. But many simple robot navigation problems do not require such sophisticated solutions. For instance, you would not want to equip a mail-boy robot with a 3-D image-understanding system when a simple buried cable or track might do the job. Remember: the application dictates the system. At this time, let's take a brief look at some simple nonvision techniques that can be employed for robot position and proximity sensing.

Mechanical Position and Proximity Sensors

Probably the simplest way for a robot to detect a given position is by using a mechanical limit switch like those shown in Figure 7-12. These switches can be placed on various parts of the robot so that an interrupt is generated to the robot CPU when the switch makes contact with an object. They can also be used on surrounding fixtures to define the limits of robot travel.

The limit switch is connected to the CPU using a simple pull-up resistor circuit like the one shown in Figure 7-13. When the switch is opened or closed, the circuit generates a logic transition that can be easily detected by a digital control circuit (computer). A disadvantage of mechanical switch sensors is that to be activated they must come in physical contact with an object.

Inductive Position and Proximity Sensors

Inductive and capacitive sensors are both used for noncontact position and proximity sensing. As you are probably aware, an inductor is constructed by coil-

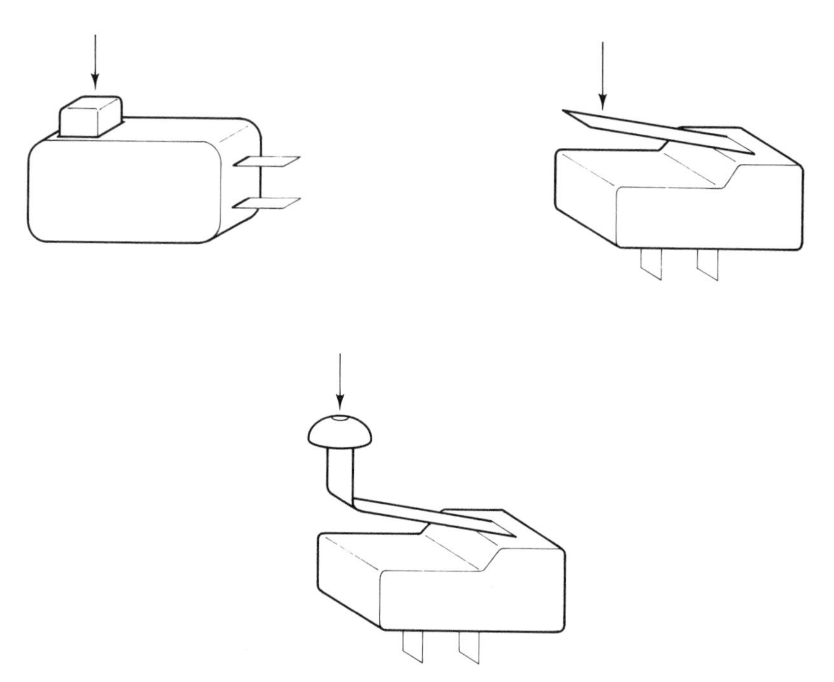

Figure 7-12 Some typical limit switches.

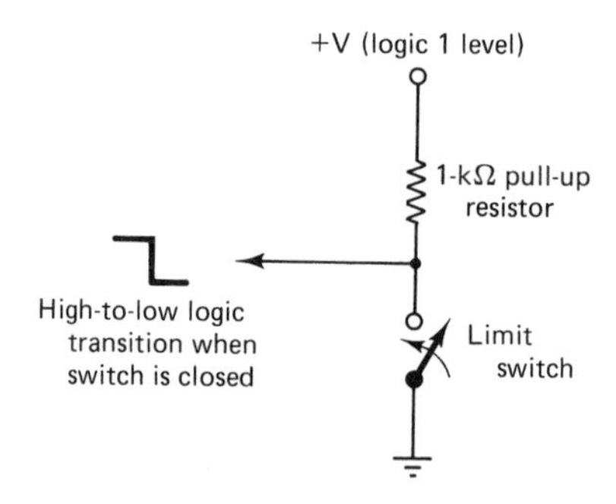

Figure 7-13 A pull-up resistor circuit is used to generate a logic transition when the limit switch is activated.

ing a wire around a ferrous core. The inductance of the inductor is determined using the following physical relationship:

$$L = \frac{\mu N^2 A}{l}$$

where L = the inductance in henrys
μ = a constant, called permeability, for a given inductor core
N = number of coil turns
A = cross-sectional area of the core
l = effective length of the core.

For any given inductor with a movable core, the only variable in this equation is the effective length, l, of the inductor core. The effective length is that part of the core that comes in contact with the magnetic field created around the inductor coil. Consequently, the inductance of an inductor can be changed by moving the core in and out of the magnetic field created by an ac excitation voltage applied to the coil. This idea is illustrated in Figure 7-14.

Now, suppose the inductor core is attached to a moving part of a robot, and the inductor coil is stationary. Then any movement can be measured by a change in inductance. If the inductor is connected as part of an ac bridge circuit, like those shown in Figure 7-14, the change in inductance is converted to a proportional change in the amplitude of the bridge output. This output can be rectified and converted to a digital value for position sensing and control purposes.

For proximity sensing, the inductor is placed in a resonant *LC* oscillator circuit. The magnetic field of the inductor creates an active zone that, if penetrated by a metal object such as a robot manipulator, will change the inductance of the circuit, resulting in a change in the circuit output.

A very common inductor device used for extremely accurate position sensing is the ***linear variable differential transformer (LVDT)*** shown in Figure 7-15. The primary transformer winding of the LVDT is excited with an ac source, and a voltage is induced by transformer action into two secondary windings. The two secondary windings are wound in opposite directions so that the two induced secondary voltages are 180° out of phase. When the core is located precisely between the two secondary windings, the combined output, V_{out}, is zero, since the two secondary voltages cancel each other. However, if the core is moved in either

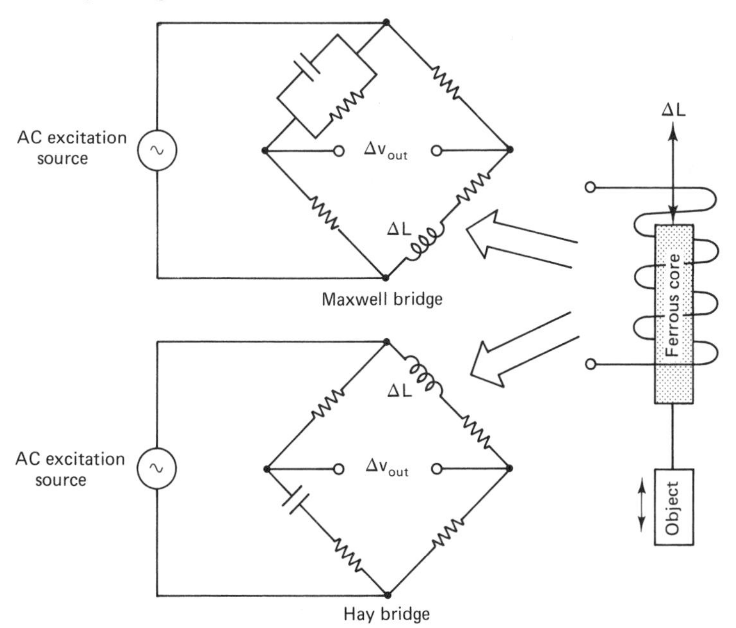

Figure 7-14 Inductive position sensor and associated ac bridge circuits.

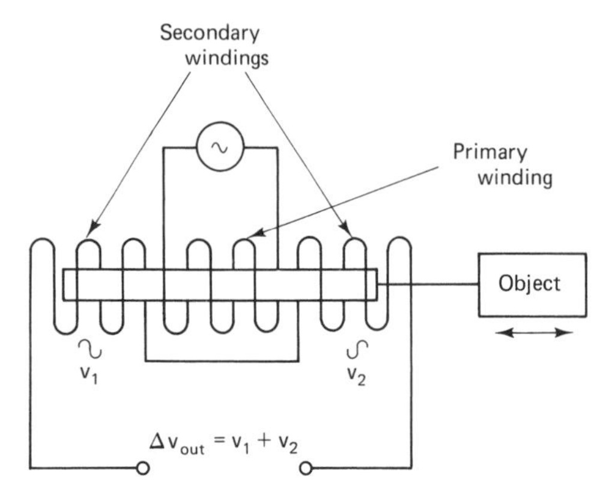

Figure 7-15 Linear variable differential transformer, a special type of inductive position sensor.

direction, the output amplitude increases in proportion to the amount of movement, and the phase of the output indicates in which direction the core has moved.

Thus, the LVDT produces a voltage output whose amplitude is a function of the amount of core movement, and phase is dependent on the direction of core movement. This output can be rectified and converted to a digital value for position sensing and control purposes.

The LVDT is very popular for use in industrial position sensing because it is extremely linear and accurate. Commercial LVDTs can resolve position to within 1 or 2 μm. They are also available in extremely small packages. The movable core must be attached to the moving object.

Capacitive Position and Proximity Sensors

The physical relationship for a capacitor is

$$C = \epsilon \frac{A}{d}$$

where C = the capacitance in farads

ϵ = a constant, called permittivity, that depends on the dielectric material

A = the surface area of one plate

d = distance between the plates

Three ways to use a capacitor to sense position are illustrated in Figure 7-16. For linear position sensing, the moving object is connected to a movable plate of the capacitor. As the object moves, the distance between the plates or the effective plate area of the capacitor is changed. In either case, the result is a change in capacitance due to the preceding relationship. For angular position sensing, a rotating object is connected to the shaft of a variable capacitor. The capacitance of this capacitor changes when its shaft is rotated. Any of these capacitive position sensors can be placed in an ac bridge like those shown to produce a voltage output that is proportional to the amount of object movement. The bridge output must be rectified and converted to digital for control purposes.

To sense proximity, two plates of a capacitor are spaced widely apart in an

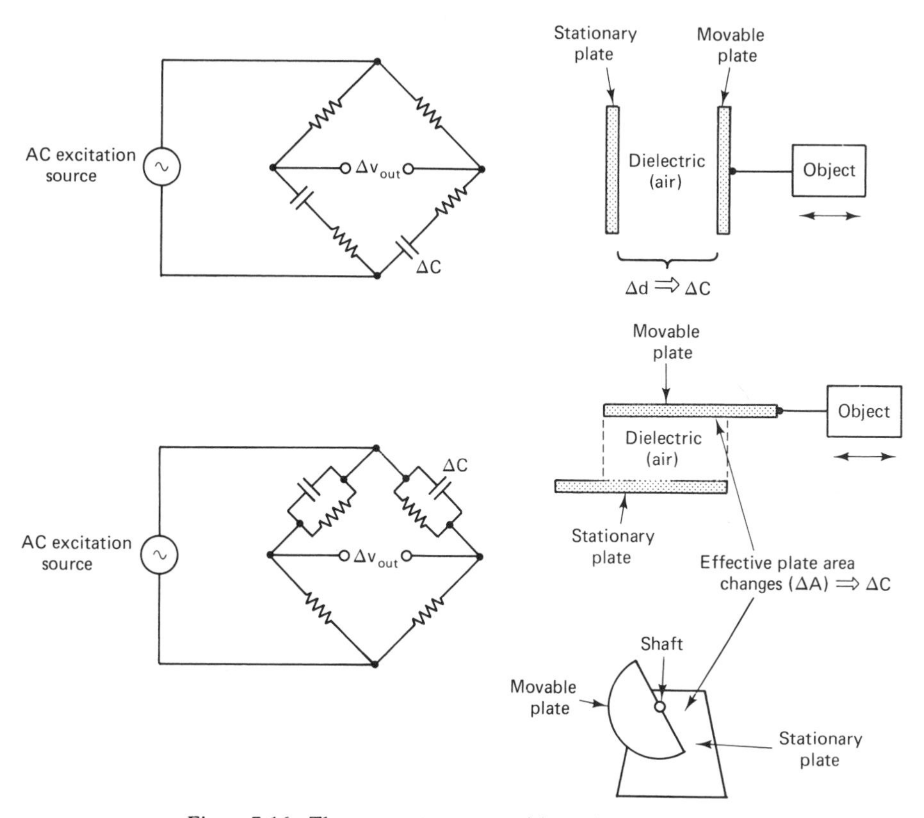

Figure 7-16 Three ways to sense position using capacitance.

RC oscillating circuit. The active zone is between the widely spaced capacitor plates. If an object penetrates this zone, it changes the capacitor dielectric constant, which changes the capacitance, resulting in a change in the output amplitude of the oscillator. The object can be any material that is insulated from ground.

Magnetic Sensors

Magnetic sensors, such as ***reed switches*** and ***Hall-effect devices***, are used for close proximity or position sensing. Both of these devices can be used to generate a logic transition when a magnetic field is detected. They are relatively inexpensive and readily available.

The magnetic reed switch, shown in Figure 7-17(a), consists of two magnetic reed contacts enclosed in a glass tube. The reed contacts close in the presence of a magnetic field and, when connected in a pull-up resistor circuit as shown in Figure 7-17(b), can be used to generate a logic transition to a robot control circuit.

A Hall-effect device is a solid-state component about the size of a small signal transistor, as shown in Figure 7-18. Since they are solid state, they are more reliable and rugged than the glass magnetic reed switch. When connected in a pull-up re-

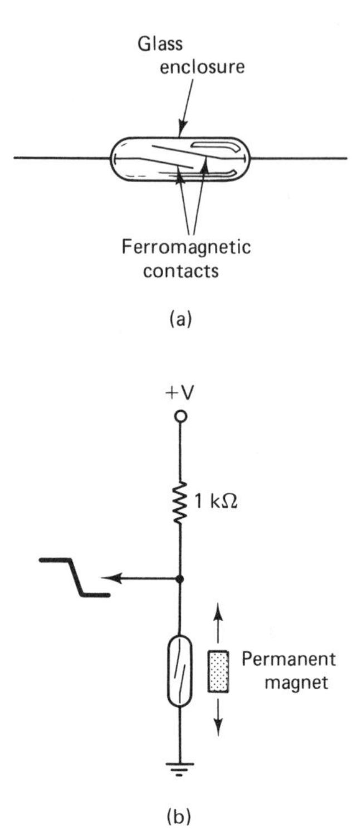

Figure 7-17 (a) Magnetic reed switch, and (b) associated pull-up resistor circuit.

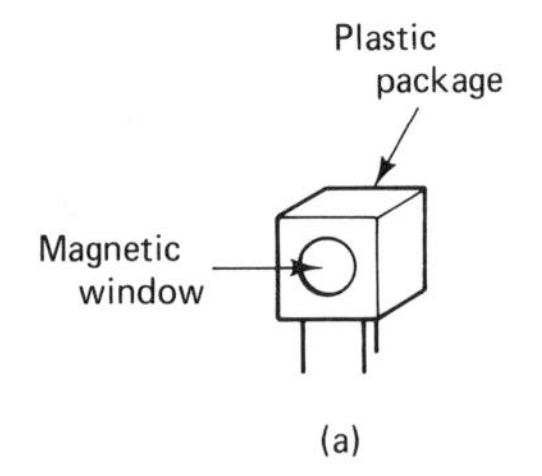

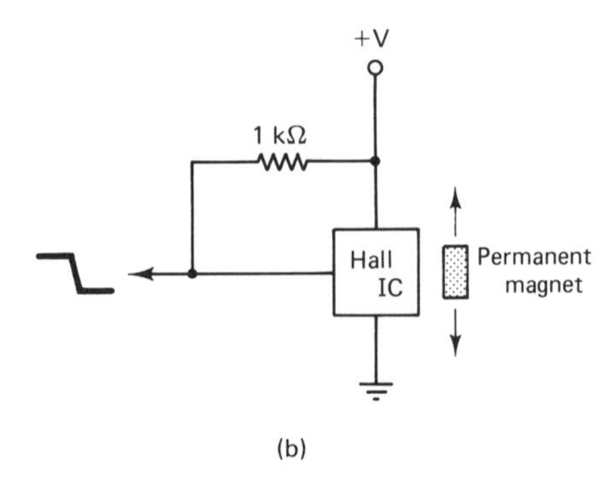

Figure 7-18 (a) Hall-effect device, and (b) its corresponding pull-up resistor circuit.

sistor circuit as shown, the Hall-effect device will generate a logic transition when activated by a magnetic field.

Figure 7-19 shows how both magnetic reed switches and Hall-effect devices are used to sense linear and angular position. When a magnet is attached to a moving object, such as a robot gripper, the magnetic sensors can be used to detect its position. The sensors are positioned precisely so that the gripper position can be accurately determined as the sensors are sequentially activated by the presence of the magnetic field. Closer spacing of the sensors provides increased position resolution.

Single reed switch or Hall-effect devices are also used for very close proximity sensing. The device is activated when a magnet attached to a robot manipulator or gripper comes in close proximity to it. The disadvantage of using these devices is that they can only be used for very close sensing. Their sensing distance is only from ⅛ to ½ inch using standard permanent magnets.

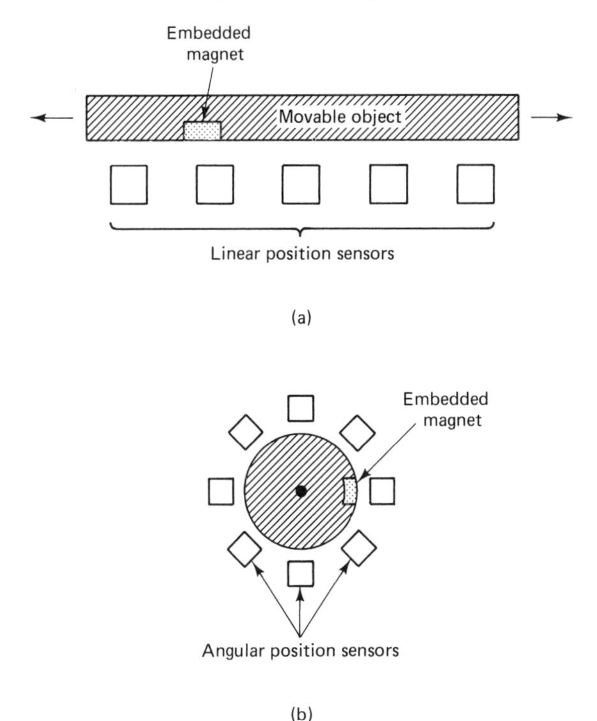

Figure 7-19 Both (a) linear and (b) angular position can be sensed using magnetic devices such as magnetic reed switches or Hall-effect devices.

Optical Position and Proximity Sensors

There are numerous types of optical sensing devices, from simple photodiodes to sophisticated vision systems. Let's complement our knowledge of vision systems here by looking at three simple devices that can be used for position and/or proximity sensing. They are the ***phototransistor, optical interrupter,*** and ***optical reflector.***

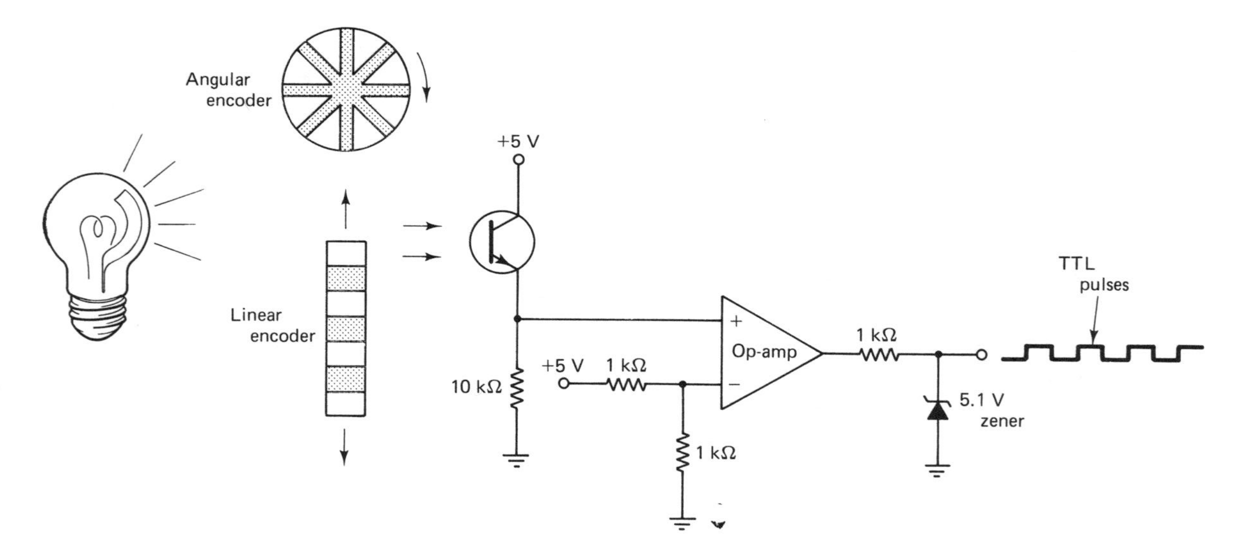

Figure 7-20 Linear and angular position sensing using a phototransistor.

The simple phototransistor can be used with an optical encoder to detect both linear and angular position, as shown in Figure 7-20. Here the optical encoder makes or breaks the light seen by the phototransistor. If the phototransistor is connected in a simple op-amp comparator circuit as shown, the circuit generates digital pulses that can be counted to determine position. The optical encoder must be constructed with precise encoder increments so that each count represents a given linear or angular distance.

The same circuit (Figure 7-20) can be used for proximity sensing. When an object enters or leaves a beam of light projected into the phototransistor, the circuit generates a logic transition that can be detected by a digital control circuit.

A phototransistor and associated infrared light source can be purchased in a single package called an optical interrupter. The optical interrupter, shown in Figure 7-21, contains a slot that separates the light source from the phototransistor. Slot widths range from 0.1 to 0.375 inch.

A linear or angular optical encoder wheel is passed through the slot, as shown, to measure position. Again, a counter circuit must be used to count the output pulses of the interrupter circuit to determine position. The interrupter can also be used as a proximity sensor by attaching a thin opaque strip to a moving object. A logic transition is generated when the strip enters or leaves the slot. Optical interrupters are often used in place of mechanical limit switches for proximity sensing. They are not subject to the maintenance problems of mechanical switches, since they are solid-state integrated devices.

The optical reflector module shown in Figure 7-22 is like the optical inter-

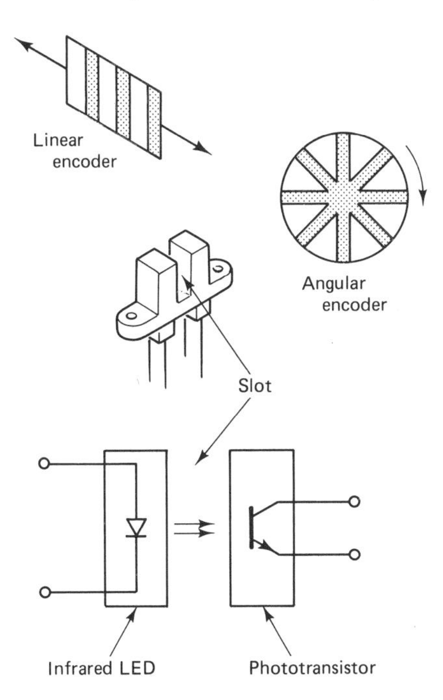

Figure 7-21 An optical interrupter contains an infrared light source and phototransistor in a single package.

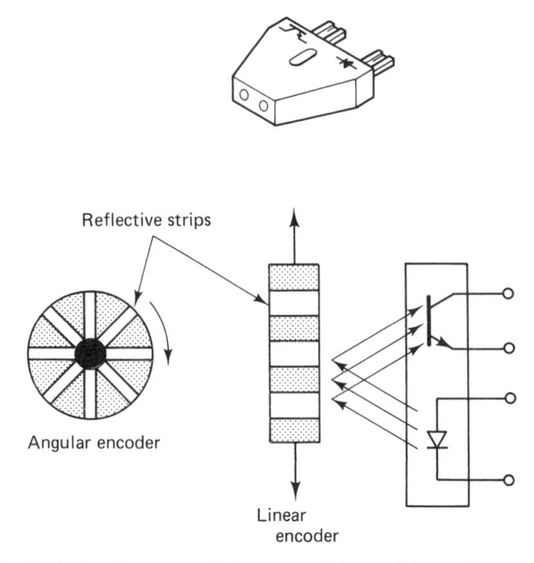

Figure 7-22 Optical reflector modules are used for position and proximity sensing.

rupter, except that both the light source and phototransistor are mounted side by side in the same plane. To measure position with this device, the optical encoder must contain reflective strips, as shown. Pulses are generated from the optical reflector circuit as the reflective encoder strips pass across the face of the device, causing the light source to be reflected back into the adjacent phototransistor. Again, the pulses must be counted to determine position. These devices can also be used as proximity sensors by attaching a reflective strip to a moving object. A signal transition is generated when the reflector module detects the strip.

This completes my discussion of range finding and navigation. There are many different ways to use the ideas and devices discussed here, depending on the particular application. You now have the basics required to begin tackling just about any ranging, navigation, position, or proximity-sensing problem.

SUMMARY

Range data must be combined with 2-D images to create the 3-D information necessary for real-world navigation. Two fundamental techniques used for ranging are triangulation and time-of-flight. Triangulation can be passive or active. Passive triangulation involves two stationary imaging devices that produce stereo images. Range is a function of the disparity, or pixel displacement, between the two images. The biggest problem with passive triangulation is the correspondence problem. With active triangulation, the scene is scanned by a moving spot or stripe of light that is reflected by objects into a stationary camera. The reflected image points are detected and their displacement used to calculate range. Active triangulation can also employ a nonstructured light source and a moving camera.

Time-of-flight, or TOF, ranging involves the calculation of the time it takes for a signal to reach and return from an object. Various types of signals are used, including electromagnetic, light, and sound. The width of the signal beam determines the amount of object resolution, while the speed of the signal determines the depth resolution. Laser ranging provides the best object resolution, but the poorest depth resolution. Conversely, ultrasonic ranging generates the best depth resolution, but poorest object resolution.

Regardless of the type of ranging system, the resulting range information is used to create gray-scale range images, or depth maps. Range images are stored as matrices and, when combined with intensity images, provide the basic low-level knowledge required for 3-D image understanding.

Depending on the application, simple position and proximity sensors can be used to complement or replace sophisticated vision systems. These devices include mechanical limit switches, inductive sensors, capacitive sensors, magnetic sensors, and optical sensors.

PROBLEMS

7-1 How does passive triangulation differ from active triangulation?

7-2 Define disparity as related to stereo images.

7-3 A stereo vision system employs two TV cameras whose focal length is 10 cm and that are located 20 cm apart. Two corresponding pixel points are located 1 cm from their respective lens centers. Calculate the range to an object if the disparity between the two stereo images is 2 cm.

7-4 In what part of the scene is the object point in Problem 7-3 located?

7-5 The range value calculated in Problem 7-3 is the distance from the object point to ________________.

7-6 Suppose the correspondence between the two image points in Problem 7-3 could only be resolved to 0.2 cm. How does this affect the accuracy of the range measurement?

7-7 How do stereo vision systems attempt to reduce the correspondence problem?

7-8 Name at least two forms of active triangulation.

7-9 Suppose the spot in Figure 7-4 is projected at an angle of 30°. What is the range from the camera lens to an object point that produces a reflected image point 0.5 cm from the center of the image?

7-10 Explain how object boundaries and regions can be determined from light stripe images.

7-11 A burst of ultrasound takes 5 ms to reach and return from an object. What is the range, in feet, of the object?

7-12 What is the range of the object if laser light is substituted for ultrasound in Problem 7-11?

7-13 State a disadvantage for each type of TOF ranging signal.

7-14 Name two types of transducers used in ultrasonic ranging systems.

7-15 What frequency range is typical for a commercial ultrasonic ranging system?

7-16 A commercial ultrasonic range finder is used with a 100-kHz, 8-bit counter. What is the maximum range of the system?

7-17 An object at a range of 1 ft would produce a count of ________________ in the counter in Problem 7-16.

7-18 What is the maximum possible resolution of the system in Problem 7-16?

7-19 How does a range image differ from an intensity image?

7-20 List at least three types of noncontact proximity sensors.

7-21 Explain how capacitance can be used for proximity sensing.

7-22 Explain how Hall-effect devices are used for position sensing.

7-23 Describe the difference between an optical interrupter and an optical reflector.

7-24 Two Hall-effect devices are located 1 ft apart. A robot moves a magnetic part in a straight line from one device to the next. A pulse from the first Hall-effect device clears a 1-kHz, 8-bit counter. A pulse from the second Hall-effect device stops the counter at a count of 0110 1001. At what velocity, in feet per second, has the object been moved by the robot?

8 Tactile Sensing

Tactile, or touch, sensing is required when intelligent robots must perform delicate assembly operations. During the assembly operation, an industrial robot must be capable of recognizing parts, determining their position and orientation, and sensing any problems encountered during assembly from the interaction of parts and tools. Many assembly tasks that are visually difficult are simple with touch.

Imagine trying to thread a nut on a bolt in a dark room using only your sense of touch. Let's assume that the bolt is being held in a rigid position, and you are to perform the operation with one hand. The first thing you must do is feel around for the bolt and determine its position and orientation. Then you must find the nut, pick it up, and move it to the end of the bolt. You will then probably fumble around rocking the nut and turning it in a reverse direction until it seats itself onto the first thread. (Try it!) Once seated on the first thread, you must turn the nut clockwise (assuming it is a right-hand thread), applying the appropriate torque until it is firmly seated against the object being bolted. Of course, you must monitor the turning force, or torque, to make sure that you are not cross-threading the nut.

This example should give you a feel for some of the properties that must be incorporated into a tactile sensing system. Just as your eyes process the light images of a visual scene, your hand with its joints and skin must generate the information required for tactile sensing. In this chapter, you will first learn about those properties of tactile sensing that are important to robotics. After this, you will learn about the three fundamental sensing operations that must be performed by a tactile-sensing system: sensing ***joint forces***, sensing ***touch***, and sensing ***slip***.

Like vision, tactile-sensing technology is still in its early stages of development. With this in mind, I have written this chapter to introduce you to current basic principles of tactile sensing. Although the technology will evolve to produce better tactile-sensing systems, the basic concepts presented here will remain constant throughout the industry.

8-1 TACTILE-SENSING SYSTEM

Before you begin learning the design details of tactile sensing, you must understand how tactile sensing fits in with other robot sensing operations. In addition, you must understand why tactile sensing is required to perform many operations that cannot be performed using other types of sensory data. Once you have this overall perspective, you will understand how tactile-sensing systems can be developed and integrated with other types of sensors to satisfy a given application task.

This section will begin with a short discussion of contact versus noncontact sensing, followed by a discussion of the need for tactile sensing in industrial robotics. Then you will be introduced to the sensing operations required in a tactile system in preparation for detailed discussions of those operations in the remainder of the chapter.

Contact versus Noncontact Sensing

Robot sensing systems fall into two fundamental categories: ***noncontact sensing*** and ***contact sensing.*** A noncontact sensor measures the response of light, sound, magnetic, or electromagnetic energy to the environment surrounding a robot. On the other hand, a contact sensor measures the response of some form of physical contact between the robot and its environment. Most of the vision, ranging, and proximity sensors discussed in the last two chapters fall into the noncontact sensing category. Tactile sensing, however, is a form of contact sensing. A tactile-sensing system must measure things like pressure, force, torque, and electrical quantities that result from direct physical contact between the robot and its environment.

Need for Tactile Sensing

You might be wondering why a robot needs a sense of touch, when vision seems like a more capable alternative. It turns out that many times tactile sensing is more desirable than vision, especially for industrial operations. Here's why: First, tactile sensing is relatively simple and therefore more economical than vision. Tactile systems employ simple sensing devices such as switches, strain gauges, and pressure transducers. The data generated by these devices are relatively easy to condition, store, and analyze as compared to visual data. Even with the fastest computers, image analysis and understanding still takes minutes as opposed to seconds. This often makes vision impractical for real-time control of a robot. Furthermore, large amounts of memory are required to store visual information as opposed to tactile information.

Second, tactile-sensing data provide information about the contact between the workpiece and the robot end-effector. Such information is critical when assembling close-tolerance parts or performing industrial operations like drilling, grinding, deburring, and welding, just to mention a few. Vision systems cannot supply all the information needed to control such operations. For instance, a vision system cannot determine whether an object is hard or soft, whether or not it is being grasped properly, or whether there is any interference between the fit of two parts. Could a vision system detect a cross-threading problem in a nut and bolt threading operation?

Finally, tactile sensing allows a robot to analyze environments that cannot be seen due to inadequate illumination or object obstruction. Think about what you do when entering a dark room. Don't you first feel for the light switch in order to turn on the light so you can see? Here your sense of touch complements your sense of vision. Tactile sensing can also complement vision to reduce perspective distortion. Vision systems suffer from perspective distortion, since a given situation is often interpreted differently when seen from different visual perspectives.

Tactile-sensing Operations

Think about the nut and bolt operation for a minute. What things must be sensed in order to perform the operation, and how can these human sensing abilities be duplicated in a machine? First, you must use your sense of *touch* to recognize the nut from the bolt and determine their relative positions and orientations. In addition, you must apply just enough gripping *force* to assure that the nut does not *slip* out of your grasp. To do this, you must detect any slippage and adjust your gripping force accordingly. You also must detect any slip while turning the nut on the bolt and adjust your grasping force to compensate for the slippage. While threading the nut, you must monitor the amount of tension, or *force*, being exerted on the various muscles in your hand, wrist, and arm. This force information allows you to control the amount of torque applied during the threading operation. Thus, three basic things must be sensed while performing the nut and bolt operation: ***touch*, *slip*,** and ***force*.**

It is highly unlikely that the tactile-sensing abilities of a robot will ever match those of a human being, nor do they need to. Simple object recognition, position, and orientation are possible with far less tactile-sensing abilities than you and I possess. However, using the human hand, wrist, and arm as a model for a robot, a tactile-sensing system reduces to the following three fundamental sensing operations:

1. ***Joint forces:*** sensing the force applied to the robot hand, wrist, and arm (manipulator) joints.
2. ***Touch:*** sensing the pressures applied to various points on the hand, or gripper, surface.
3. ***Slip:*** sensing any movement of the object while it is being grasped.

These three major sensing operations are illustrated by the robot manipulator pictured in Figure 8-1. Here force must be sensed at the gripper, wrist, and robot arm joints. Touch and slip sensing are accomplished by locating tactile sensors around the robot gripper. The next three sections of this chapter are devoted to a discussion of how tactile-sensing systems like the one in Figure 8-1 can be designed to incorporate the three sensing operations.

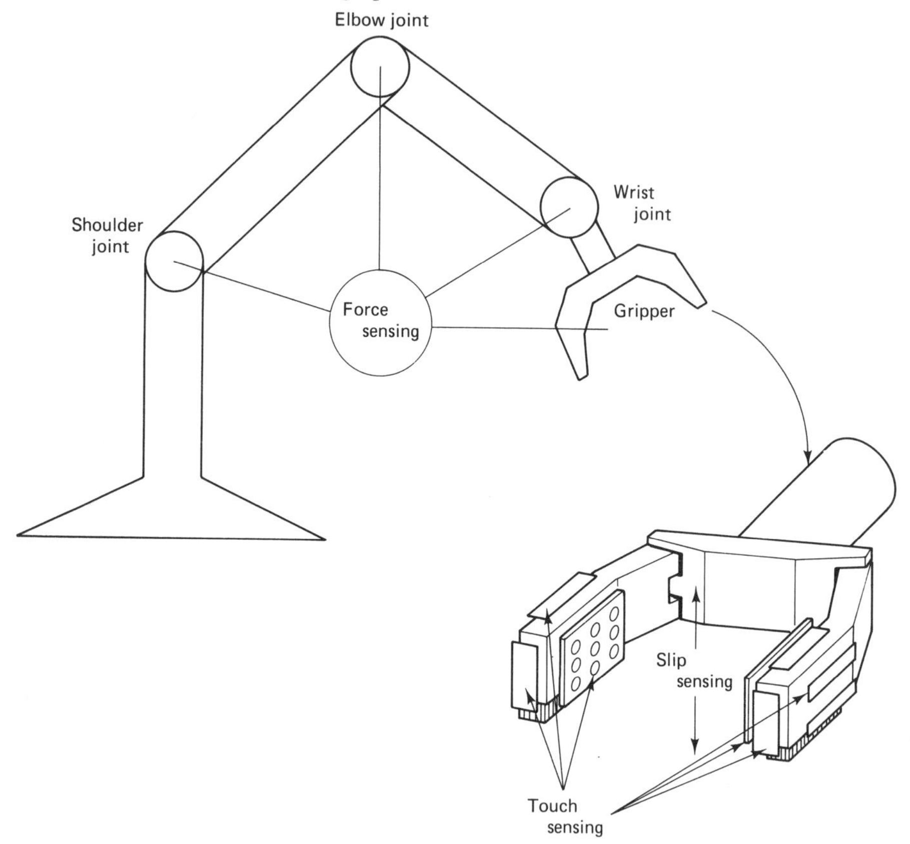

Figure 8-1 A complete tactile sensing system includes joint force sensing, touch sensing, and slip sensing.

8-2 SENSING JOINT FORCES

Any robot manipulator interacts with the environment around it through the exchange of forces with surrounding objects. Consequently, if you can measure these forces accurately, you can program a computer to control the manipulator accordingly. There are basically three ways to measure the forces acting on a robot manipulator: (1) by measuring the forces acting on the arm joints of the manipulator, (2) by measuring the forces acting on the wrist of the manipulator, or (3) by measuring the force exerted by the manipulator on the surrounding objects.

Arm Joint Force Sensing

A problem that hinders any type of force measurement is that you cannot measure force directly. You can only measure the indirect result of force. When a robot manipulator is performing a given task, the exchange of forces with surrounding objects is reflected in the robot arm joints. These forces are seen as different loads to the electric, hydraulic, or pneumatic manipulator control systems. For example, if electric motors are used to control the joint movements, forces acting on a given joint change the motor load, which varies the amount of armature current required by the motor. Thus, the force acting on a joint driven by a dc motor can be measured indirectly by measuring the armature current. Likewise, a force acting on a hydraulic-driven joint can be determined by measuring the hydraulic back pressure. Air pressure can be used to determine the force applied to a pneumatic-driven joint.

The advantage of measuring arm joint forces indirectly is that a separate system of force sensors is not required. The joint forces are simply determined by measuring load variables that already exist in the system. The major disadvantage of sensing arm joint forces in a manipulator is that the resulting force measurements do not always provide an accurate indication of the exchange of forces between the robot end-effector and its surrounding objects. To get accurate force information, you must account for things like joint friction, the load of the arm itself, and the inertial forces created through arm movement. These problems can be reduced by measuring forces at the wrist, rather than the arm joints.

Wrist Force Sensing

The wrist of a robot is the last link of the manipulator between its arm and end-effector. All the forces acting on the wrist assembly can be reduced to the six fundamental forces shown in Figure 8-2. Notice that three of the forces are lateral and three are rotational. The three lateral forces, labeled F_x, F_y, and F_z, relate to the manipulator ***lift, sweep,*** and ***reach,*** respectively. The three rotational forces, or torques, are labeled T_x, T_y, and T_z, and are called ***yaw, pitch,*** and ***roll,*** respectively.

The most common means of measuring these forces is by using ***strain gauge*** sensing elements. These devices are extremely rugged, reliable, and, most importantly, inexpensive. It is worthwhile taking a closer look at strain gauges, since they are so common in the industrial robot environment.

Strain Gauges

First, remember that you cannot measure force directly. You must measure the result of the force acting on an object. If the object is solid, it will be deformed as a result of the applied force. This resulting deformation is called ***strain.*** Strain is defined as the change in length of an object divided by its original length, or

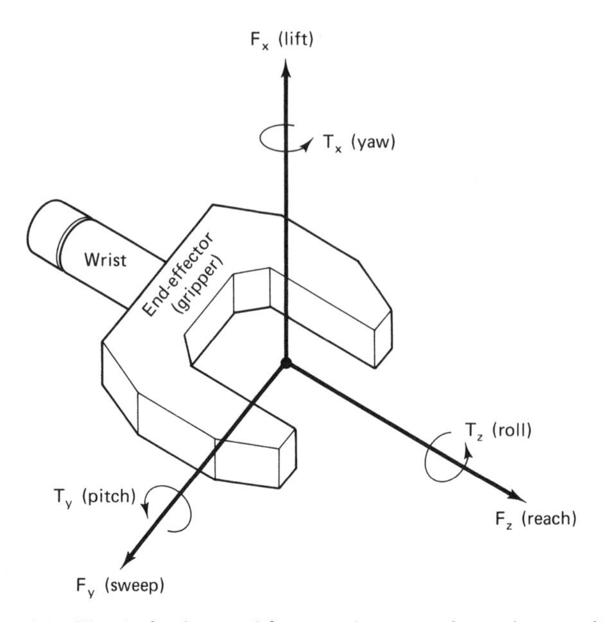

Figure 8-2 The six fundamental forces acting on a robot wrist assembly.

$$\text{Strain} = \frac{\Delta l}{l}$$

where Δl = change in the object's length caused by the applied force
l = original length of the object

Example 8-1

The length of a robot gripper finger is 10 cm. Suppose a reach force is applied to the end of the finger that causes it to compress by 0.1 mm. Calculate the resulting deformation, or strain.

Solution

Strain is found by dividing the change in length by the original length. Since the 10-cm robot finger is compressed 0.1 mm, the resulting strain is

$$\text{Strain} = \frac{\Delta l}{l}$$

$$= \frac{0.1 \text{ mm}}{10 \text{ cm}}$$

$$= \frac{100 \times 10^{-6} \text{ m}}{0.1 \text{ m}}$$

$$= 1000 \times 10^{-6}$$

Notice that the above strain has no units, since meters cancels meters. In fact, strain values are usually expressed in micro, or μ, quantities, where μ stands for 10^{-6}. Using this notation, the above strain would be expressed as 1000 micros, or 1000 μ.

A ***strain gauge*** is a device that measures the amount of strain that results from a force being applied to an object. There are piezoelectric, semiconductor, and resistive strain gauges. Piezoelectric strain gauges generate a small voltage output that is proportional to strain due to the piezoelectric principle. These devices are used to measure relatively large dynamic forces such as acceleration and vibration. Semiconductor strain gauges operate on the principle that the resistance of a semiconductor changes as the result of strain. These devices are very sensitive and durable and have low hysteresis. However, they are nonlinear and extremely sensitive to temperature changes. The most common type of strain gauge is the resistive strain gauge shown in Figure 8-3.

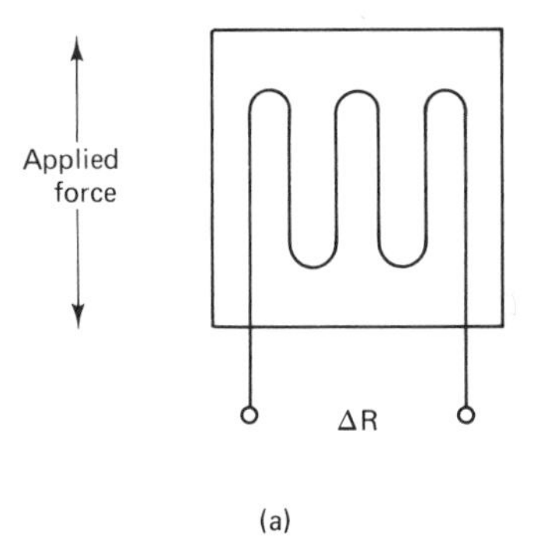

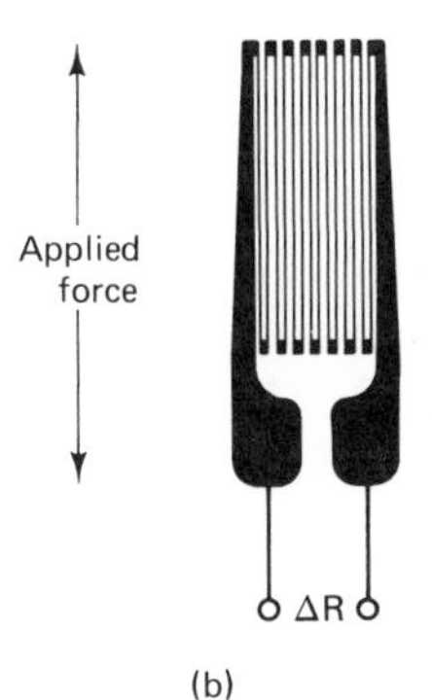

Figure 8-3 Resistive strain gauges: (a) wire-bonded strain gauge; (b) foil strain gauge.

The gauge in Figure 8-3(a) is called a ***wire-bonded*** strain gauge, since it consists of a small wire coil cemented to a thin paper or plastic base. The ***foil*** strain gauge shown in Figure 8-3(b) gets its name from the fact that a conductive foil is photoetched to a nonconductive substrate. Observe that in both cases the strain gauge consists of several conductor loops arranged in an accordion-type fashion.

Strain gauges measure forces that are applied at a 90° angle to the length of the coil turns as shown in the figure. Any force acting in this direction changes the overall length of the conductor, resulting in a proportional change in conductor resistance due to the following relationship:

$$R = \frac{\rho l}{A}$$

where ρ = a constant, called ***resistivity***, related to the conductor material
l = length of the conductor
A = cross-sectional area of the conductor

Since the conductor resistivity is constant for a given strain gauge, this relationship says that any change to the conductor's length or cross-sectional area caused by a force acting 90° to the conductor coil causes a proportional change in resistance. However, forces acting in the same direction as the length of the coil result in an accordion type of action and do not cause any resistance change.

It can be shown that the amount of resistance change of a conductor due to a strain is

$$\Delta R = 2R_{\text{nominal}} \times \text{strain}$$

where ΔR = change in resistance
R_{nominal} = unstrained resistance of the conductor
strain = $\Delta l/l$, in micros

Example 8-2

Suppose that a force is applied to a strain gauge that has an unstrained resistance of 1000 ohms (Ω) and the applied force results in a strain of 1000 μ. Calculate the change in resistance of the conductor.

Solution

Using the preceding equation, the change in conductor resistance must be

$$\begin{aligned} \Delta R &= 2R_{\text{nominal}} \times \text{strain} \\ &= 2(1000\ \Omega) \times (1000\ \mu) \\ &= 2\ \Omega \end{aligned}$$

Strain gauges are available with nominal, unstrained, resistance values anywhere between 50 and 5000 Ω. Furthermore, most strain gauges are rated by a sensitivity factor called a ***gauge factor (GF).*** Gauge factor is defined as follows:

$$\text{GF} = \frac{\Delta R/R_{\text{nominal}}}{\text{strain}}$$

where ΔR = change in resistance due to strain
R_{nominal} = unstrained resistance of the gauge
strain = $\Delta l/l$, in micros

Gauge factors for commercial resistance-type strain gauges usually range from 2 to 10, depending on the material used for the resistive strain element. Semiconductor strain gauge factors range from 50 to 150.

Example 8-3

Calculate the gauge factor of the strain gauge in Example 8-2.

Solution

Using the values in Example 8-2, the gauge factor of this strain gauge is

$$\mathrm{GF} = \frac{\Delta R/R_{\mathrm{nominal}}}{\mathrm{strain}}$$

$$= \frac{(2\ \Omega)/(1000\ \Omega)}{1000\ \mu}$$

$$= 2$$

Example 8-4

A semiconductor strain gauge has an unstrained resistance of 250 Ω and a gauge factor of 125. Calculate the change in resistance that results from a strain of 1000 μ to this gauge.

Solution

$$\mathrm{GF} = \frac{\Delta R/R_{\mathrm{nominal}}}{\mathrm{strain}}$$

Then

$$\Delta R = \mathrm{GF} \times \mathrm{strain} \times R_{\mathrm{nominal}}$$

$$= 125 \times 1000\ \mu \times 250\ \Omega$$

$$= 31.25\ \Omega$$

When using a strain gauge to measure the force acting on an object, the gauge must be attached to the object as shown in Figure 8-4. The two strain gauge leads are then connected as part of a potentiometer circuit or dc Wheatstone bridge, as shown. In either case, any resistance change due to an applied force results in a change in output voltage from the potentiometer or bridge circuits. This analog voltage output must then be converted to a digital value by an A/D converter for use by the robot controller. I should mention that you must condition the Wheatstone bridge output signal before it can be converted. This is accomplished with the simple op-amp differential amplifier circuit, called an ***instrumentation amplifier***, shown in Figure 8-5.

One problem that is often encountered when using strain gauges is tempera-

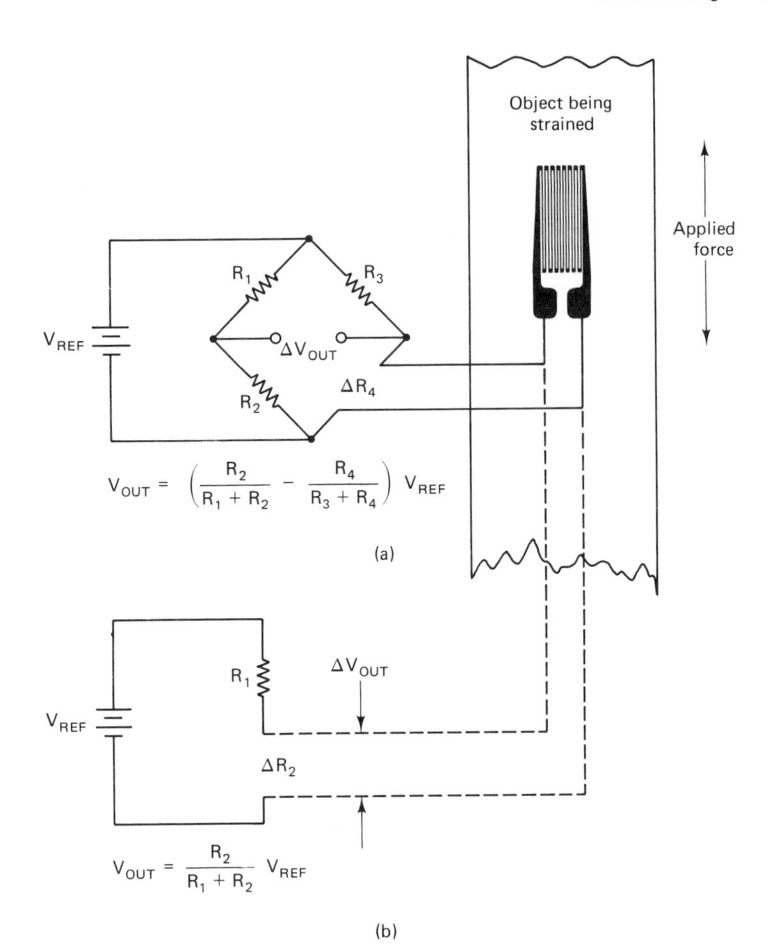

Figure 8-4 A resistive strain gauge must be connected into (a) a Wheatstone bridge or (b) a potentiometer circuit.

ture. Changes in temperature also cause the resistance of a strain gauge to change. This might result in erroneous force data, especially when the robot operates between extreme temperature environments. To compensate for temperature changes, a ***dummy*** strain gauge is placed next to the ***active*** strain gauge, as shown in Figure 8-6. Notice that the dummy gauge is oriented 90° in relation to the active gauge. Thus, any forces affecting the active gauge do not affect the dummy gauge, and vice versa. Temperature changes, however, affect the resistance of both gauges equally. Since the two gauges are connected into opposite sides of the bridge circuit, any common changes due to temperature are canceled out. Only resistance changes due to an applied force on one of the gauges causes the bridge output to change. Commercial strain gauges are available with two matched resistive elements for temperature-compensation purposes. These devices have two matched foil elements photoetched onto the same substrate, but oriented at 90° angles to each other.

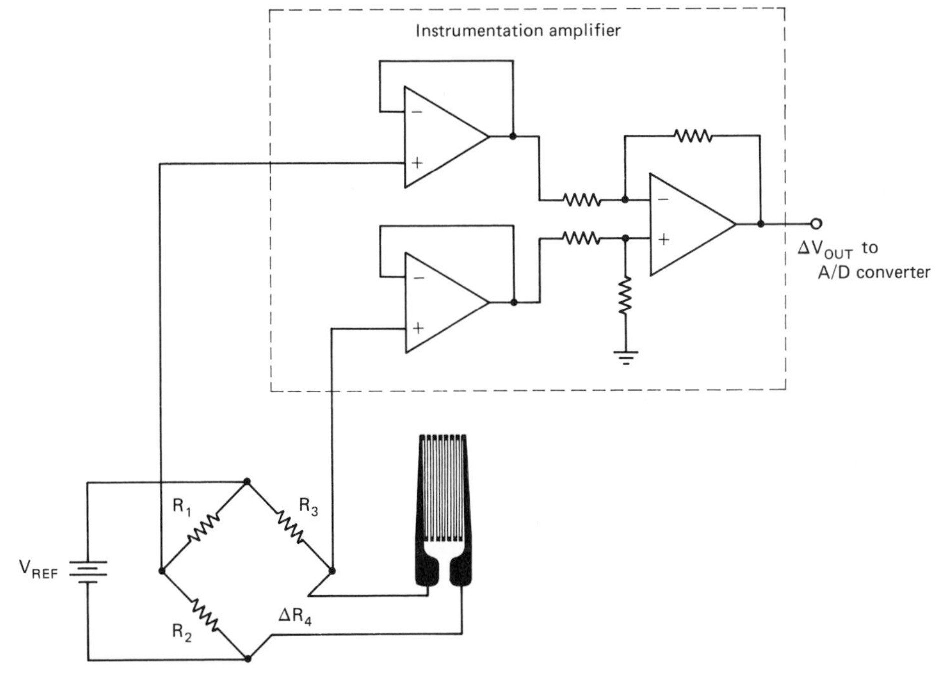

Figure 8-5 The output of a Wheatstone bridge must be conditioned using an instrumentation amplifier circuit.

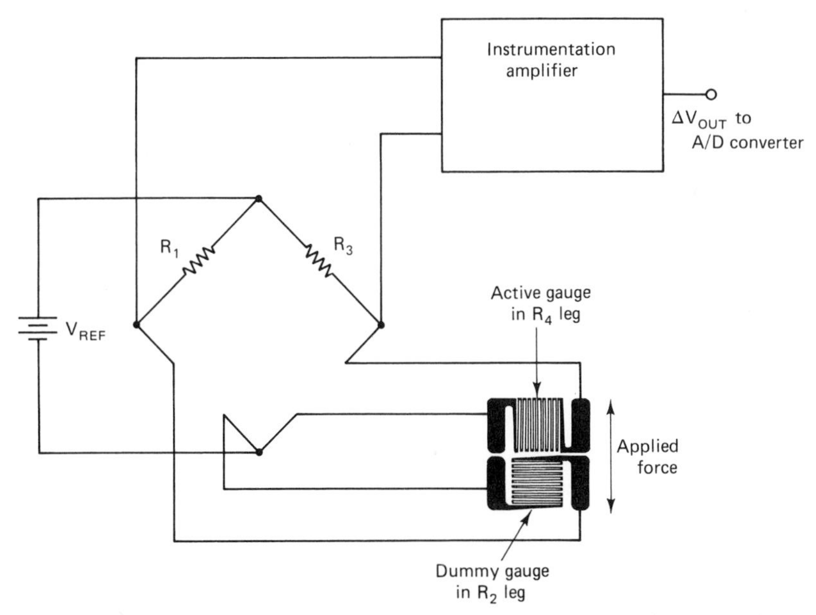

Figure 8-6 Temperature compensation requires an active and a dummy strain gauge oriented 90° to each other.

Gripping Force Sensor Module. A simple strain gauge module that can be used to sense gripping forces is shown in Figure 8-7. Here the strain gauge module acts as one of the fingers of a robot gripper. Strain gauges 2 and 3 are attached to the inside of the finger, while strain gauges 1 and 4 are attached to the outside of the finger. Notice that the sensing module has been milled thin at the strain gauge points to allow it to flex when gripping an object.

Here's how it works: When an object is grasped, any gripping force causes strain gauges 2 and 3 to stretch and gauges 1 and 4 to compress. Thus, by the preceding resistance equation, the resistance of gauges 2 and 3 increases, while the

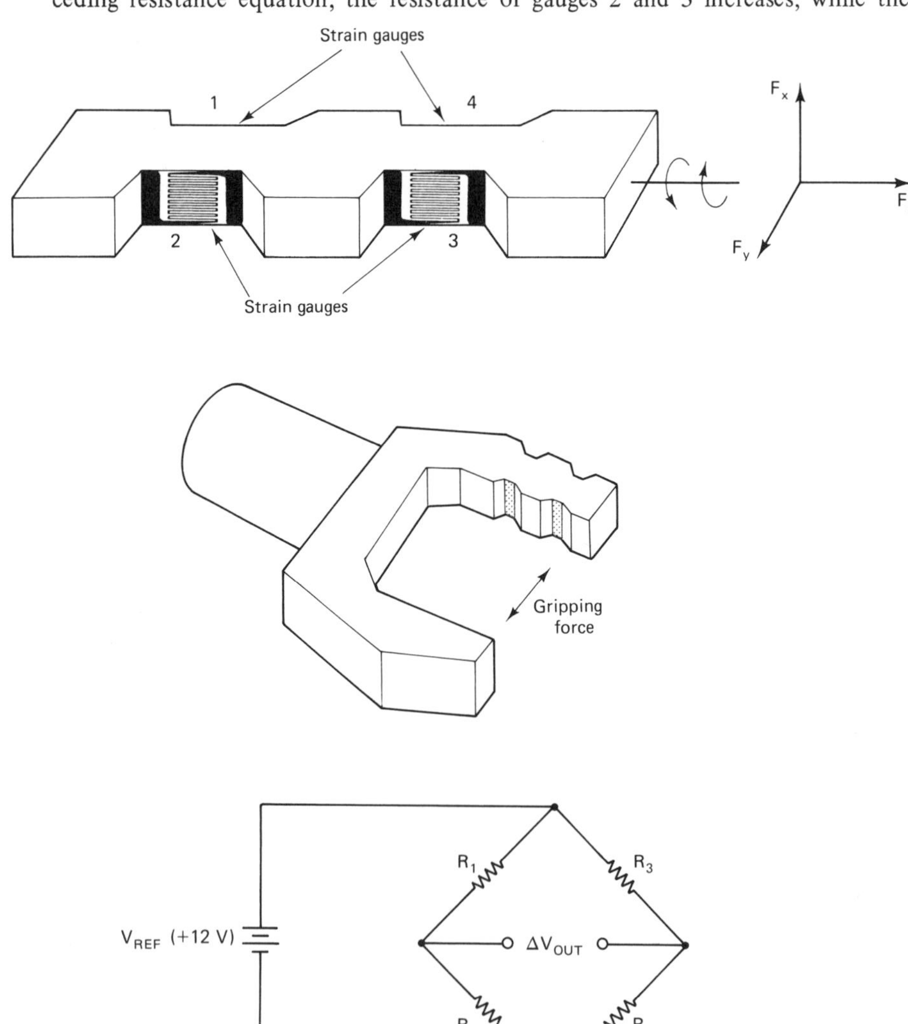

Figure 8-7 Strain gauge gripping force sensor.

resistance of gauges 1 and 4 decreases. If the four strain gauges are connected as part of a Wheatstone bridge, as shown, the voltage out of the bridge will increase in direct proportion to the grasping force.

Example 8-5

The voltage out of the Wheatstone bridge, V_{out}, in Figure 8-7 is

$$V_2 - V_4$$

where

$$V_2 = \frac{R_2}{R_2 + R_1} V_{REF}$$

$$V_4 = \frac{R_4}{R_4 + R_3} V_{REF}$$

Suppose the strain gauge from Example 8-3 is used to make the force-sensing module.

(a) What is the bridge output voltage when no gripping force is applied?

(b) What output voltage is produced for a gripping force that causes a strain of 2000 μ?

(c) What output voltage is produced for a reach force, F_z, applied to the end of the gripper finger that results in a 1000-μ compression strain on the finger?

Solution

(a) The strain gauge used in Example 8-3 has a nominal unstrained resistance of 1000 Ω. Thus,

$$V_2 = \frac{1000\ \Omega}{1000\ \Omega + 1000\ \Omega} 12\ \text{V}$$
$$= 6\ \text{V}$$

$$V_4 = \frac{1000\ \Omega}{1000\ \Omega + 1000\ \Omega} 12\ \text{V}$$
$$= 6\ \text{V}$$

As a result, when no gripping force is applied, the bridge is balanced, and

$$V_{out} = V_2 - V_4$$
$$= 6\ \text{V} - 6\ \text{V}$$
$$= 0\ \text{V}$$

(b) When a gripping force causes a strain of 2000 μ, the change in each strain gauge resistance is

$$\Delta R = 2R_{nominal} \times \text{strain}$$
$$= 2(1000\ \Omega) \times 2000\ \mu$$
$$= 4\ \Omega$$

Consequently, the resistance of strain gauges 2 and 3 goes up by 4 Ω, while the resistance of gauges 1 and 4 goes down by 4 Ω. This results in the following strain gauge resistances:

$$R_1 = 996\ \Omega$$

$$R_4 = 996\ \Omega$$

$$R_2 = 1004\ \Omega$$

$$R_3 = 1004\ \Omega$$

Substituting these values into the voltage equations, you get

$$V_2 = \frac{1004\ \Omega}{1004\ \Omega + 996\ \Omega}\ 12\ \text{V}$$

$$= 6.024\ \text{V}$$

$$V_4 = \frac{996\ \Omega}{996\ \Omega + 1004\ \Omega}\ 12\ \text{V}$$

$$= 5.976\ \text{V}$$

This gives an output voltage of

$$V_{\text{out}} = V_2 - V_4$$

$$= 6.024\ \text{V} - 5.976\ \text{V}$$

$$= 0.048\ \text{V}$$

(c) Ideally, a direct reach force, F_z, will cause all four strain gauges to compress by the same amount. This causes all the resistances in the bridge to decrease by the same amount, resulting in a balanced condition, or 0-V output. You conclude, therefore, that this particular gripper design cannot be used to measure reach forces.

Wrist Force Sensing Systems. The gripper sensing module just discussed will only measure gripping forces. What about the six fundamental forces that need to be measured at the wrist joint? Suppose you connect several of the strain gauge modules together as shown in Figure 8-8. Now you have a system that will measure gripping force as well as lift, reach, sweep, roll, pitch, and yaw. Notice how these six forces are sensed due to the location and orientation of the sensing modules in the wrist assembly.

Each gauge module in the wrist assembly must be connected into a separate Wheatstone bridge circuit. The individual bridge outputs are conditioned with an instrumentation amplifier and multiplexed into an A/D converter for conversion to digital and use by the robot controller. These force feedback signals provide the basic information required for an intelligent robot to sense the interaction of its end-effector with surrounding objects and successfully control a manipulator operation.

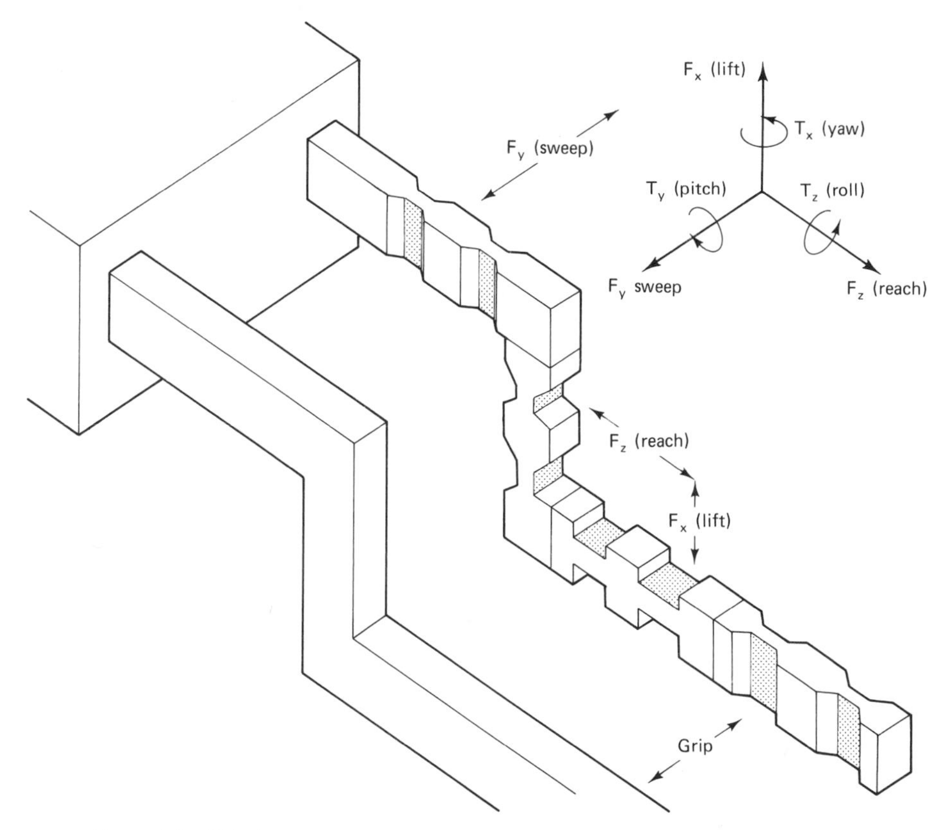

Figure 8-8 Wrist force sensing system that measures all six fundamental forces.

Another type of wrist force sensing system is shown in Figure 8-9. This system also employs strain gauges, but the gauges are attached to a hollow tube that forms the wrist assembly. Here, a hollow aluminum tube about 3 inches in diameter is milled to produce eight narrow beams where 16 strain gauges are attached, one to each side of each beam. The small neck at each end of each beam bends or twists under a load and strains the two gauges at the other end of the beam. This causes the resistance of one strain gauge to increase and the other to decrease. If the two strain gauges are connected into a voltage divider circuit as shown, the circuit output will be proportional to the amount of force applied to the beam. The eight beams are located so that the six fundamental wrist forces can be sensed.

Pedestal Force Sensing

A common law in physics states that for every force there is an equal and opposite force. Using this idea, the forces acting on a robot manipulator do not have to be measured at the manipulator. Rather, they can be measured on the

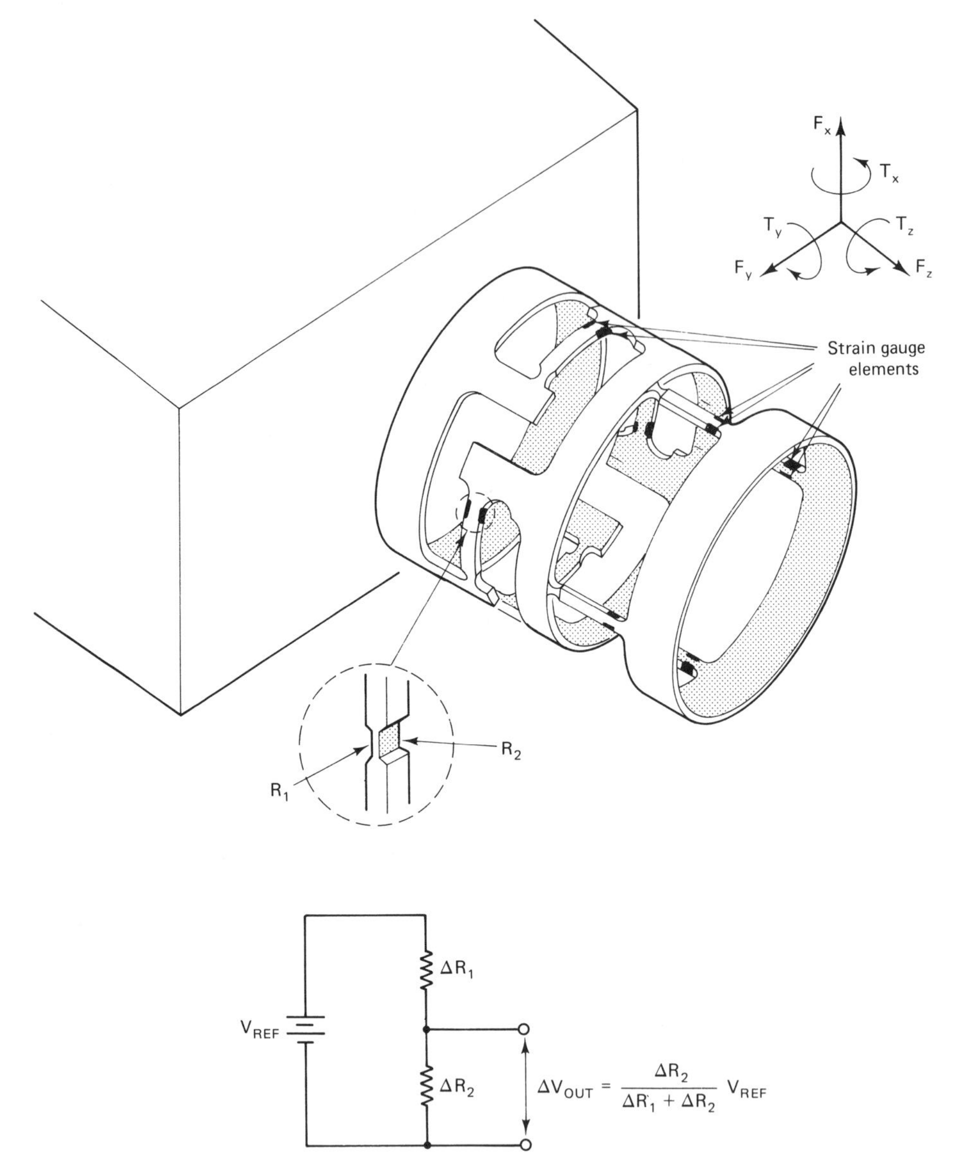

Figure 8-9 Strain gauge wrist sensor.

objects that are being manipulated. With pedestal force sensing, the object workpieces are firmly attached to a pedestal sensing support structure like the one shown in Figure 8-10. This particular pedestal sensor consists of three metal plates, separated by eight strain gauge beams. Four strain gauge beams connect the top plate to the middle plate, and four connect the middle plate to the bottom plate, as shown. The top plate beams have their strain gauges oriented so that they will

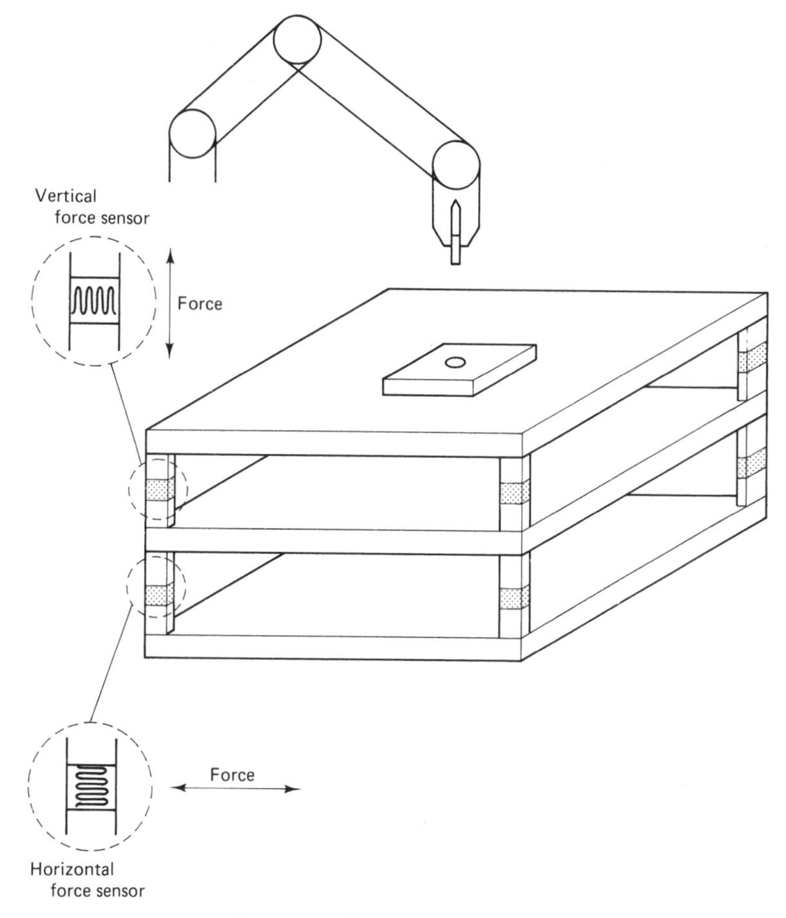

Figure 8-10 Pedestal sensing system.

detect any downward, or vertical, forces. The strain gauges on the bottom plate beams are oriented to detect any lateral, or horizontal, forces.

The four vertical force sensors form a Wheatstone bridge, whose output indicates any downward force. Likewise, the four horizontal sensors are part of a Wheatstone bridge circuit whose output indicates any lateral force. The force information generated by this type of pedestal sensor is adequate to perform simple assembly operations, such as placing a peg in a hole. However, more complicated operations are difficult due to the lack of sufficient force information.

8-3 SENSING TOUCH

Force sensing at the arm joints and wrist provides coarse data that can be used to control the position and orientation of the end-effector, or hand, relative to surrounding workpieces. However, a sense of touch must be used to generate the finer data required to describe the interaction between the hand and workpiece. In other words, most of the action is at the interface between the hand and workpiece. Consequently, a sense of touch is required in addition to arm and/or wrist force sensing to get a complete picture of the operation.

What is touch, or tactile, sensing? If you think of it in terms of your human sense of touch, it is a combination of force and temperature sensing. The tiny nerve endings in your skin transmit impulses to your brain that allow you to feel force and temperature. From this you can recognize objects and determine their position and orientation. Ideally, a robot touch sensor should duplicate the function of the human skin. That is, it must provide enough information to recognize objects and determine their position and orientation. To do this, robot touch sensors must be small, rugged, sensitive, and flexible. Let's begin this section by expanding on some of these ideal properties of touch, or tactile, sensing. Then we will explore several types of tactile sensors that can be used to implement various degrees of touch sensing.

Properties of a Tactile Sensor

Although it might be desirable that a tactile sensor have all the sensing properties of our human skin, it is not practical or even required for many application tasks. Again, the application must dictate the system. For instance, most applications related to product assembly do not require the tactile system to sense temperature. For these applications, the task reduces to force sensing.

Let's assume that the ideal goal of a tactile sensor is to identify objects and determine their position and orientation using force sensing in and around the robot hand. With this goal in mind, the tactile sensor must have six ideal properties:

1. ***An array:*** The tactile sensor must consist of an array of small force or pressure sensing elements.
2. ***Skinlike:*** The sensor array must be a skinlike material that is thin, flexible, and durable under harsh industrial environments.
3. ***Fast response:*** The sensing elements must have a fast response time, on the order of 1 millisecond (ms).
4. ***Sensitive:*** The sensing elements must have a sensitivity of at least 1 gram and be able to sense forces up to 1 kilogram.
5. ***Low hysteresis:*** The sensing elements must have very low hysteresis.
6. ***Smart:*** The sensing system must be inherently smart.

An Array. To identify objects and determine their orientation, the tactile sensor must consist of an array of sensing elements spaced about 0.1 inch apart. In other words, a 1-inch^2 tactile sensor should contain 10×10, or 100, individual sensing elements. This will satisfy most industrial tasks. Although finer resolution might be desirable, it is not always essential, especially if arm joint and/or wrist force sensing are used in conjunction with the tactile sensor. In fact, many application tasks can be satisfied using less than ten sensing elements per inch. Again, the application will dictate the size and resolution of the sensor array.

Skinlike. Ideally, the surface of the tactile sensor should be as flexible and durable as human skin. Flexibility is required for the sensor to conform to the object's shape. Durability is required to withstand the repeated handling of thousands of objects. In addition, many specialized applications require that the sensing surface withstand harsh environments. Such environments might include extreme temperature, high electrical interference, and corrosive liquids and gases, just to mention a few.

Other ideal properties of the skinlike sensor surface include the ability to sense temperature and slippage. Temperature sensing could aid in object identification. The ability to sense slip is required to adjust the grasping force so that hard and heavy objects, as well as soft and fragile objects, can be handled.

Fast Response. A response time of around 1 ms is ideal to coordinate the tactile data with other sensory data. Such a response allows the robot to compete with the human sense of touch in performing real-time operations.

Sensitive. The dynamic range of the tactile sensing elements must be at least 1000:1. To satisfy most applications, the sensing range should be between 1 gram and 1 kilogram. Ideally, the sensor output should be linear within this range. However, this is not a necessary requirement, since software lookup tables can be used to adjust for nonlinear data.

Low Hysteresis. Hysteresis relates to the lag of the effect in the sensor when the force acting on it is changed. In other words, the sensor must respond in the same way for a decrease in force as it does for an increase in force over the same range. If the sensor output increases with a given increase in force, then decreases for a like amount of decrease in force, the decreasing force/output transfer function must retrace the increasing transfer function.

Smart. The tactile sensing system must be a smart system. That is, it must employ on-board signal conditioning and sensory data processing. Single-chip microcomputer tactile-sensing cells made possible by VLSI technology should collect, condition, partially analyze, and transmit tactile sensory data to the robot controller. This greatly simplifies the connections between the robot hand and controller. In addition, the controller is not tied up performing low-level data analysis. Rather, the controller supervises the overall system and makes the final control decisions based on information received via the various smart robot sensing systems.

Of course, the properties listed are ideal and won't all be realized for several years. At present, there are no existing tactile sensors that satisfy all these properties. However, as the technology improves, tactile sensing will begin to employ more and more of these ideal sensor characteristics. Keep these properties in mind for the rest of this section, where you will be introduced to various tactile-sensing techniques.

The sensing elements that make up the tactile sensor can be divided into two broad categories, depending on their output: ***binary tactile sensors*** and ***analog tactile sensors.*** As the names imply, a binary tactile sensor produces a two-state output, and an analog tactile sensor generates a continuous output that is proportional to an applied force. The following material is devoted to a discussion of various sensors that fit into these two general categories.

Binary Tactile Sensors

Binary tactile sensors are the simplest type of touch sensors. The sensing element is basically a switch that produces a two-state on/off output when it comes in contact with an object. The simplest binary sensing element is the limit switch, which was discussed in the last chapter under proximity sensing. The limit switch idea can be extended into a more sensitive element called a ***whisker sensor***. Whisker sensors are also a form of proximity sensor and, therefore, can be used to determine object position. However, for true tactile sensing, binary switches must be arranged into a keyboard-like matrix in order to generate contact data over a given area, much like a binary pixel matrix provides noncontact data of an object's image.

Limit Switches. Limit switches and their associated circuits were introduced in Chapter 7 under proximity sensing. I do not intend to repeat that discussion here; however, I would like to suggest several ways in which limit switches can be used to aid in the touch-sensing process. First, a simple touch sensor can be constructed by using two limit switches, placing one on the inner side of each gripper finger as shown in Figure 8-11. This arrangement can be used to determine the

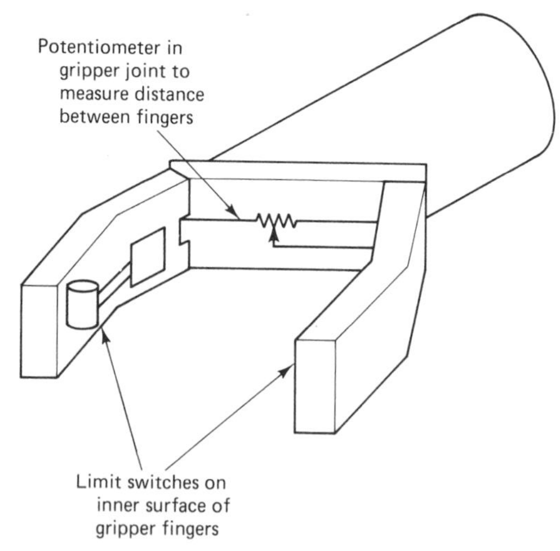

Figure 8-11 Limit switches make a simple touch sensor.

presence or absence of an object, as well as the thickness of the object. Here's how: The gripper is placed over an object and begins to close on the object. A potentiometer is placed in the gripper joint as shown to measure the distance between the fingers. When the two limit switches are both actuated, the gripper stops closing and the potentiometer output is proportional to the distance between the fingers. Of course, if the output is zero, there is no object present.

Binary switches can also be placed around the outside of the gripper fingers as shown in Figure 8-12. Here the switches are used as external sensing plates to detect when the outside surfaces of the gripper make contact with an object. These, combined with the limit switches on the inner surface of the fingers, allow the robot to determine object position and perform simple pick-and-place operations. However, it is difficult to recognize objects and determine their orientation using such a crude system.

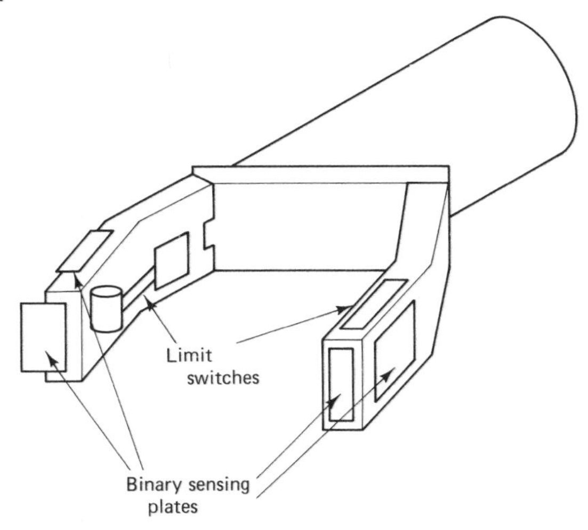

Figure 8-12 Binary sensing plates can be added to the outer surfaces of the gripper fingers.

Whisker Sensors. Most tactile sensing systems, like the limit switch system, are very rigid and require a relatively high contact force to actuate the limit switch sensing elements. On the other hand, many assembly parts are light and will move under the slightest touch. It is easy to disturb the position and orientation of such parts when searching for objects using a clumsy gripper assembly like the one just described.

An alternative idea is to add a superlight whisker sensor on the end of the robot gripper, as shown in Figure 8-13. The whisker sensor is used to probe a given area before actually attempting to grasp an object. This allows the robot manipulator to come in contact with an object without disturbing it. The probing information is then analyzed to determine the object's position and orientation and to control the manipulator and gripper movements accordingly. Tactile-sensing elements inside the gripper can then be used when working with the object.

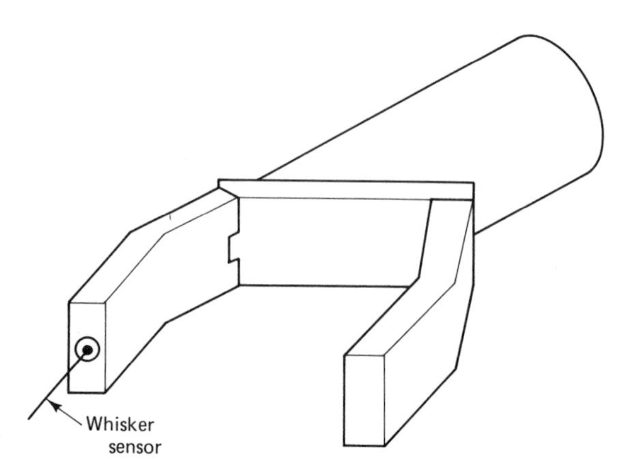

Figure 8-13 A whisker sensor added to the end of a robot finger can be used to search for objects without disturbing them.

A simple pneumatic whisker sensing system is shown in Figure 8-14. Here the whisker assembly is placed inside the robot finger. The system requires both a vacuum source and an air-pressure source. These combine to allow the whisker to be extended and retracted during an assembly operation. Semiconductor pressure sensors (to be discussed shortly) are used to detect the slightest deflection of the whisker and transmit a logic signal to the robot controller.

The whisker sensing system is simple, economical, and effective in determining the position and orientation of assembly parts. In addition, such a system can be used to track an edge, groove, or seam. This can aid in object identification and control the robot during a production process such as seam welding.

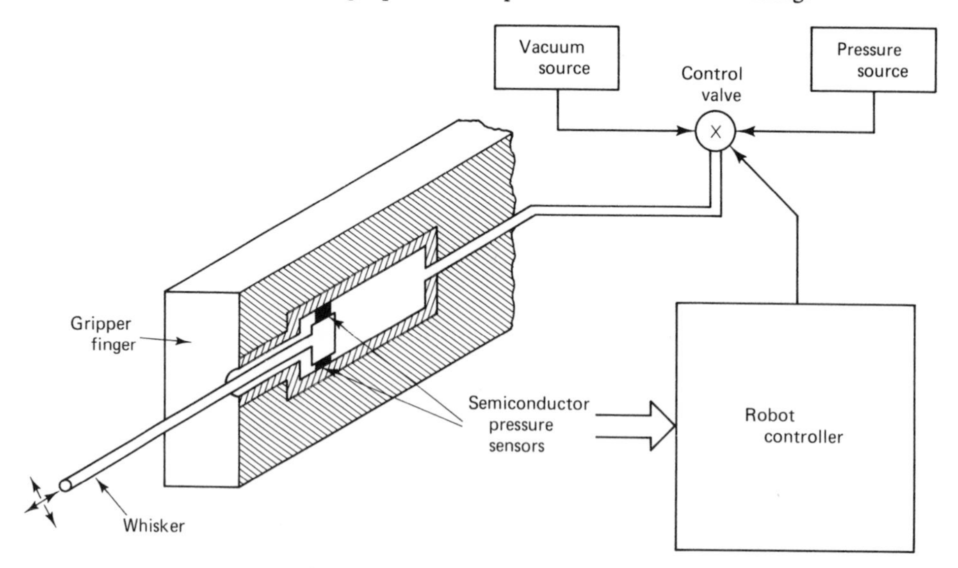

Figure 8-14 Pneumatic whisker sensing system.

Binary Switch Arrays. Binary switches can be placed in an array and mounted on the inner surface of a robot gripper to form an ***area tactile sensor***, as shown in Figure 8-15. Here, as the gripper closes on an object, only those switches that are in contact with the object are actuated to produce a logic 1. Those switches not in contact with the object are not actuated and produce a logic 0. The result is a binary tactile image that can be used for object recognition. In addition, any slippage of the object within the gripper can be detected by examining several tactile images over a given time period.

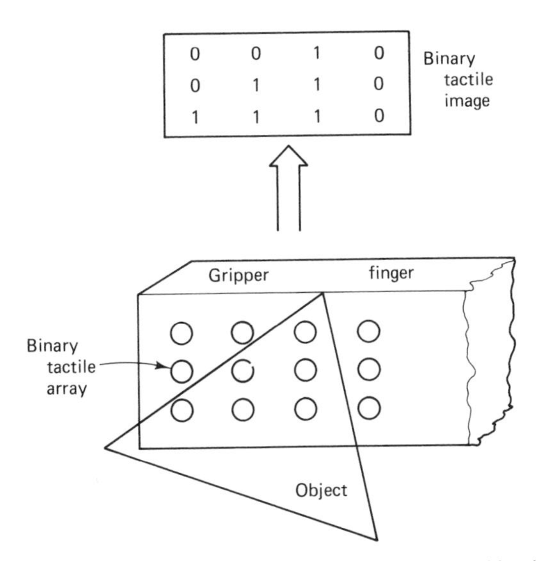

Figure 8-15 An area sensor can be used to aid in object identification.

You might be wondering how such an array sensor can be constructed. One approach is to use an array of snap-action pressure switches, as shown in Figure 8-16. Here a thin steel plate is punched to form an array of small spherical domes. The plate is connected to a voltage supply by a pull-down resistor circuit and sandwiched between a flexible insulating gripping surface and a high-pressure air supply. When a given switch is not in contact with an object, the high air pressure keeps the switch open, resulting in a logic 0 at that point. However, when a switch comes in contact with an object, the contact force causes the switch to overcome the air pressure and close, resulting in a logic 1 at the switch point. The switch matrix is multiplexed and decoded just like a computer keyboard to generate the binary tactile image. The sensitivity of the system is adjusted by adjusting the air pressure.

Another approach is to integrate an array of semiconductor pressure sensors into the gripper and use a thresholding technique to generate a binary image. This idea will be discussed under analog tactile sensors, since semiconductor pressure-sensing elements are analog devices.

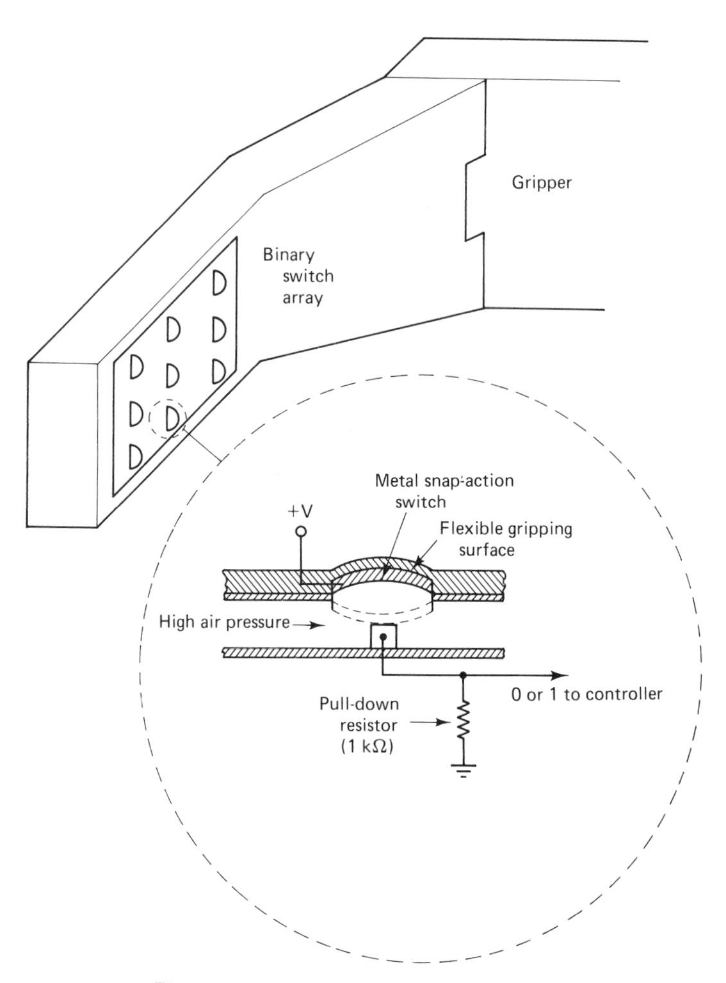

Figure 8-16 Binary snap-switch tactile array.

Analog Tactile Sensors

An analog tactile sensor differs from a binary sensor in that the analog sensing element produces a voltage or current output that is proportional to the amount of force sensed by the element. In most cases, the analog sensing element is a resistive element, whose resistance varies with an applied force. The change in resistance is seen as a change in voltage or current by Ohm's law. Let's take a brief look at several analog tactile-sensing elements so that you can get a feel for how these devices work.

Conductive Sensing Elements. There is a group of flexible conductive materials that change their resistive properties under pressure. Two such materials for use in tactile sensing are ***conductive rubber*** and ***conductive foam.***

Conductive Rubber. Most conductive-rubber tactile-sensing elements are made from silicon rubber that has been mixed with metallic compounds. The rubber allows them to be flexible, or ***compliant***, and the amount of metallic compound in the mix controls the conductive properties of the element. The conduction of the rubber changes as it is compressed by an external force, or pressure. Typical conduction curves are shown in Figure 8-17. These curves indicate resistivity in ohms per centimeter versus pressure in pounds per square inch, or psi, for two typical conductive-rubber compounds. First, notice that resistance is inversely proportional to the amount of applied pressure. The resistivity of compound A decreases rapidly as pressure increases from 0 to about 4 psi, and then levels off at about 100 Ω/cm. The resistivity of compound B decreases almost exponentially with applied pressure. In fact, if you were to graph the response of compound B on log paper, you would get a straight line. Another observation is that compound B covers a broader range of pressures, but is not as sensitive as compound A.

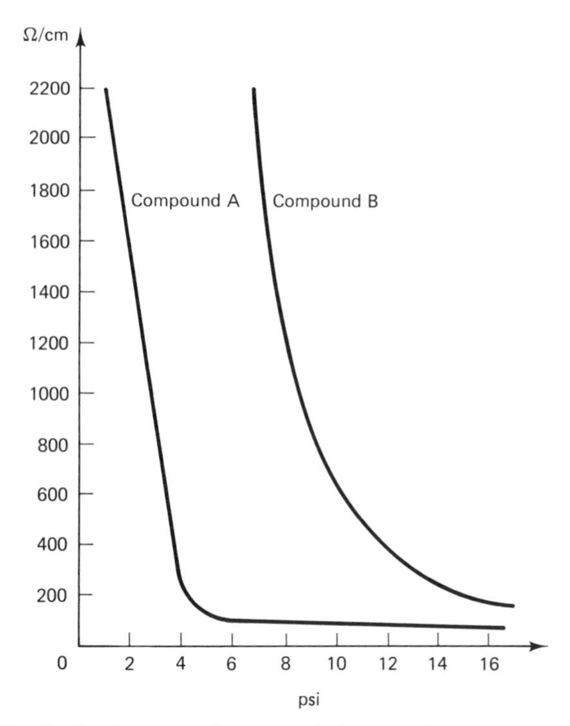

Figure 8-17 Conduction curves for two typical conductive-rubber compounds.

One way to use conductive rubber in a tactile sensor is to place a sheet of the rubber over a PC board that has been etched to provide an array of concentric ring pairs, as shown in Figure 8-18. Each pair of concentric rings forms the electrodes of a variable resistor, as illustrated by the exploded view in the figure. Since the conductive rubber is placed against the ring pair, it creates a resistance between

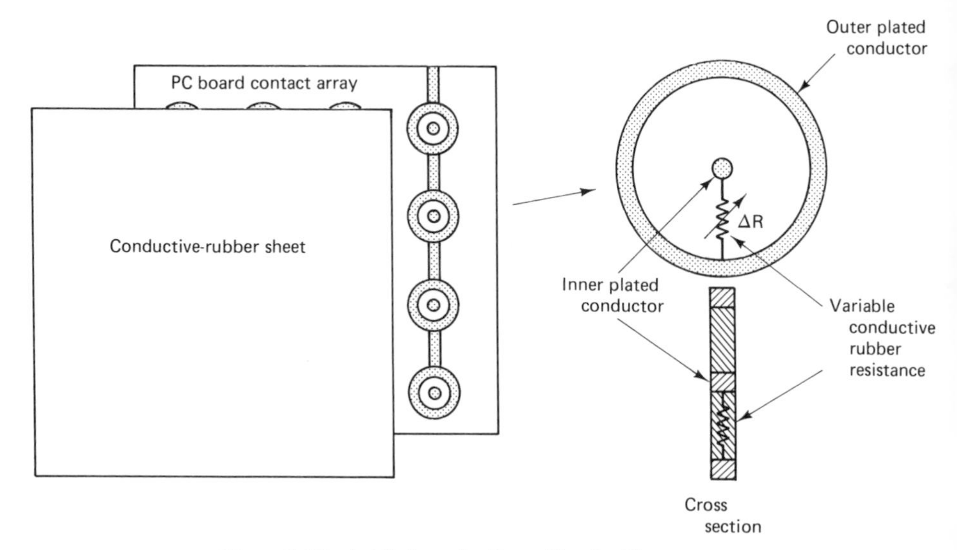

Figure 8-18 A 4 × 4 conductive-rubber tactile array sensor.

the outer ring and the center ring that varies with the amount of pressure applied to the rubber at that point in the array.

Notice that the outer rings are linked together on the PC board to form the columns of the array. The center rings of each element in a given column are connected to diodes, which are part of an external decoding circuit. A schematic diagram of the decoding circuit for *one* of the columns is shown in Figure 8-19. Observe that the four outer rings of the sensor column are connected together and to the collector of a column-select transistor. This common connection is provided by the PC board. The column-select decoding logic selects one of the four columns by applying a logic 1 to the base of the respective column-select transistor. Only the column 0 circuit is shown here. This entire circuit must be repeated for columns 1, 2, and 3.

Now, once a column is selected, current in a given row flows from the +5-V supply down through a fixed 1-kΩ resistance, through the diode to the inner ring connection at the cathode of the diode. Current then flows through the pressure-sensitive conductive rubber to the common outer ring connection, and then through the transistor to ground. Thus, there is a separate current path through each sensor in the selected column. The amount of current in each path depends on the variable conductive-rubber resistance.

Notice that each current path forms a voltage divider between the conductive-rubber element and the fixed 1-kΩ resistance. As a result, the voltage developed across the conductive-rubber element is directly proportional to its resistance. The individual conductive-rubber voltages are selected using a 4:1 analog multiplexer and passed to an A/D converter.

In summary, one of four columns is selected by the column-select logic. Then

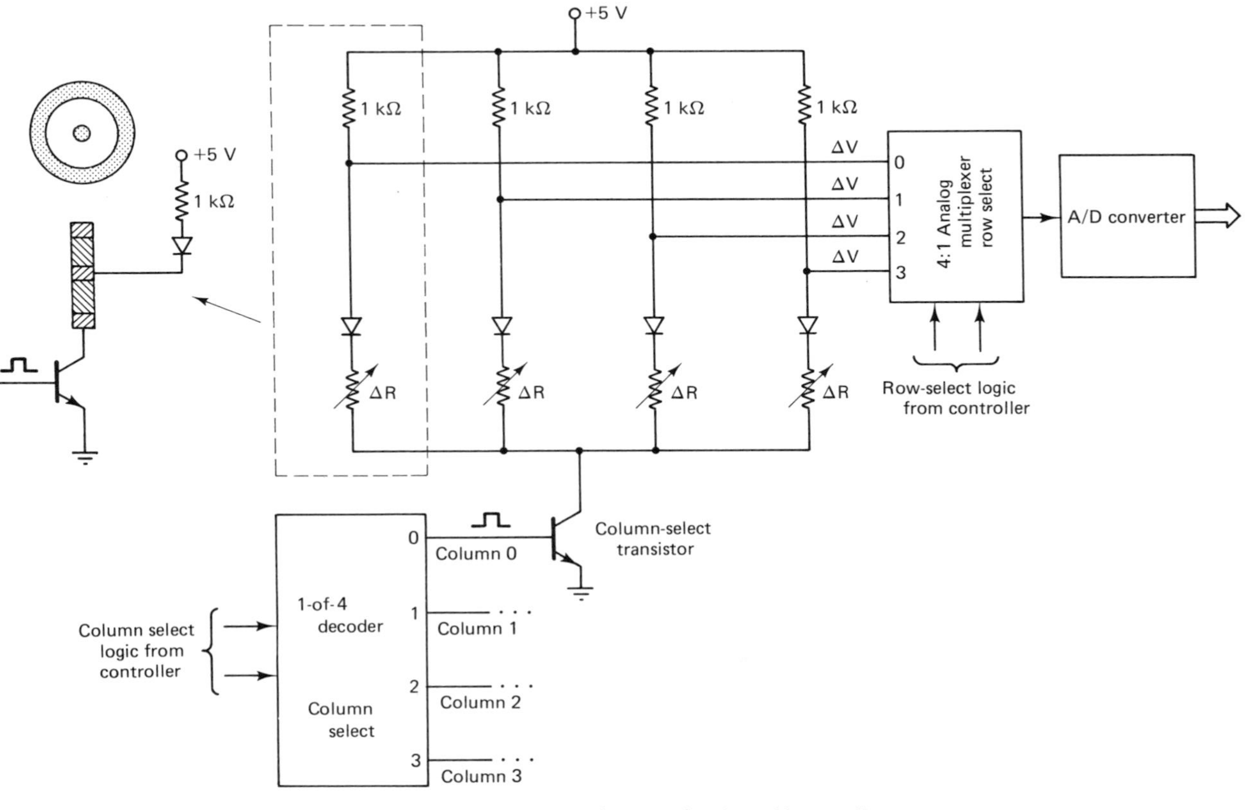

Figure 8-19 Decoding circuit for a conductive-rubber tactile array.

the voltages developed across the four sensors within that column are multiplexed into an A/D converter, one at a time, via an analog multiplexer. This process is then repeated for each of the other three columns in the array. The converted tactile data can then be stored in the form of a ***tactile matrix***, which can be analyzed and interpreted in much the same way as a gray-scale picture matrix in a vision system.

Another way to make a tactile array sensor from conductive rubber is shown in Figure 8-20. Here the array is created by overlapping rows and columns of convex conductive-rubber cords. Notice that the cords are shaped like the letter D. The convex portions of any two given cords come in contact with each other to form a row/column junction within the array. The resistance at the junction of a row and column is proportional to the amount of compression at that point.

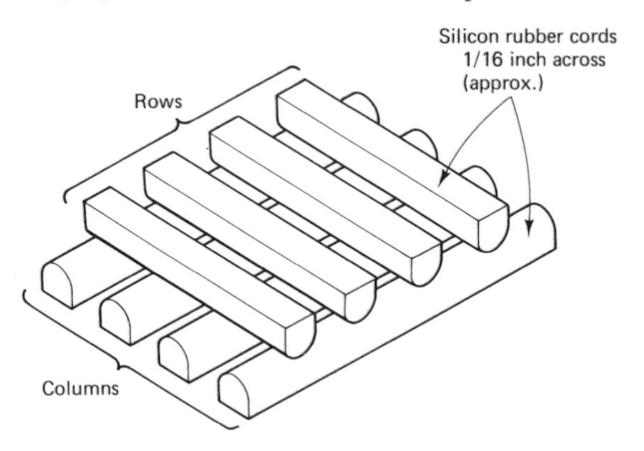

Figure 8-20 Area tactile sensor made by overlapping convex D-shaped conductive-rubber cords.

Look at the exploded view of a row/column junction in Figure 8-21(a). The resistance between the row and column cords is relatively high when no pressure is applied to the junction. However, as the junction is compressed, the surface contact area between the row and column cords increases. This increases the number of electrical paths between the two cords and decreases the resistance at their junction.

The diagram in Figure 8-21(c) shows how a voltage divider is formed at the junction. One cord is connected to a +5-V supply via a 1-kΩ fixed resistor and the other cord is connected to ground. The contact area between the two cords forms the lower half of the voltage divider using the variable contact resistance. Any change in this resistance due to compression between the cords is seen as a change in voltage across the junction.

The array is decoded using a circuit like the one in Figure 8-22. A given column is selected using the 4 : 1 analog multiplexer. A given row within the column is then selected when the row select decoder applies a logic 1 to the base of the respective row-select transistor. This grounds the selected row through the op-amp and row-select transistor. As a result, a current path is created from the +5-V supply through the fixed 1-kΩ resistor, through the variable rubber contact resistance,

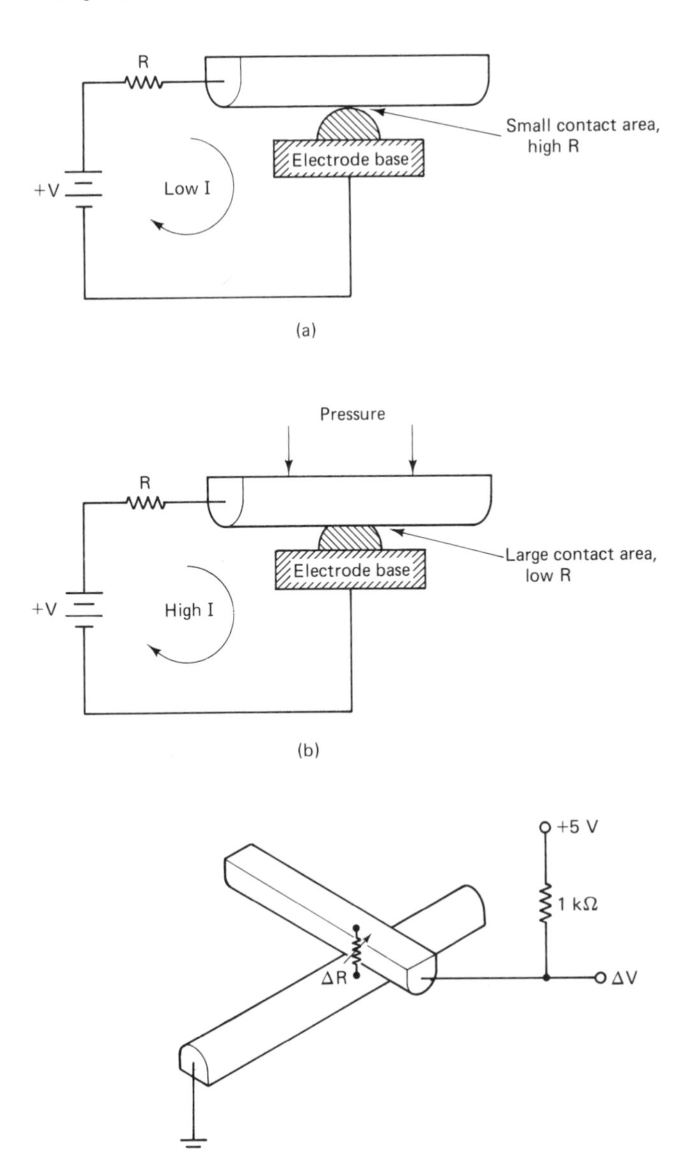

Figure 8-21 Conductive-rubber cord pressure-sensing element.

through the transistor to ground. Thus, the voltage across the selected row/column junction is seen at the bottom of the pull-up resistance and applied to the A/D converter for conversion to digital. The columns and rows are sequentially selected in this manner by the row- and column-select logic until the entire array has been scanned.

You are probably wondering why the selected column output is fed back by

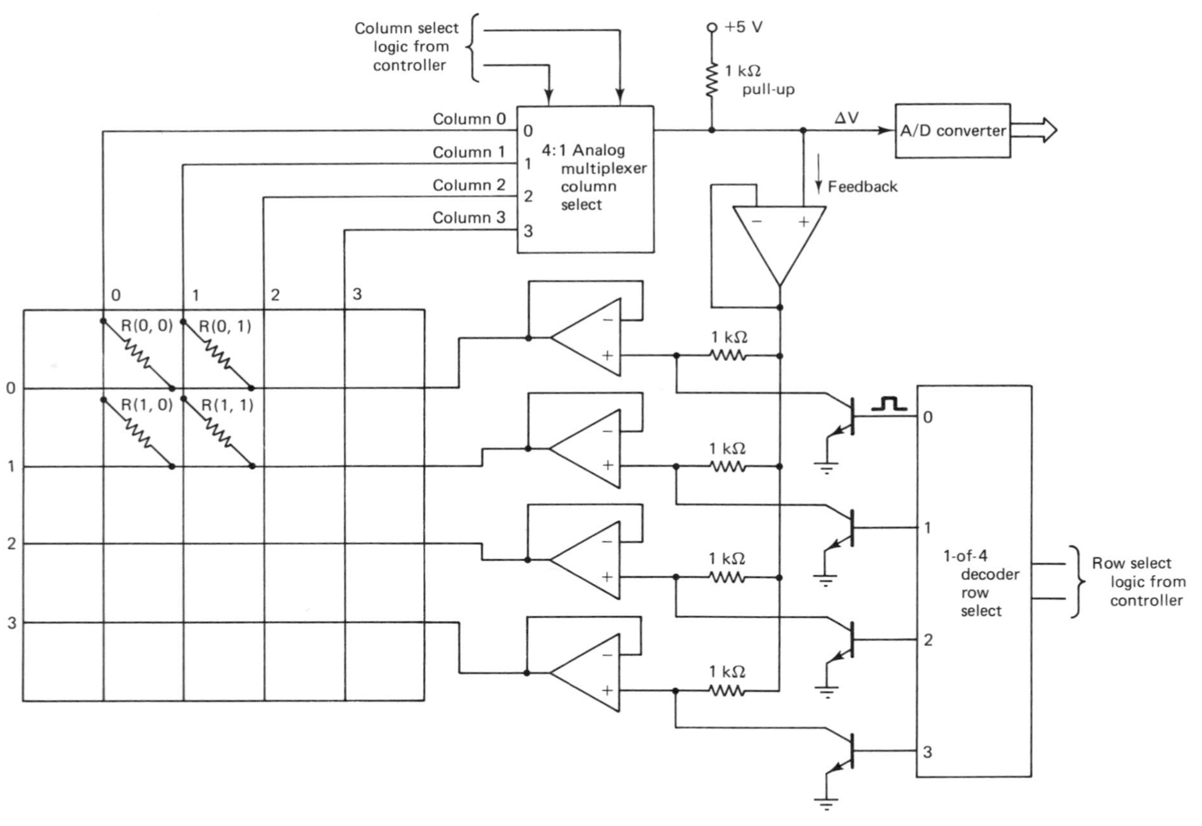

Figure 8-22 Decoding circuit for the conductive-rubber cord tactile array.

voltage-follower op-amps to all the unselected rows in the array. If this were not done, there would be multiple current paths within the array through junction points other than the one that has been selected. For instance, suppose that the junction at row 0 and column 0 is selected. The desirable current path is then through the row 0/column 0 junction resistance, $R(0, 0)$. Now, if the remaining rows (1, 2, and 3) were simply left open, or floating, there are alternate paths for the current to flow. One such path is down through $R(1, 0)$, across row 1 and up through $R(1, 1)$, across the selected row 0, through the transistor to ground. In fact, there are several such current paths within the array. This does not allow you to isolate the selected row/column resistance.

The solution to the problem is to feed back the selected column output to all unselected rows so that there is no difference in potential across the selected column and the unselected rows. If both sides of an unselected row/column junction are at the same potential, there can be no current flow through that junction. Thus, the only current path is through the selected row/column junction.

The voltage-follower op-amps are used in the feedback circuit because of their very high impedance characteristic. This eliminates any loading of the output by reducing the amount of feedback current.

Conductive Foam. You have probably seen conductive foam, since it is commonly used to prevent static electricity from damaging MOS integrated circuits. The foam is impregnated with carbon so that it acts as a conductor to dissipate any static charges. Like conductive rubber, conductive foam has a resistance property that decreases under compression. If the foam is placed between two metal contacts, as shown in Figure 8-23, the resistance between the contacts decreases as the foam is compressed. Of course, the actual resistance between the contacts also depends on the thickness of the foam.

The associated graph shows how the resistance of the foam changes under compression. Notice that the sensor is extremely sensitive and fairly linear. A compression change of only 0.15 inch (in.) produces a resistance change of almost 19 kΩ. Also, the sensor is especially sensitive and linear below 0.020 in. of compression.

A tactile array sensor can be constructed by sandwiching a layer of foam between rows and columns of conductors, as shown in Figure 8-24. One set of parallel conductors is placed on top of the foam to form the columns of the array. Another set of parallel conductors is placed beneath the foam and perpendicular to the top set of conductors to form the array rows. This 4 × 4 array uses ¼-in. foam and provides 16 intersection, or junction, points. The row and column wires must be small, around 32 gauge, so that the array is compliant. Spacing the wires ½ in. apart results in a 2 in. by 2 in. tactile array. Of course, larger arrays and increased resolution can be obtained using more rows and columns. For instance, a 4 in. by 4 in. array can be constructed using 16 rows and 16 columns and spacing the wires ¼ in. apart. Such an array would provide 256 junction points, resulting in better resolution of the tactile image. In any event, the row and column conductors must be

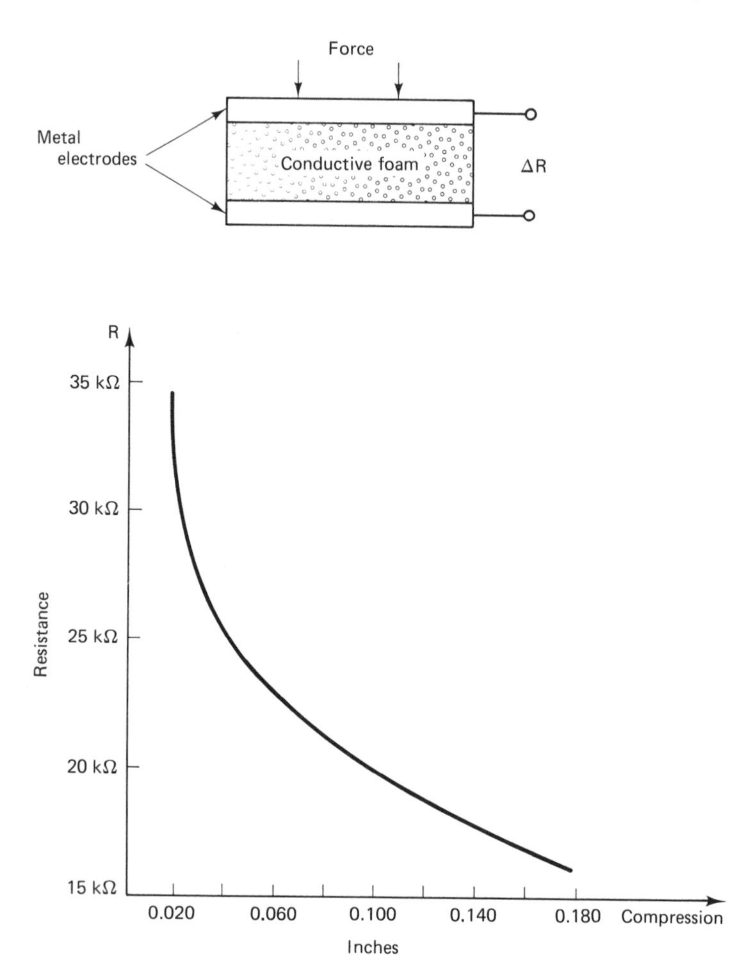

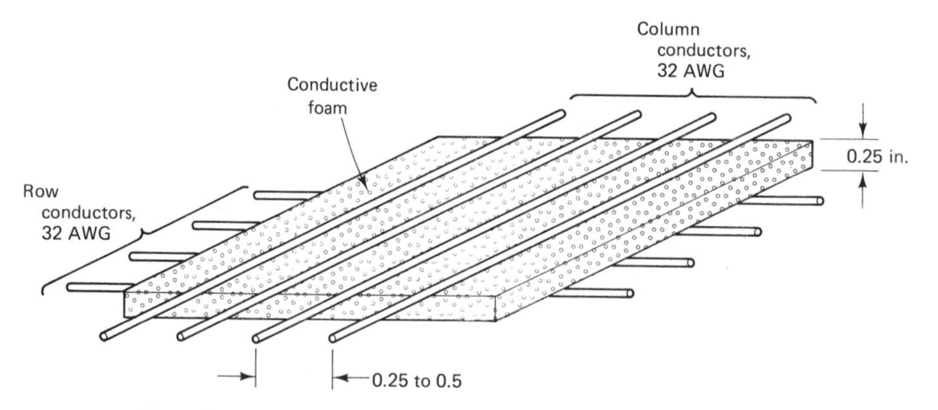

Figure 8-24 A 4 × 4 tactile sensing array made from conductive foam.

spaced far enough apart so that the resistance at one point in the array is not affected by changes in resistance at an adjacent point in the array.

The conductive-foam array must be scanned similar to the conductive-rubber cord array discussed earlier. Like the rubber cord array, the selected intersection points in the foam array must be isolated from the unselected points to prevent false resistance readings due to multiple current paths within the array. The decoding circuit discussed earlier (Figure 8-22) can also be applied to the conductive-foam array. Other circuits using various row and column multiplexer arrangements are also possible. Regardless of the type of decoding circuit used, the basic idea is the same: you must create a difference in potential across the selected row/column intersection, while maintaining no difference in potential across the unselected row/column intersections. Thus, current can only flow through the selected intersection. The current flow is proportional to the foam resistance at the intersection, which produces a proportional voltage that can be converted to a digital value.

Semiconductor Sensors. Semiconductor pressure sensors are also used for tactile sensing. Their advantages include high sensitivity, reliability, good linearity, low hysteresis, and relatively low cost. In addition, they are manufactured using standard IC technology and do not require extensive signal conditioning, since they often include on-board signal-conditioning circuits.

A typical semiconductor sensing element is shown in Figure 8-25. The element consists of an *N*-type semiconductor material that has been etched to form a vacuum cavity. Over the top of the cavity is a very thin (approximately 0.001 in.) pressure diaphragm that deflects as pressure is applied to the sensor. Four ***piezoresistive*** elements are formed on the top, or pressure side, of the diaphragm by diffusing

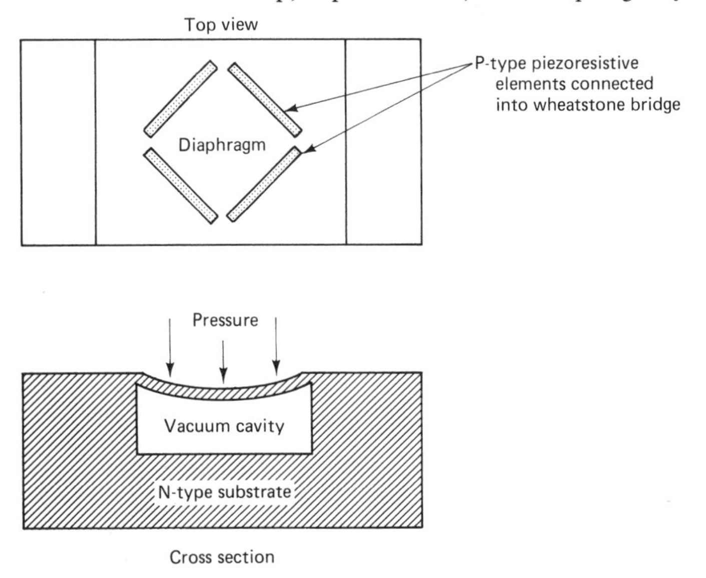

Figure 8-25 Typical piezoresistive semiconductor pressure-sensing element.

a *P*-type semiconductor material into the diaphragm, or by depositing thick-film resistors onto the diaphragm. The four resistive elements act like tiny strain gauges that change resistance as the diaphragm deflects under pressure.

The four piezoresistive elements are internally connected to form a Wheatstone bridge circuit. This increases the sensitivity of the device and provides for built-in temperature compensation, since the element pairs are connected into opposite sides of the bridge. Some pressure-sensing ICs also include differential-amplifier signal-conditioning circuits in the same chip. This allows the chip output to be applied directly to the input of an A/D converter.

Using VLSI technology, tactile sensors could be developed by integrating an array of semiconductor pressure sensors within the same IC substrate. All the necessary signal conditioning, decoding logic, and conversion circuitry could also be integrated within the IC. Moreover, the IC could contain the computer logic and memory required to analyze and temporarily store the tactile data. This would free the robot controller of the low-level data-analysis task, leaving it available to perform the higher-level intelligent tasks required for recognition and control.

As compared to conductive rubber and foam sensors, semiconductor sensors are extremely sensitive, have a fast response, are more immune to noise, and have low hysteresis. As a result, they look very promising for future applications. With present technology, their biggest disadvantage is noncompliancy. Unlike conductive rubber and foam, it is difficult to make a flexible artificial skin out of semiconductor sensors.

Pressure-sensitive Paint. A possible solution to the noncompliant nature of the semiconductor sensor is found in pressure-sensitive paint, or ***pressistor.*** The paint is made by mixing piezoresistive semiconductor powders with an organic material. The combination produces a liquid that can be painted onto electrode arrays or impregnated into pourous foam to produce a tactile sensor. Once the paint cures, its resistance changes in proportion to an applied force.

A typical force-sensing element made from pressure-sensitive paint is shown in Figure 8-26. Here the paint is placed between two metal electrodes. The associated graph shows how the element resistance changes inversely with an applied force. Notice its linearity and extreme sensitivity. A force of only 2 ounces causes the element to change its resistance by over 200 MΩ.

8-4 SENSING SLIP

Think about what you do to control the amount of force that you use to grasp an object. If you are familiar with the object, you have a feel for its hardness and weight and know just about how much grasping force to apply without damaging or dropping the object. However, if you are not familiar with the object, you will

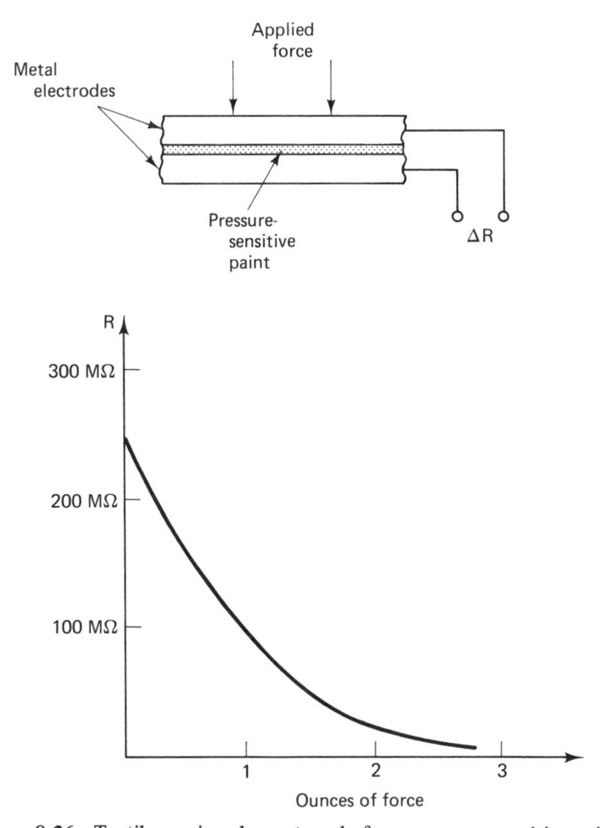

Figure 8-26 Tactile sensing element made from pressure-sensitive paint.

most likely apply just enough grasping force to prevent damage to the object and assure that it will not *slip* out of your grasp.

The same is true of a robot gripping system. If the hardness and weight of an object are known, gripping force sensors can be used to sense when the optimum amount of gripping force has been applied. However, when handling unknown objects, the optimum gripping force cannot be predetermined. As a result, usually too much or too little gripping force is applied. Too much force causes damage to the object, while too little force results in the object being dropped.

One way for a robot to handle unknown objects is to adjust the amount of gripping force according to the amount of slip that is detected between the object and gripper. As a result, the robot compensates for objects of different weights using slip sensing, just as you and I do. The question then is how can a robot gripper be designed to sense slip. There are basically two ways: (1) by sensing any ***vibration*** between the object and gripper, or (2) by sensing any movement, or ***displacement***, between the object and gripper. The remainder of this section is devoted to a discussion of these two slip-sensing techniques.

Vibration Sensing

Slip between an object and robot gripper can be sensed by using a record-player-type stylus like the one shown in Figure 8-27. A sapphire needle is attached to a piezoelectric crystal as shown. The stylus assembly is placed within the robot gripper so that the needle protrudes slightly beyond the inner surface of the gripper finger. Consequently, the stylus will come in contact with the surface of any object being grasped.

Once an object is grasped, the stylus will trace the surface texture of the object. Any slippage causes a vibration of the stylus, resulting in a small voltage output from the piezoelectric crystal. The output is amplified and conditioned so that it can be detected by the robot controller. The controller can then apply additional gripping force in small increments until no more slip is detected. The signals shown in Figure 8-27 represent the original and conditioned output of the piezoelectric slip sensor. Notice that the analog sensor output is converted to a digital signal transition for detection by the robot controller.

The advantage of using such a device for sensing slip is that it is very sensitive

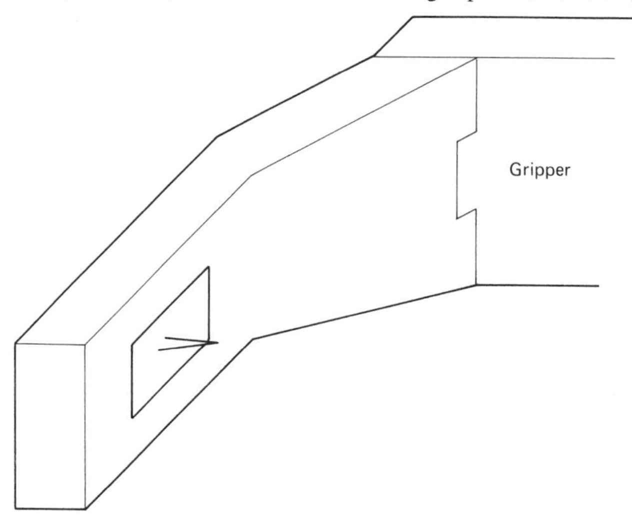

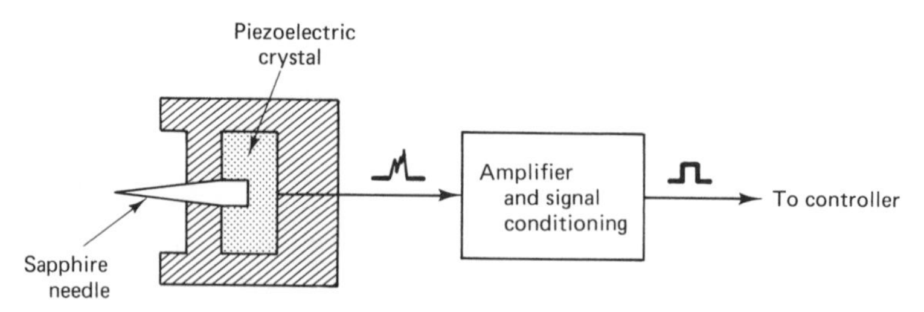

Figure 8-27 Stylus-type slip sensor.

and immune to noise. The disadvantage is that the sensor only provides an indication that some slip has occurred. The exact amount of slippage cannot be determined from the sensor output.

In addition to sensing slip to determine grasping force, such a system could also be used for product inspection to inspect an object for surface defects. For this application, the gripper would not actually grasp the object. Rather, the robot would open the gripper, place the stylus in contact with the object surface, and move the stylus across the surface to generate an analog output proportional to the surface texture. The analog output is then conditioned and converted to digital for interpretation by the robot controller. The same surface texture information could also be used in conjunction with visual information to identify unknown objects.

Displacement Sensing

Slip results from movement, or displacement, between the object and robot gripper. Thus, slip can be detected by sensing any object movement once it is grasped. This can be accomplished in two ways: (1) by comparing tactile images from an array sensor over a given period of time, or (2) by placing a sensor in the robot gripper that moves in direct proportion to any object movement.

Successive tactile images created by an array sensor and stored in memory can be compared to detect object movement. Any displacement from one image to the next is an indication of slip. The amount of displacement is directly proportional to the amount of slip. This type of slip detection can be built into the tactile-sensing software used with the array sensor. Of course, this complicates the system software and might result in an object being dropped before the slip is detected and the grasping force adjusted.

A simpler technique to detect slippage is to place a rubber roller within the gripper, as shown in Figure 8-28. The roller surface must have a high coefficient of

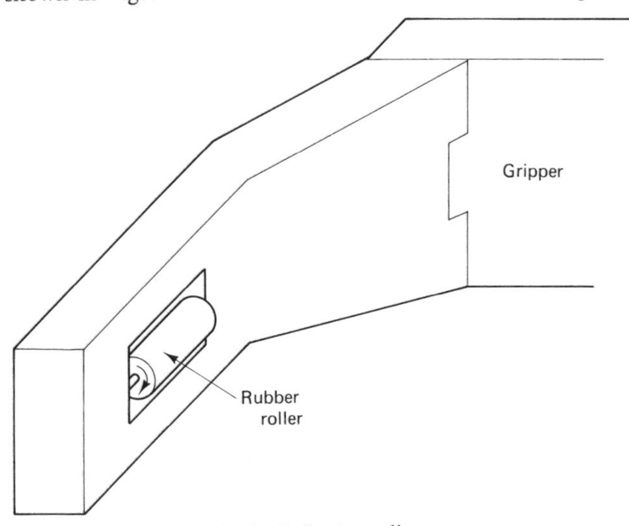

Figure 8-28 Roller-type slip sensor.

friction so that any object movement results in a proportional roller movement. The roller movement can be detected using either magnetic or optical techniques. In Figure 8-29(a), a small magnet is embedded within the roller. A Hall-effect device is then used to detect the roller movement. In Figure 8-29(b), the roller shaft is connected to an optical encoder that passes through the slot of an optical interrupter to detect movement. The number of magnets embedded in the roller in Figure 8-29(a) or number of slits in the optical encoder of Figure 8-29(b) determines the sensitivity and resolution of the system.

The advantage of the roller-type slip sensor is that the *amount* of object movement can be determined by counting the number of sensor pulses. In addition, the velocity of movement can be calculated by measuring the amount of time between one sensor pulse and the next. The slip velocity can be used to control the speed of the gripper motor. For instance, if the slip velocity is relatively high, the gripper must close faster in order to grasp the object before it is dropped.

Example 8-6

The shaft of a roller-type slip sensor is attached to an optical encoder wheel that has 10 slits spaced 0.1 in. apart. An optical interrupter generates a digital pulse as

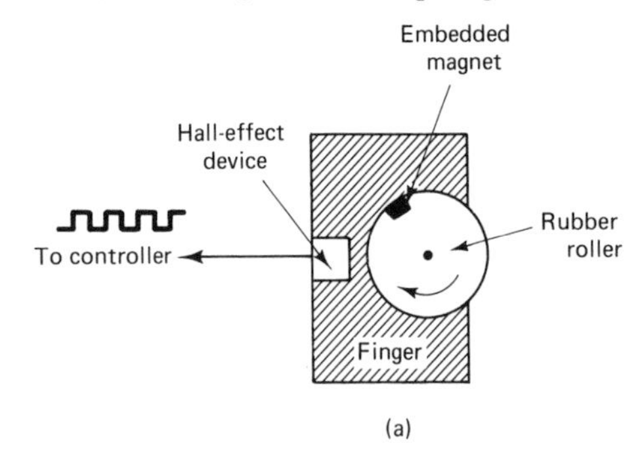

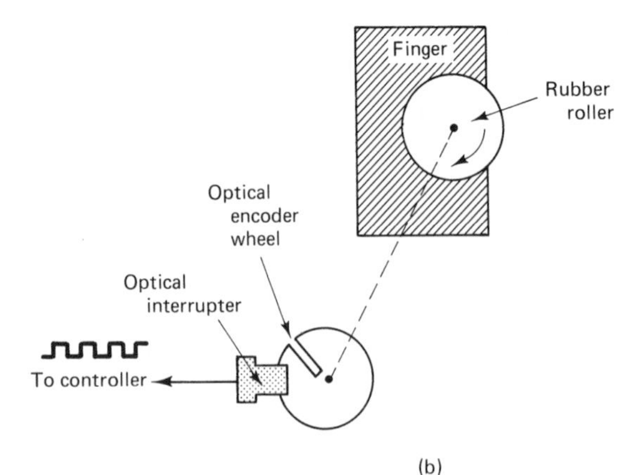

Figure 8-29 Two roller-type slip sensors: (a) magnetic and (b) optical.

an interrupt input to the robot controller each time a slit is detected. The robot controller uses the interrupt to increment an 8-bit counter. Suppose a timer is started when the first slip pulse is detected and the counter is read after 0.5 s.

(a) What is the maximum amount of slip that can be measured with this system?

(b) How much slip has occurred if the counter contains the hexadecimal value 1C?

(c) How fast is the object slipping?

Solution

(a) Using an 8-bit counter, the maximum amount of slip that can be measured is 255 × 0.1 in., or 25.1 in.

(b) For each count, the object has slipped 0.1 in. Thus, for a count of 1C, or 28 decimal, the object has slipped 28 × 0.1 in., or 2.8 in.

(c) Since the object has slipped 2.8 in. in 0.5 s, the slip velocity is 2.8 in./0.5 s = 5.6 in./s.

As you might guess, the roller sensor is not as sensitive as the stylus sensor discussed previously. Also, it cannot be used to measure the surface texture of an object.

Software

The flow chart in Figure 8-30 summarizes the software required to support slip sensing. Basically, the idea is to detect when the inner surface of the gripper comes in contact with an object. This can be accomplished using a simple limit switch. The gripper then closes on the object using a minimum amount of force as sensed by a gripper force sensor. The robot then attempts to pick up the object. If any slip is detected, the slip sensor interrupts the robot controller, which increases the gripping force slightly. If more slip is detected, the controller continues to increase the gripping force in steps, as shown at the bottom of the figure, until the object has been firmly grasped and no more slip occurs.

SUMMARY

Tactile, or touch, sensing is required especially when intelligent robots must perform delicate assembly operations. Tactile sensing allows a robot to determine object position and orientation, as well as to sense any problems encountered during assembly from the interaction of parts and tools. Furthermore, tactile sensing aids in the identification of unknown objects.

A complete tactile-sensing system must perform three fundamental sensing operations: joint force sensing, touch sensing, and slip sensing. Joint force sensing involves the sensing of hand, wrist, and arm joints. Touch sensing involves sensing the pressures applied to various points on the hand, or gripper, surface. Slip sensing requires the robot to sense any movement of the object while it is being grasped.

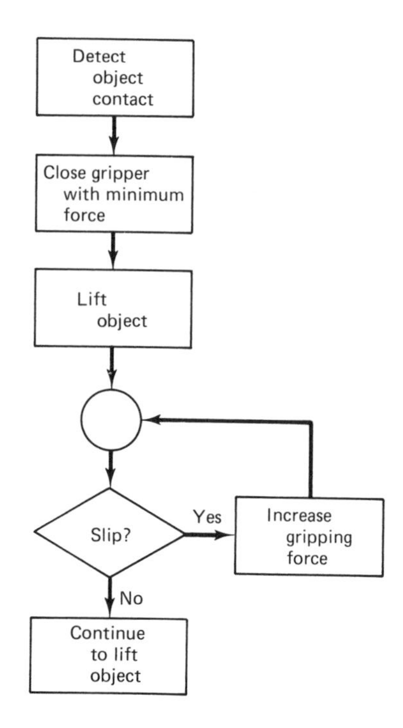

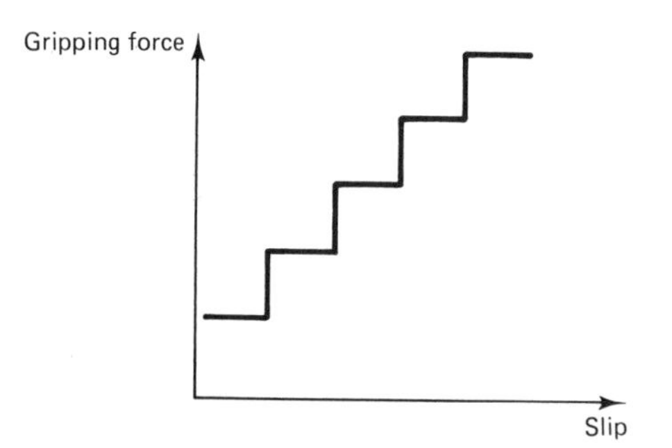

Figure 8-30 Flowchart summary of slip-sensing software.

Arm joint forces can be sensed without any special sensing devices by measuring the resulting loads to the manipulator control systems. The six fundamental wrist joint forces (lift, sweep, reach, yaw, pitch, and roll) are usually sensed using various strain gauge arrangements in the robot wrist assembly. A strain gauge is a force-sensing element whose resistance changes in proportion to the amount of force applied to the element.

Touch-sensing elements should be configured in an array that is skin-like, sensitive, responsive, smart, and has low hysteresis. There are two general categories of touch sensors: binary and analog. A binary touch sensor produces a two-state output, while an analog touch sensor generates a continuous output that is proportional to an applied force. Binary touch sensors include limit switches, whisker sensors, and binary switch arrays. Analog touch sensors include conductive rubber, conductive foam, semiconductor pressure sensors, and pressure-sensitive paint.

Slip sensing is required for a robot to create the optimum amount of grasping force applied to an object. This prevents damage to the object and allows the object to be lifted without being dropped. Two methods used to detect slip are (1) to sense any vibration between the object and gripper and (2) to sense any movement, or displacement, between the object and gripper. In both methods, the idea is to increase the gripping force in steps as slip is detected until the object has been firmly grasped and no more slip occurs.

PROBLEMS

8-1 Name the three fundamental sensing operations of a tactile system.

8-2 List three ways to measure forces acting on a robot manipulator.

8-3 How can arm joint forces in an electric manipulator control system be determined without using any separate force sensors?

8-4 What are the six fundamental forces acting on a robot wrist assembly?

8-5 A strain gauge sensing element is 5 cm long. A force applied to the element causes it to compress by 0.2 mm. Calculate the resulting strain of the sense element.

8-6 What is the resistance of the strained element in Problem 8-5 if its unstrained resistance is 2500 Ω?

8-7 Calculate the gauge factor for the strain gauge used in Problems 8-5 and 8-6.

8-8 The same force (Problem 8-5) is applied to a semiconductor strain gauge that has a GF of 100 and an unstrained resistance of 500 Ω. How much does the resistance of the semiconductor element change as a result of this force?

8-9 Suppose four resistive strain gauges (Problems 8-5, 8-6 and 8-7) are used to construct a gripping-force sensing module like the one shown in Figure 8-7. The strain gauges are connected into a Wheatstone bridge that uses a reference voltage of 12 V. Calculate the bridge output resulting from a gripping force strain of 4000 micros.

8-10 What is the bridge output in Problem 8-9 if the resistive strain gauges are replaced with semiconductor strain gauges from Problem 8-8?

8-11 List the ideal properties of a touch sensor.

8-12 The two broad categories of touch sensors are ________________ and ________________.

8-13 A tactile sensor that can be used to probe objects without disturbing them is the ________________.

8-14 Explain how slip can be detected using a binary tactile array.

8-15 List at least three analog tactile-sensing elements.

8-16 Explain, in general, how conductive analog tactile-sensing arrays are decoded.

8-17 Describe the construction of a typical semiconductor pressure sensor.

8-18 Why is slip sensing important in a tactile system?

8-19 Name two general ways to provide slip sensing in a tactile system.

8-20 A roller-type slip sensor is embedded with 12 permanent magnets spaced 0.25 in. apart, and a Hall-effect device is used to detect the roller movement. How far has an object slipped if the Hall-effect device generates 10 pulses in 0.5 s?

8-21 What is the slip velocity of the object in Problem 8-20?

8-22 Describe the general software required to support slip sensing.

References

Artificial Intelligence

Barr, Avron, and Edward A. Feigenbaum, eds., *The Handbook of Artificial Intelligence, Volume I.* Stanford, CA: HeurisTech Press, and William Kaufmann, Inc., Los Altos, CA, 1981.

Graham, Neill, *Artificial Intelligence.* Blue Ridge Summit, PA: TAB Books, 1979.

Lewis, J. W., and F. S. Lynch, "GETREE: A Knowledge Management Tool," *IEEE Proceedings: Trends & Applications 1983.* Silver Spring, MD: IEEE Computer Society Press, 1983, pp. 273–277.

Raphael, Bertram, *The Thinking Computer.* San Francisco: W. H. Freeman, 1976.

Stolfo, Salvadore, J., "Knowledge Engineering: Theory and Practice," *IEEE Proceedings: Trends & Applications 1983.* Silver Spring, MD: IEEE Computer Society Press, 1983, pp. 97–104.

Winston, Patrick H., *Artificial Intelligence.* Reading, MA: Addison-Wesley, 1979.

Speech Synthesis

Ciarcia, Steve, "Build the Microvox Text-to-Speech Synthesizer," *BYTE Magazine*, October 1982, pp. 40–64.

Ciarcia, Steve, "Use ADPCM for Highly Intelligible Speech Synthesis," *BYTE Magazine*, June 1983, pp. 35–48.

Heath/Zenith Educational Products, *Robotics and Industrial Electronics.* Benton Harbor, MI: The Heath/Zenith Co., Inc., 1983.

Hoot, John E., "Voice Lab," *BYTE Magazine*, July 1983, pp. 186–206.

Sclater, Neil, *Introduction to Electronic Speech Synthesis.* Indianapolis, IN: Howard W. Sams, 1983.

Witten, Ian H., *Principles of Computer Speech.* New York: Academic Press, 1982.

Speech Recognition

Barr, Avron, and Edward A. Feigenbaum, eds., *The Handbook of Artificial Intelligence, Volume I.* Stanford, CA: HeurisTech Press, and William Kaufmann, Inc., Los Altos, CA, 1981.

Bronson, Edward C., and Leah Siegel, "Distributed Processing for Speech Understanding," *IEEE Proceedings: Trends & Applications 1983.* Silver Spring, MD: IEEE Computer Society Press, 1983, pp. 126–132.

Hutchins, Michael W., and Lee K. Dusek, "Advanced ICs Spawn Practical Speech Recognition," *Computer Design*, May 1984, pp. 133–139.

White, George M., "Speech Recognition: An Idea Whose Time Is Coming," *BYTE Magazine*, January 1984, pp. 213–222.

Zue, Victor W., and Daniel P. Huttenlocher, "Computer Recognition of Isolated Words from Large Vocabularies," *IEEE Proceedings: Trends & Applications 1983.* Silver Spring, MD: IEEE Computer Society Press, 1983, pp. 121–125.

Vision

Ballard, Dana H., and Christopher M. Brown, *Computer Vision.* Englewood Cliffs, NJ: Prentice-Hall, Inc., 1982.

Cohen, Paul R., and Edward A. Feigenbaum, *The Handbook of Artificial Intelligence, Vol. 3.* Stanford, CA: HeurisTech Press, and William Kaufmann, Inc., Los Altos, CA, 1982.

Nevatia, Ramakant, *Machine Perception*, Englewood Cliffs, NJ: Prentice-Hall, Inc., 1982.

West, James K., "Machine Vision in the Real World of Manufacturing," *Computer Design*, April 1983, pp. 89–96.

Winston, Patrick H., *Artificial Intelligence.* Reading, MA: Addison-Wesley, 1979.

Range Finding and Navigation

Ahuja, N., et al., "Three Dimensional Robot Vision," *Industrial Applications of Machine Vision.* Silver Spring, MD: IEEE Computer Society Press, 1982, pp. 206–213.

Ballard, Dana H., and Christopher M. Brown, *Computer Vision.* Englewood Cliffs, NJ: Prentice-Hall, Inc.

Cohen, Paul R., and Edward A. Feigenbaum, *The Handbook of Artificial Intelligence, Vol. 3.* Stanford, CA: HeurisTech Press, and William Kaufmann, Inc., Los Altos, CA, 1982.

Hall, E. L., et al., "Curved Surface Measurement and Recognition for Robot Vision," *Industrial Application of Machine Vision.* Silver Spring, MD: IEEE Computer Society Press, 1982, pp. 187–199.

Nevatia, Ramakant, *Machine Perception.* Englewood Cliffs, NJ: Prentice-Hall, 1982.

Staugaard, Andrew C., *Microprocessor Applications.* Benton Harbor, MI: The Heath/Zenith Co., 1983.

Tactile Sensing

Harmon, Leon D., "Touch-Sensing Technology: A Review," *SME Technical Report*, MSR80-03, 1980.

Masuda, Ryosuke, et al., "Slip Sensor of Industrial Robot and Its Application," *Electrical Engineering in Japan*, vol. 96, no. 5, 1976, pp. 129–136.

Purbrick, John A., "A Force Transducer Employing Conductive Silicone Rubber," Artificial Intelligence Laboratory, Massachusetts Institute of Technology, Cambridge, MA, 1981.

Rosen, Charles A., and David Nitzan, "Use of Sensors in Programmable Automation," *Computer*, December 1977, pp. 13–23.

Snyder, W. E., and J. St. Clair, "Conductive Elastomers as Sensor for Industrial Parts Handling Equipment," *IEEE Transaction on Instruments and Measure*, vol. IM-27, no. 1, March 1978, pp. 94–99.

Ueda, Minoru, et al., "Tactile Sensors for an Industrial Robot to Detect Slip," source unknown.

Wang, S. S. M., and P. M. Will, "Sensors for Computer Controlled Mechanical Assembly," *Industrial Robot*, March 1978, pp. 9–18.

Answers to Chapter Problems

Chapter 1

1-1 The two common denominators of an intelligent robot are sensory perception and intelligence.

1-2 The five categories of sensory perception that can be incorporated into a robot are vision, tactile sensing, range finding and navigation, speech synthesis, and speech recognition.

1-3 A vision system can be used to build the knowledge base of a robot by acquiring information about its surroundings, storing this information in memory, and using the knowledge gained for future problem-solving tasks.

1-4 The science of computer vision is called visual machine perception.

1-5 A typical robot light-detection system utilizes an LDR or phototransistor circuit to translate light levels to a proportional analog voltage level. The analog voltage is then converted with an A/D converter and read by the computer (see Figure 1-3).

1-6 A robot vision system that uses a TV camera as the visual sensing device scans a scene with the TV camera(s). The TV camera output is converted to a digital code by an A/D converter and stored in memory as a digital image (see Figure 1-4).

1-7 Adding a sense of touch to a robot is called tactile sensing.

1-8 Compliance is the ability of a robot to adjust for external interference forces during an assembly task.

1-9 The three fundamental tasks that must be performed by a computer vision system are image transformation, image analysis, and image understanding.

1-10 The speech-synthesis technique that allows an unlimited vocabulary is called phoneme speech synthesis.

1-11 The two techniques presently employed in speech-recognition systems are isolated-word recognition and connected-speech understanding.

1-12 Speaker-independent systems use the connected-speech understanding speech-recognition technique.

1-13 Artificial intelligence is the common denominator for most of the components required in an intelligent robot since AI techniques must be used with vision systems, tactile sensing, navigation, and speech recognition. Without AI, sensory perception and integration in a robot would be impossible.

1-14 A sensing cell is a single chip VLSI processor that is dedicated to a given sensing task, usually associated with tactile sensing.

1-15 The hardware aspect of knowledge engineering involves the networking of a multiprocessor system. The software component of knowledge engineering involves the application of hierarchical planning and knowledge representation.

1-16 The two categories of knowledge representation are general knowledge and expert knowledge.

1-17

Route	Mileage
A-C-B	6
A-C-D-E-B	8
A-D-E-B	7
A-D-C-B	9

1-18 See Figure P1-18.

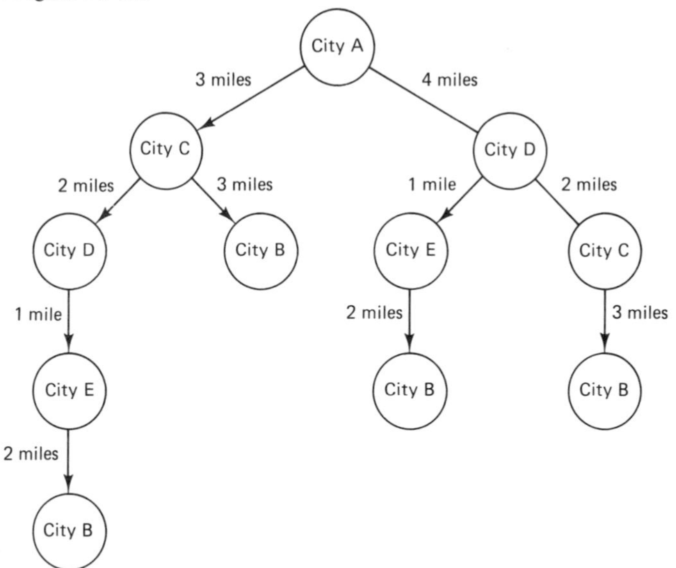

Figure P1-18 Solution to Problem 1-18.

Chapter 2

2-1 The human brain exhibits a digital character since the presence or absence of nerve impulses can be viewed as an on/off, or two-state, operation.

2-2 Synapses are neuron junction points and become excited when the sum of several nerve impulses exceeds a given threshold level. Depending on the threshold level, a synapse can respond like an OR or an AND gate. If the synapse excites when only one nerve impulse is present, it functions like a simple OR gate. If it responds to a threshold level equal to that of all of its input nerve impulses, it functions like an AND gate.

2-3 The human brain is inventive, creative, perceptive, capable of abstract and inductive thinking, forgetful, inaccurate, slow, and becomes fatigued.

2-4 Digital computers are fast, accurate, consistent, inflexible, not creative, and not capable of abstract thinking or inductive reasoning.

2-5 According to Marvin Minsky, artificial intelligence "is the science of making machines do things that would require intelligence if done by men."

2-6 State-space networks divide a problem into states, operators, and goals.

2-7 Problem-reduction representation graphs divide a problem into subproblems.

2-8 Forward reasoning begins with an initial state and progresses through a series of intermediate states to a goal state. Backward reasoning begins with the goal state and progresses through a series of subproblems to primitive problems whose solution solves the goal state problem. Forward reasoning is generally used for state-space searches, while backward reasoning is used for problem-reduction searches.

2-9 The overall problem-solving approach used in AI reduces to representing the problem as states, or models, within the computer, operators to manipulate the states, and a control strategy that applies the operators to produce a problem solution.

2-10 The tree-searching procedure should accomplish three tasks: (1) always find a solution if one exists; (2) always find the best solution; (3) always find the most efficient solution.

2-11 (a) $X \rightarrow A \rightarrow B \rightarrow X$
$X \rightarrow A \rightarrow B \rightarrow D$
$X \rightarrow A \rightarrow B \rightarrow Y$
$X \rightarrow A \rightarrow C \rightarrow E$
$X \rightarrow A \rightarrow C \rightarrow Y$
$X \rightarrow B \rightarrow D$
$X \rightarrow B \rightarrow A \rightarrow X$
$X \rightarrow B \rightarrow Y$
$X \rightarrow B \rightarrow A \rightarrow C \rightarrow E$
$X \rightarrow B \rightarrow A \rightarrow C \rightarrow Y$

(b) See Figure P2-11(b).

2-12 (a) Using breadth-first searching, the minimum number of paths is four. Using depth-first searching, the minimum number of paths is two.

(b) Using breadth-first searching, the maximum number of paths is 12. Using

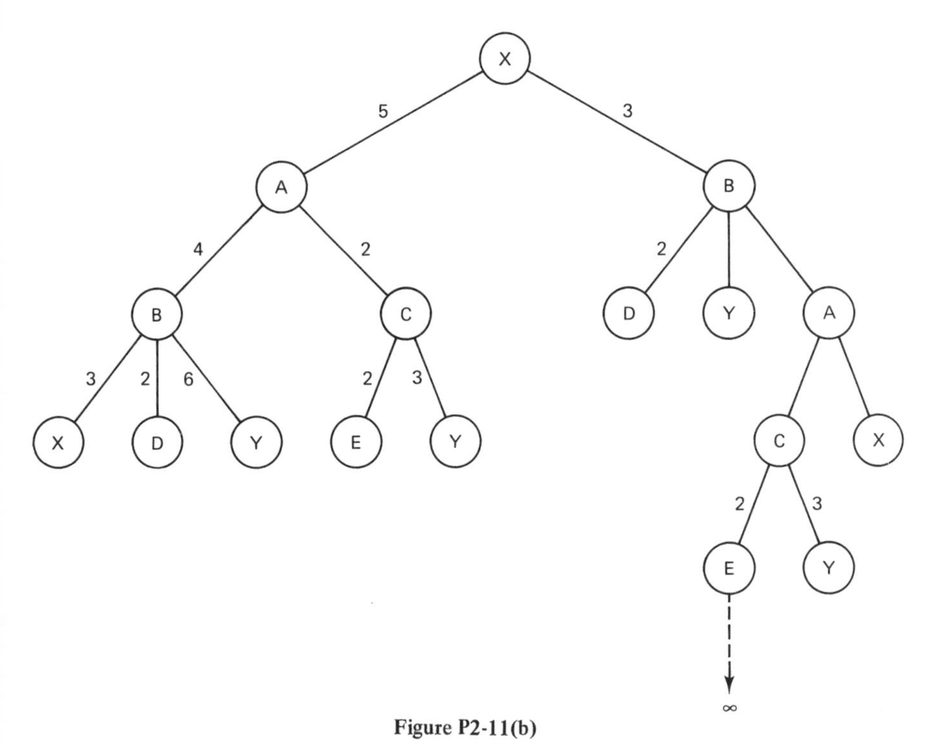

Figure P2-11(b)

depth-first searching, the maximum number of paths is infinite through node E.

2-13 **(a)** The minimum-distance path from point X to point Y is $X \rightarrow B \rightarrow Y$, with a total distance of nine feet.

(b) A progressive deepening search strategy would find the minimum-distance path in this problem most efficiently with minimum risk.

2-14 See Figure P2-14.

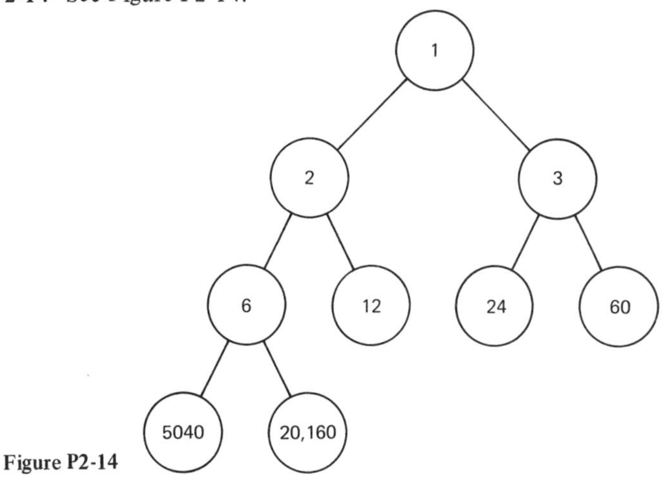

Figure P2-14

2-15 Assume that the discs are labeled *A, B, C, D* from the smallest to the largest disc. Assume also that the pegs are labeled 1, 2, and 3. The object is to move the four discs from peg 1 to peg 3 using the rules given in the text. The four-disc problem can be broken down into three subproblems, one of which is a primitive problem. See Figure P2-15.

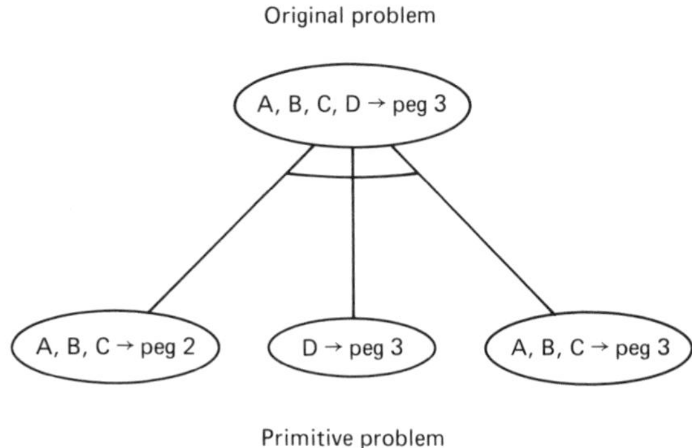

Figure P2-15

2-16 See Figure P2-16.

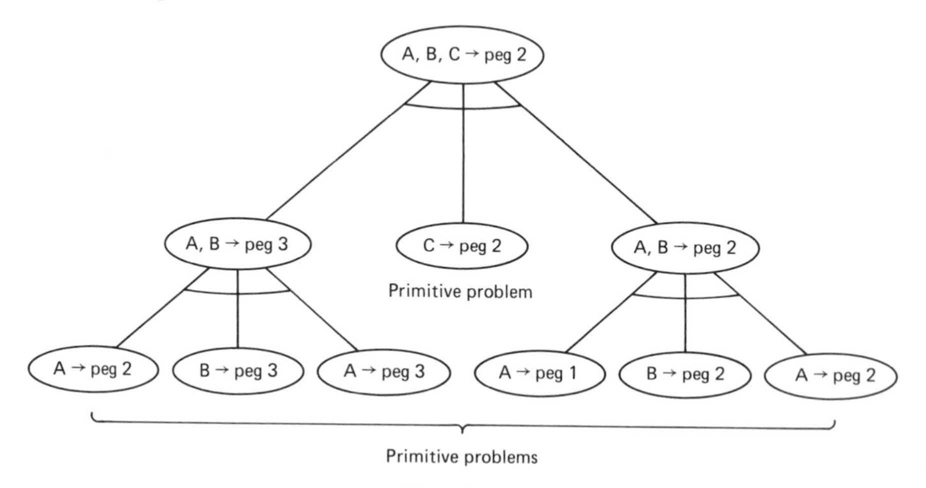

Figure P2-16

2-17 See Figure P2-17.

2-18 The answers to this question depend on your individual search, which depends on a "flip of the coin."

2-19 Let the heuristic evaluation function be the number of tiles, including the space, that are not in their proper goal state position for any given state. Associate this function with the cost required to get to the given state. If you then follow the lowest-cost path, you will always find the most efficient solution path.

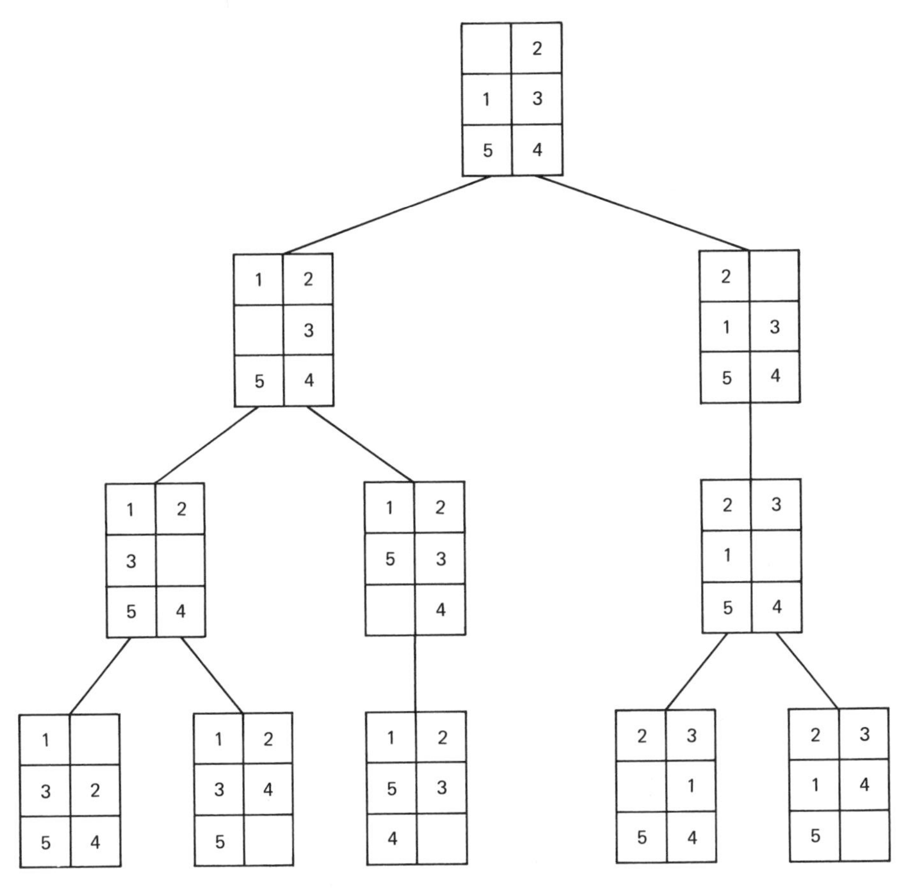

Figure P2-17

Chapter 3

3-1 Propositional logic represents knowledge using statements, called propositions, that are either true or false. Predicate logic extends the idea of propositional logic by using variables, quantifiers, and functions within the logic statement. Propositional logic is a subset of predicate logic.

3-2 The truth table for both logic expressions is

A	B	$\overline{A \vee B}$, $\bar{A} \wedge \bar{B}$
T	T	F
T	F	F
F	T	F
F	F	T

Since the truth table is the same for both expressions, the equivalency is proved.

3-3 The truth table for $A\bar{B} \vee \bar{A}B$ is

A	B	$A\bar{B} \vee \bar{A}B$
T	T	F
T	F	T
F	T	T
F	F	F

The equivalency is proved since this truth table is the same as the exclusive-OR connective defined in the problem.

3-4 Modus ponens.

3-5 The statement can be disproved by observing the implies truth table. Notice that, if B is true and $A \rightarrow B$ is true, A can be either true or false. Thus, you cannot say that A will always be true using the given premises. The statement in Problem 3-5 is sometimes called the modus moron rule, for obvious reasons.

3-6 Resolution.

3-7 **(a)** BOY (DAVID)
(b) BOY (DAVID) $\wedge$ SICK (DAVID) or SICK (BOY (DAVID))
(c) LESS (x, y)
(d) $(\forall x)$[MAN$(x) \rightarrow$ MORTAL(x)]
(e) PART $(x) \wedge$ BIN $(x) \rightarrow (\forall x)$[RED$(x) \wedge$ CIRCULAR(x)]

3-8 Step 1: The statement can be reduced to predicate notation as follows:
Premise 1: ROBOT (BOB)
Premise 2: $(\exists x)$[ROBOT $(x) \rightarrow$ HAS MANIPULATOR (x)]
Conclusion: HAS MANIPULATOR (BOB)
Step 2: Negate the conclusion: $\overline{\text{HAS}}$ MANIPULATOR (BOB)
Step 3: Reduce the premises to clauses:
Premise 1: ROBOT (BOB) is already a clause.
Premise 2: $(\forall x)$[$\overline{\text{ROBOT}}(x) \vee$ HAS MANIPULATOR(x)]
Step 4: Use unification to substitute BOB for x in premise 2.
$\overline{\text{ROBOT}}$(BOB) $\vee$ HAS MANIPULATOR(BOB)
Step 5: You now have
Premise 1: ROBOT(BOB)
Unified Premise 2: $\overline{\text{ROBOT}}$(BOB) $\vee$ HAS MANIPULATOR(BOB)
Negated conclusion: $\overline{\text{HAS}}$ MANIPULATOR(BOB)
Applying the resolution principle to premises 1 and 2, you get
Premises 1 and 2: HAS MANIPULATOR(BOB)
Negated conclusion: $\overline{\text{HAS}}$ MANIPULATOR(BOB)
Applying the resolution principle again to the negated conclusion, you get the empty set. Thus, the negated conclusion produced a contradiction, and the conclusion "Bob has a manipulator" must be true by the resolution principle.

3-9 IF an animal has hair and gives milk, THEN it is a mammal.
IF an animal eats meat, THEN it is a carnivore.
IF an animal is a mammal AND a carnivore AND brownish-yellow AND has stripes, THEN it is a tiger.
IF an animal has claws AND forward eyes AND pointed teeth, THEN it is a carnivore.

3-10 Using a backward-chaining reasoning strategy, the production rules in Problem 3-9 would be activated in the following order:

1. IF an animal is a mammal AND a carnivore AND brownish-yellow AND has stripes, THEN it is a tiger.
2. IF an animal has hair AND gives milk, THEN it is a mammal.
3. IF an animal eats meat, THEN it is a carnivore.
4. IF an animal has claws AND forward eyes AND pointed teeth, THEN it is a carnivore.

3-11 The production rules are executed in reverse, from conclusion to premise. The chaining ends successfully when enough conditional elements are found to support the final conclusion. The chaining ends unsuccessfully when not enough facts exist to support the conclusion or if some required fact conflicts with a known fact.

3-12 The three major parts of a production system are the context, rule base, and interpreter.

3-13 The three major phases of a production-system operating cycle are matching, conflict resolution, and action.

3-14 The major reason that production systems are popular is that human knowledge is easily represented using IF/THEN statements. In addition, backward reasoning with production systems allows verification and explanation of results and conclusions.

3-15 How was a given conclusion reached?
How was a given conclusion not reached?
Why was a given fact used in the reasoning process?
Why was a given fact not used in the reasoning process?

3-16 See Figure P3-16.

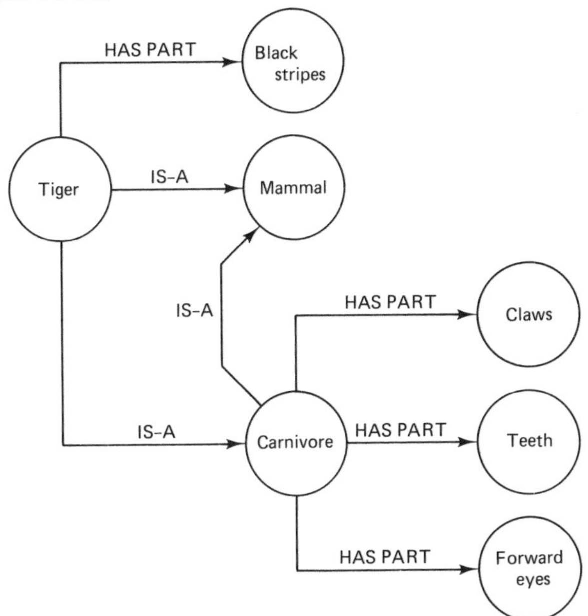

Figure P3-16

3-17 See Figure P3-17.

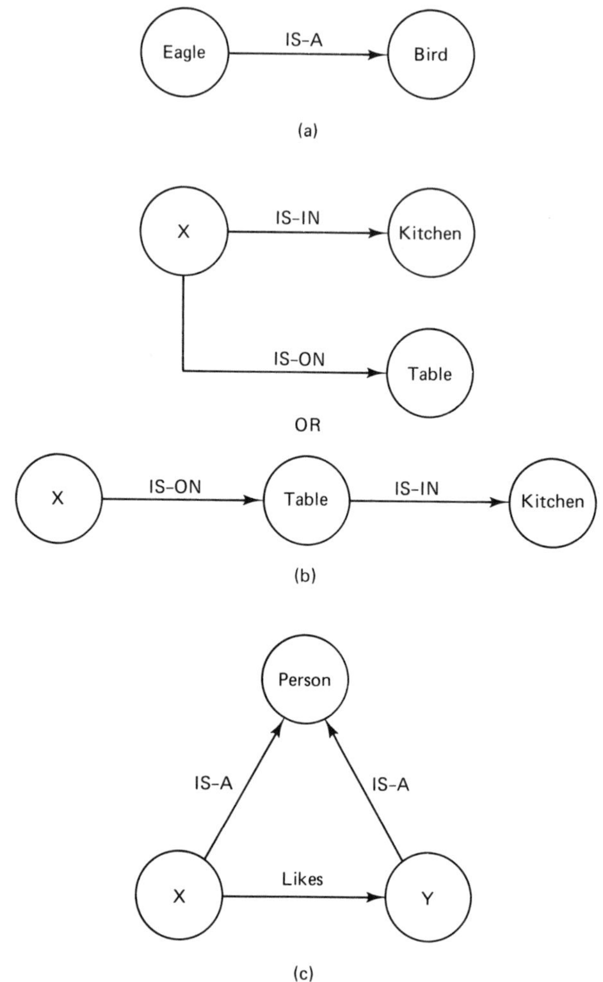

Figure P3-17

3-18 See Figure P3-18.

3-19 See Figure P3-19.

3-20 See Figure P3-20.

3-21 An expert system is a computer system that exploits the specialized knowledge of human experts to achieve high performance in a specific problem area.

3-22 Production systems.

3-23 In the knowledge-acquisition mode, a knowledge engineer works with a human expert to formalize the human expert's knowledge and integrate that knowledge into the system data base. The expert system is in the consultation mode when it is performing its function as a computerized consultant. The explana-

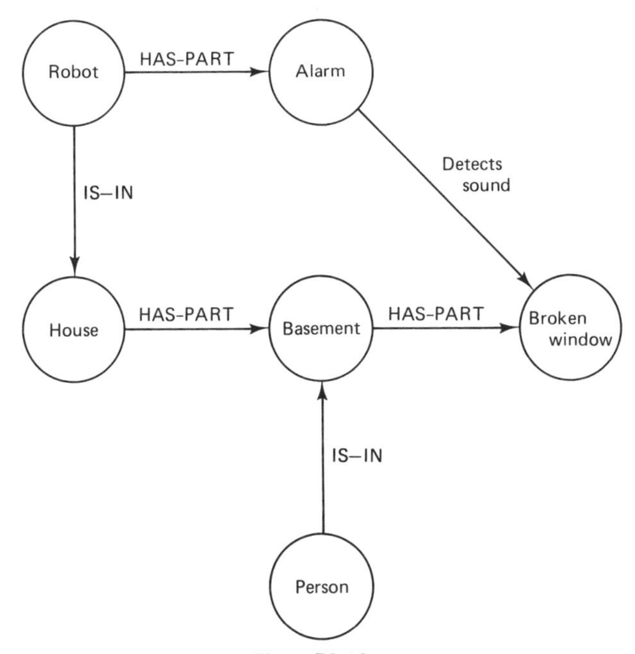

Figure P3-18

tion mode allows an expert system to explain its conclusions and reasoning process.

3-24 Why was a given fact involved?
Why was a given fact not used?
How was a given conclusion reached?
How was it that another conclusion was not reached?

3-25 See Table 3-1.

Chapter 4

4-1 Voiced sounds are created by the vibrating action of the vocal cords as air is forced through them. The sounds are then frequency modulated by the vocal cavity to produce the various voiced sounds used in speech.

4-2 A frequency range of from 300 to 3300 Hz, with a volume ratio of 1000 : 1 will produce acceptable speech.

4-3 A sound spectrograph is a frequency analyzer used by engineers to study speech. It provides a graphical display of sound frequency and intensity.

4-4 Plosives.

4-5 Voiced fricatives utilize the vocal cords, while unvoiced fricatives do not. Voiced fricatives include the following sounds: *z, zh, v, dh.* Unvoiced fricatives include *s, sh, f, th.*

4-6 Prosodic speech features convey meaning to words and phrases. These features include amplitude, pitch, and timing.

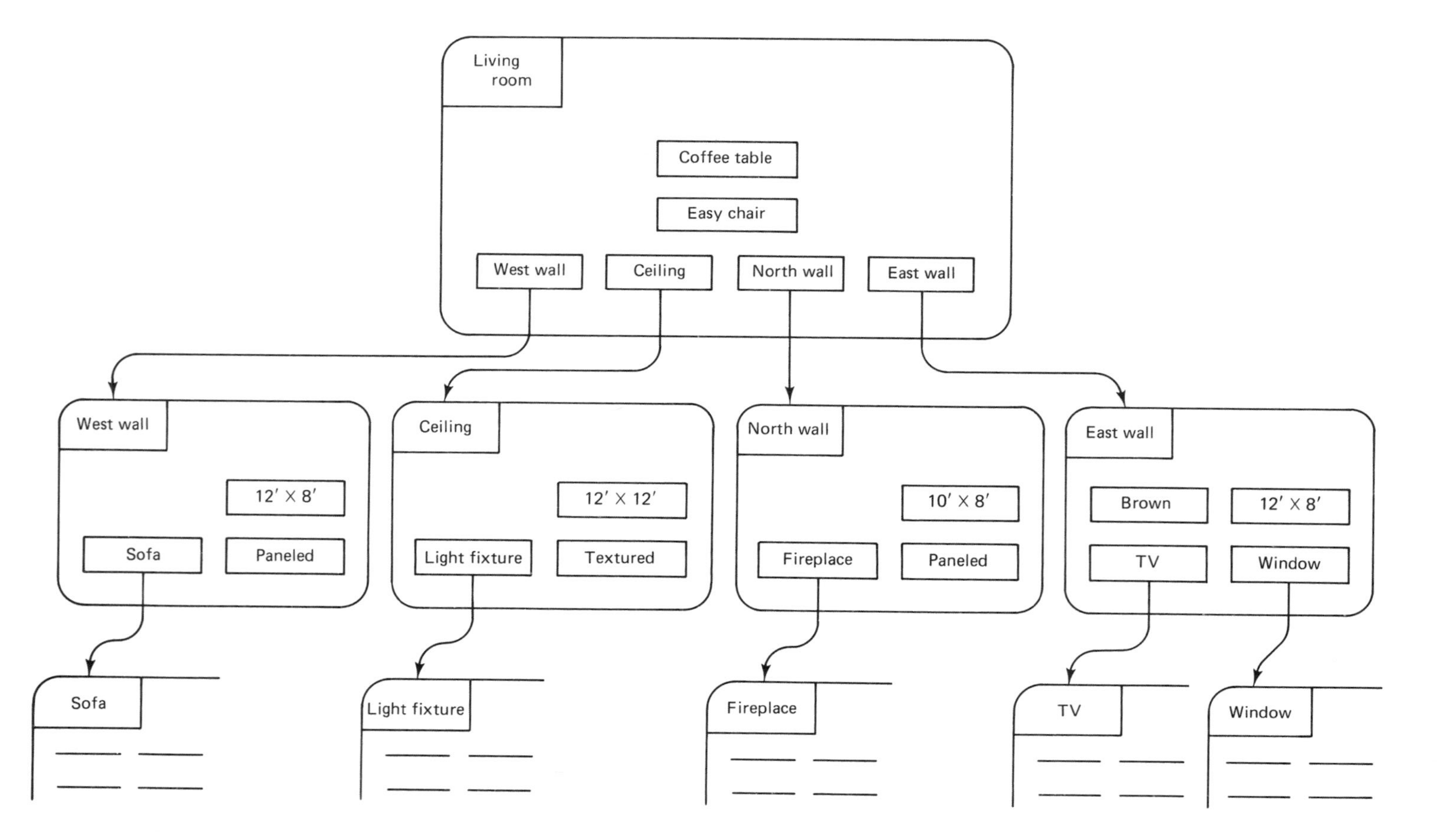

Living room
Coffee table
Easy chair
West wall
Ceiling
North wall
East wall
West wall
12′ × 8′
Sofa
Paneled
Ceiling
12′ × 12′
Light fixture
Textured
North wall
10′ × 8′
Fireplace
Paneled
East wall
Brown
12′ × 8′
TV
Window
Sofa
Light fixture
Fireplace
TV
Window

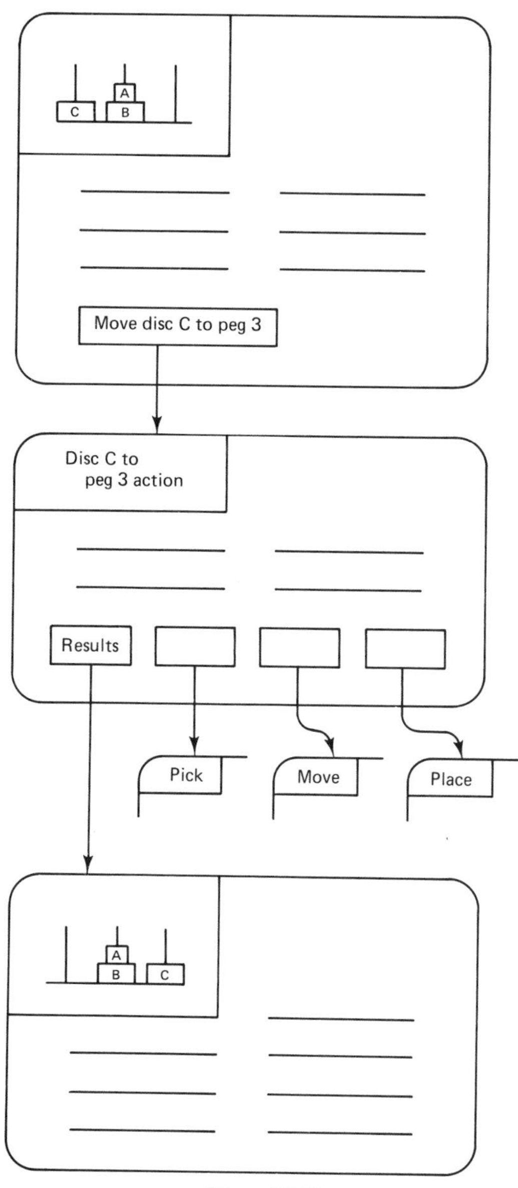

Figure P3-20

4-7 Natural speech analysis/synthesis involves the recording and subsequent playback of human speech. Artificial constructive/synthesis creates speech by stringing together the various speech sounds that are required to produce a given speech segment.

4-8 Waveform digitization.

4-9 With PCM, the analog speech waveform is sampled and converted directly into a multibit digital code by an A/D converter. The code is stored and subsequently recalled for playback. With delta modulation, only a single bit is stored for each sample. This bit, 1 or 0, represents a greater-than or less-than condition, respectively, as compared to the previous sample. An integrator is then used on the output to convert the stored bit stream to an analog signal.

Differential pulse-code modulation (DPCM) stores a multibit difference value. A bipolar D/A converter is used for playback to convert the successive difference values to an analog wave form.

Adaptive differential pulse-code modulation (ADPCM) stores a difference value that has been mathematically adjusted according to the slope of the input waveform. A bipolar D/A converter is used to convert the stored digital code to analog for playback.

4-10 (a) Data rate = 8 bits × 5000 samples/second
= 40,000 bits per second (bps)

(b) Memory in bytes = (10 s × 40,000 bps) ÷ 8 bits/byte
= 50,000 bytes

(c) Memory in bytes = (10 s × 5000 bps) ÷ 8 bits/byte
= 6250 bytes

Note: This assumes 5000 samples per second. However, delta modulation requires 32,000 samples per second, thereby requiring 40,000 bytes of memory.

(d) Memory in bytes = (10 s × 6 bits × 5000 samples/second)
÷ 8 bits/byte
= 37,500 bytes

(e) Memory in bytes = (10 s × 3 bits × 5000 samples/second)
÷ 8 bits/byte
= 12,500 bytes

4-11 The excitation source generates the voiced and unvoiced sounds, selects between the two sound sources, amplifies the sound, and passes the signal to a multistage digital filter. The digital filter modulates the excitation signal to produce the desired formant frequency spectrum. The modulated signal is then passed to a D/A converter and analog filter for conversion to an analog speech signal.

4-12 The LPC code controls the following circuit functions:

- Pitch of the voiced sounds
- Selection between voiced and unvoiced sounds
- Amplitude of the excitation signal
- Control of the digital filter by providing filter coefficients

4-13 Direct phoneme synthesis and text-to-speech phoneme synthesis.

4-14 The steps required for phoneme synthesis are as follows:

- Determine the phoneme symbol string required for the phrase
- Provide pauses

- Provide intonation
- Convert the symbol string to a code string
- Execute the string, listen, and modify accordingly

4-15

THE:	THV, UH3, UH3, UH3
UNITED:	Y1, IU, U1, N, AH1, Y1, T, I3, D
STATES:	S, T, A1, AY, Y1, T, S, S
OF:	AH, F
AMERICA:	UH2, M, EH2, EH2, R, EH3, K, UH1

4-16 PA1, THV, UH3, UH3, UH3, PA1, PA1, PA1, Y1, IU, U1, N, AH1, Y1, T, I3, D,

THE UNITED

PA1, PA1, PA1, S, T, A1, AY, Y1, T, S, S, PA1, AH, F, PA1,

STATES OF

UH2, M, EH2, EH2, R, EH3, K, UH1, PA1

AMERICA

4-17 See Figure P4-17.

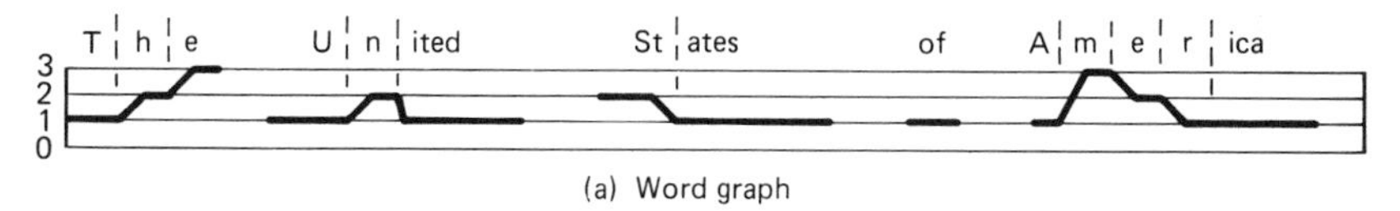

(a) Word graph

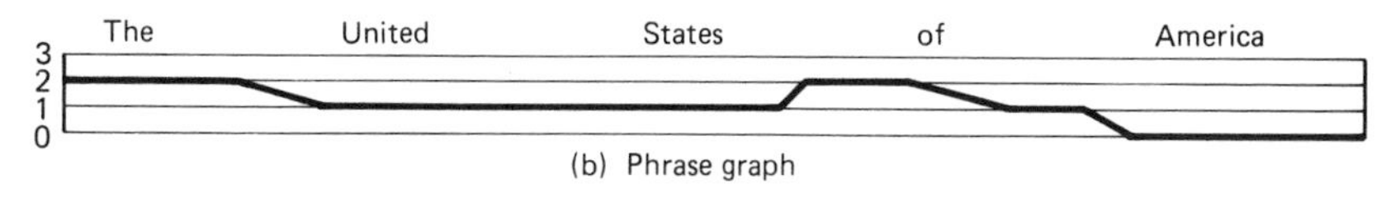

(b) Phrase graph

Figure P4-17 Intonation bar graphs.

4-18 PA1, 1/THV, 1/UH3, 2/UH3, 3/UH3, PA1, PA1, PA1,

THE

1/Y1, 1/IU, 1/U1, 2/N, 1/AH1, 1/Y1, 1/T, 1/I3, 1/D, PA1, PA1, PA1,

UNITED

2/S, 2/T, 1/A1, 1/AY, 1Y1, 1/T, 1/S, 1/S, PA1, 1/AH, 1/F, PA1,

STATES OF

1/UH2, 3/M, 1/EH2, 1/EH2, 1/R, 1/EH3, 1/K, 1/UH1, PA1

AMERICA

4-19 The three types of text-to-speech conversion are word lookup, morpheme lookup converts individual input text letters into phonemes and allophones via words for conversion to phonemes using a word dictionary. Morpheme lookup divides the input text into morphs (sounds that have meaning) for conversion to phonemes using a dictionary of about 8000 morphs. Phoneme lookup converts individual input text letters into phonemes and allophones via a phoneme dictionary.

4-20 "FICTIONAL": IF (F) THEN F
IF (I)$+:# THEN I
IF (C) THEN K
IF (TI)O THEN SH
IF I(ON) THEN UH2,N
IF #:(AL)! THEN UH,L

Chapter 5

5-1 The following criteria must be considered when designing a speech recognition or understanding system:

- Speaker dependence/independence
- Vocabulary
- Isolated word versus connected speech
- Speech recognition versus speech understanding

5-2 Industrial applications are particularly suited for speaker-dependent systems because many such applications require a limited vocabulary. Simple control words such as *yes, no, start, stop,* and the numerals 0 through 9 are sufficient for many industrial applications.

5-3 It is difficult to use isolated word recognition for connected speech because of the following:

- It is hard to detect the precise word boundaries in a connected-speech signal.
- Word signals in a connected phrase do not closely resemble the same individual word signals.
- Individual word pronunciations differ as a result of other words and punctuation within a phrase.

5-4 Speech recognition involves the use of some signal-matching process to detect the presence of a given utterance. Speech understanding involves the interpretation of the speech signals using knowledge about speech. In general, speech-understanding systems employ the principles of AI, while speech-recognition systems do not.

5-5 Using an LPC coding format, a 50-word vocabulary would require about 50 X 100 bytes, or 5000 bytes of template memory.

5-6 In a typical speech recognition system, the major hardware consists of a microphone, amplifier, bandpass filter, sample/hold circuit, and an A/D converter. Of course, an adequate CPU and memory are required in addition to these components.

5-7 The two methods employed for template matching are direct matrix comparison and dynamic programming.

5-8 Dynamic programming is superior to direct matrix comparison since it forces the best possible fit between the reference and unknown templates. As a result, end-point detection of word boundaries is not so critical.

5-9 A typical speech chip consists of an LPC speech synthesizer and LPC analyzer/code generator for speech recognition. In the speech-synthesis mode, the chip accepts sequential LPC code from the microprocessor and generates an analog speech signal to an external audio amplifier, filter, and speaker for sound reproduction. In the speech-recognition mode, the speech chip analyzes the speech signal and generates LPC code to the microprocessor for storage as memory templates.

5-10 The six broad phonetic categories used in a typical phonetic analysis system are (1) pure voiced vowel, (2) nasal, (3) voiced fricative, (4) unvoiced fricative, (5) plosive, and (6) glide.

5-11

Word	Dictionary Pronunciation	Phonetic Sound Sequence
Pick	pik	P-V-P
Place	plās	P-G-V-UF
Left	left	G-V-UF-P
Right	rīt	G-V-P
Up	up	V-P
Down	doun	P-V-N

5-12 See Figure P5-12.

5-13 The seven fundamental knowledge sources used in speech understanding are (1) pragmatic knowledge, (2) semantic knowledge, (3) syntactic knowledge, (4) lexical knowledge, (5) prosodic knowledge, (6) phonological knowledge, and (7) phonetic knowledge.

5-14 Higher-level knowledge sources are important to speech understanding because they provide expectational and constraining knowledge to the understanding task.

5-15 Knowledge can be organized in a speech-understanding system using a hierarchical, blackboard, or compiled-network representation. The hierarchical knowledge model organizes the knowledge sources in series, according to their respective levels. The blackboard model organizes the knowledge sources in parallel by means of a common data base called a blackboard. The compiled-network model organizes knowledge into one large search-tree type of network, called a speech tree.

5-16 See Figure P5-16.

5-17 A control strategy in a speech-understanding system must decide where to start, what to do next, and when to conclude the interpretation process.

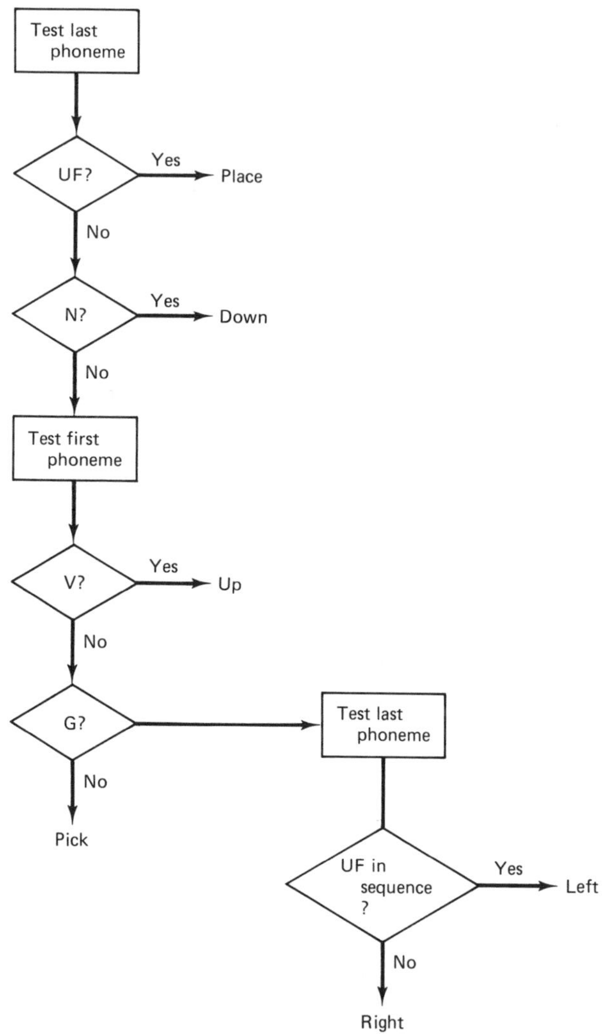

Figure P5-12 Decision algorithm flowchart.

5-18 The island-driving control strategy creates islands of identifiable words within an utterance and generates expectation branches of possible word extensions of the islands.

Chapter 6

6-1 The three main tasks that must be performed by an intelligent vision system are image transformation, image analysis, and image understanding.

6-2 In a TV camera vidicon, an electron beam scans a photoconductive image plate that generates a video signal whose amplitude is proportional to light intensity across the scan.

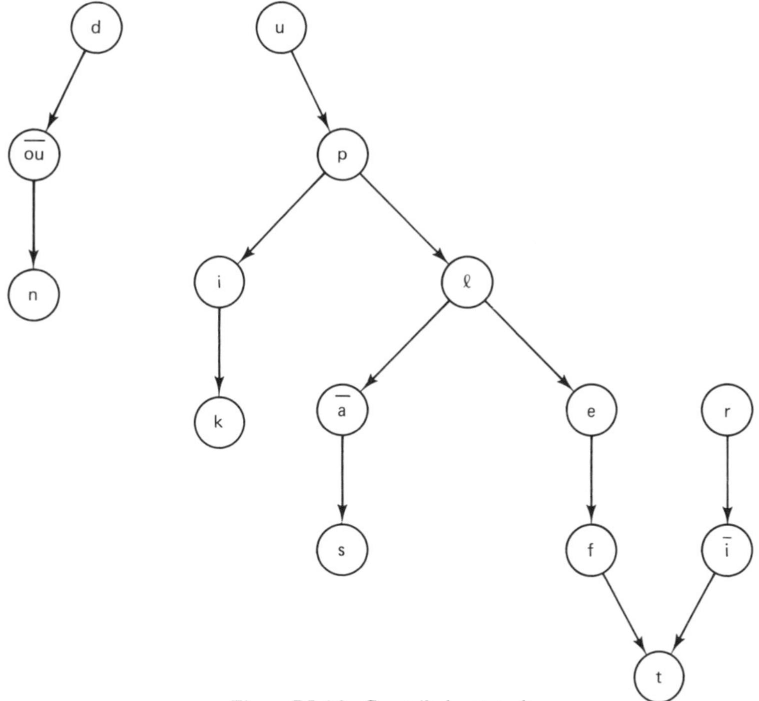

Figure P5-16 Compiled network.

6-3 A photodiode must be reversed biased when operating as an imaging device.

6-4 Imaging devices that employ a rectangular photodiode array are called diode-matrix cameras.

6-5 Light photons striking the array create a charge in the depletion region of integrated MOS devices. The individual charges are coupled in serial fashion to the output, where they are amplified, converted to digital, and stored in a frame buffer.

6-6 See Figure P6-6.

6-7 Ensemble averaging cancels out noise by averaging gray-scale pixel values between multiple picture matrices. Local averaging removes noise by replacing each gray-scale value in a single matrix with the average of several values that occur in a surrounding pixel window.

6-8 See Figure P6-8.

6-9 The Roberts cross operator approximates the first derivative, or gradient, of a digital image.

6-10 See Figure P6-10.

6-11 Three techniques that are used for finding lines from a binary matrix are tracking, model matching, and template matching.

6-12 Region splitting segments an image by dividing a scene into regions that have the same relative feature values. Region growing segments the image by merging pixels that have the same relative feature values.

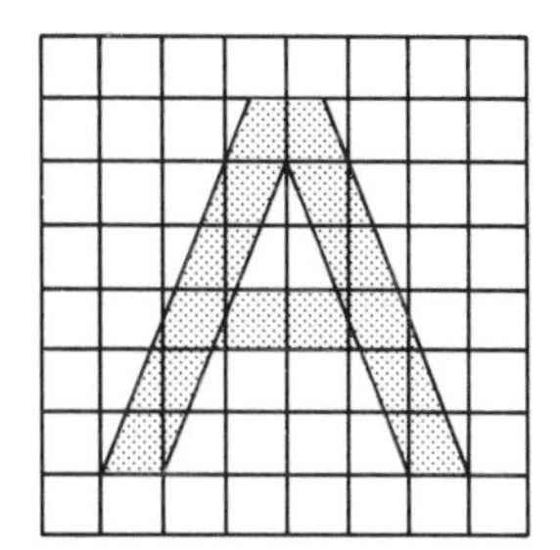

0	0	0	0	0	0	0	0
0	0	0	10	10	0	0	0
0	0	2	9	9	2	0	0
0	0	8	8	8	8	0	0
0	0	16	16	16	16	0	0
0	7	9	0	0	9	7	0
0	12	3	0	0	3	12	0
0	0	0	0	0	0	0	0

Figure P6-6

X	X	X	X	X	X	X	X
X	0	2	4	4	2	0	X
X	1	4	7	7	4	1	X
X	3	7	10	10	7	3	X
X	4	5	9	9	5	4	X
X	5	5	7	7	5	5	X
X	3	3	1	1	3	3	X
X	X	X	X	X	X	X	X

Note: Smoothed values have been rounded-off to nearest whole number.

Figure P6-8

0	0	0	1	1	0	0	0
0	0	1	1	1	1	0	0
0	0	1	1	1	1	0	0
0	1	1	1	1	1	1	0
0	1	1	1	1	1	1	0
0	0	0	0	0	0	0	0
0	0	0	0	0	0	0	0
0	0	0	0	0	0	0	0

Figure P6-10

6-13 A picture tree represents the output of the region-growing process. The leaf nodes are atomic regions, while the root node represents the entire image. Picture trees are useful during image interpretation since the region analysis can be modified as new information is introduced.

6-14 Color information can be converted to gray-scale intensity data using the brightness equation: $Y = 0.3R + 0.59G + 0.11B$, where R, G, and B are the saturation levels of red, green, and blue, respectively.

6-15 (a) Gray-level signal amplitude $= Y = 0.3R + 0.59G + 0.11B$

$$= 0.3(0.5\text{ V}) + 0.59(0.5\text{ V}) + 0.11(0.2\text{ V})$$
$$= 0.467\text{ V}$$

(b) $R\text{-}Y = 0.5\text{ V} - 0.467\text{ V} = 0.033\text{ V}$

$B\text{-}Y = 0.2\text{ V} - 0.467\text{ V} = -0.267\text{ V}$

From Figure 6-28, the color is yellow.

6-16 Texture primitives are repeated in some fashion throughout the pattern.

6-17 See Figure P6-17.

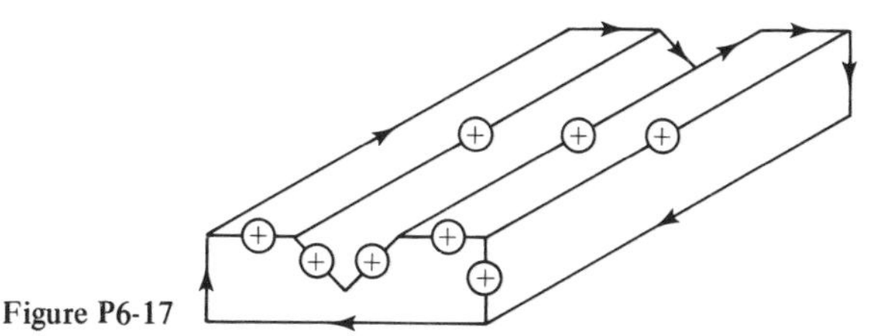

Figure P6-17

6-18 See Figure P6-18.

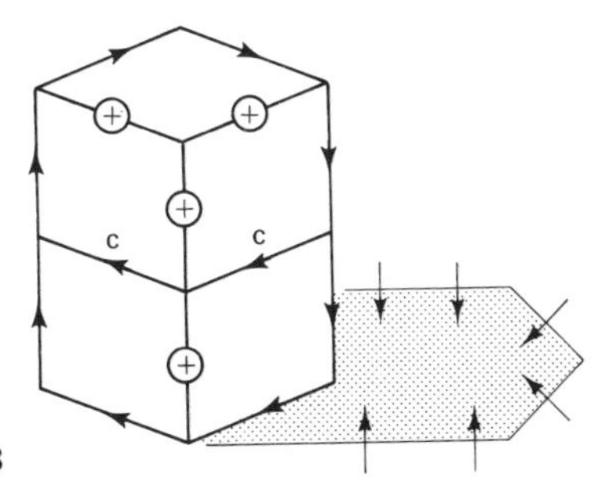

Figure P6-18

6-19 Shadows enhance image understanding, since the existence of a shadow indicates the existence of at least one other object and many times indicates that the object is being supported by another object. Cracks usually indicate the presence of several, but connected, objects.

6-20 Motion allows moving objects to be segmented from a stationary background.

6-21 Nonhuman-type sensory data that can be applied to computer vision include x-ray emissions, ultraviolet emissions, ultrasonic emissions, infrared emissions, and magnetic properties.

6-22 Expectational knowledge in a computer vision system includes expected object characteristics, positional knowledge, and commonsense knowledge.

6-23 Expectational knowledge is important in computer vision, since it is used to guide the image-understanding process.

6-24 See Figure P6-24.

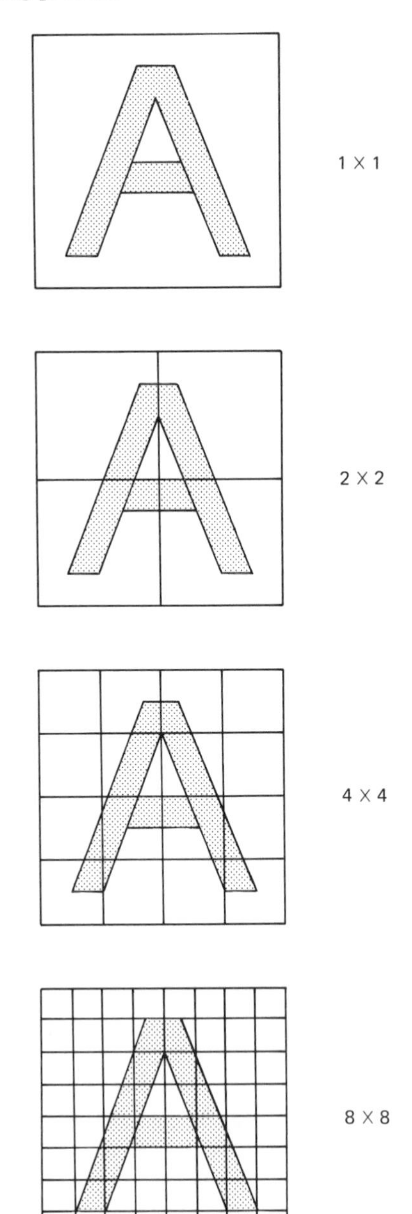

Figure P6-24

6-25 See Figure P6-25.

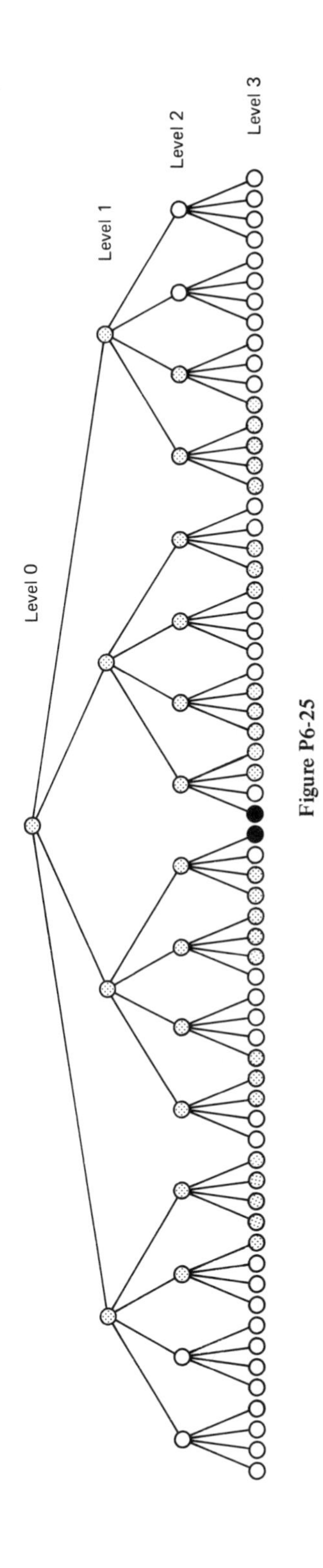

Figure P6-25

6-26 The blackboard model provides independent interaction of both high- and low-level knowledge.

6-27 The primary design considerations in an industrial vision system are cost, speed, accuracy, and reliability.

6-28 The two general applications categories that most industrial vision systems fall into are product inspection and object identification.

6-29 Difficulty level 1: Both the object position and appearance are tightly controlled.

Difficulty level 2: Either the object position or appearance are controlled, not both.

Difficulty level 3: Neither the object position nor appearance are controlled.

6-30 See Figure P6-30.

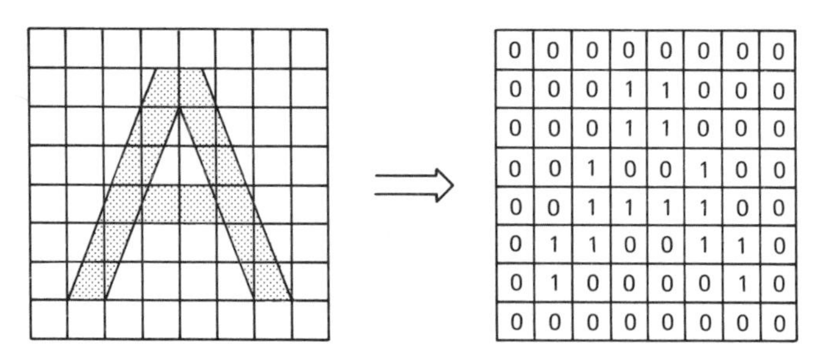

Figure P6-30

Chapter 7

7-1 Passive triangulation employs two stationary imaging devices for stereo imaging. Active triangulation involves the movement of a single imaging device or the light source, not both.

7-2 The disparity between two stereo images is the distance that the pixels are displaced from one image to the next.

7-3

$$\text{Range} = \frac{d\sqrt{f^2 + x_1^2 + x_2^2}}{|x_1 - x_2|}$$

$$d = 20 \text{ cm}$$

$$f = 10 \text{ cm}$$

$$\text{Disparity} = |x_1 - x_2| = 2 \text{ cm}$$

First, $x_1 = -1$ cm and $x_2 = +1$ cm, since any other values result in a zero disparity, or an impossible situation. (Why?) Thus,

$$r = \frac{20 \text{ cm}\sqrt{(10 \text{ cm})^2 + (-1 \text{ cm})^2 + (1 \text{ cm})^2}}{2 \text{ cm}}$$

$$= 101 \text{ cm}$$

7-4 The object point in Problem 7-3 must be located directly between the two cameras in the middle of the scene, since $x_1 = -1$ cm and $x_2 = +1$ cm.

7-5 The range value calculated in Problem 7-3 is from the object point to the center of either camera lens, since the object point is precisely in the middle of the scene.

7-6 If the correspondence between the two image pixels in Problem 7-3 can only be resolved to 0.2 cm, the disparity can range from 1.8 to 2.2 cm. This translates to a range accuracy of ±10 percent. Thus, the actual range of the object in Problem 7-3 could be anywhere from 90.9 to 111.1 cm.

7-7 By searching for similar edge or region features between the two stereo images.

7-8 Types of active triangulation include spot sensing, light striping, and camera motion.

7-9

$$f = 5 \text{ cm}$$

$$d = 30 \text{ cm}$$

$$x = 0.5 \text{ cm}$$

$$\theta_1 = 30^\circ$$

$$\theta_2 = \tan^{-1} \frac{f}{x}$$

$$= \tan^{-1} \frac{5}{0.5} = 84.3^\circ$$

$$D = d + x$$

$$= 30 \text{ cm} + 0.5 \text{ cm}$$

$$= 30.5 \text{ cm}$$

$$r = \frac{D \sin \theta_1}{\sin[180^\circ - (\theta_1 + \theta_2)]}$$

$$= \frac{(30.5 \text{ cm}) \sin 30^\circ}{\sin[180^\circ - (30^\circ + 84.3^\circ)]}$$

$$= 16.73 \text{ cm}$$

This is the range to the image point. The range to the camera lens is

$$16.73 - \sqrt{f^2 + x^2} = 16.73 - \sqrt{(5 \text{ cm})^2 + (0.5 \text{ cm})^2}$$

$$= 11.7 \text{ cm}$$

7-10 Object boundaries and regions can be determined by connecting the end points of light stripe images.

7-11

$$v = 1100 \text{ ft/s}$$
$$t = 5 \text{ ms}$$
$$r = \frac{vt}{2}$$
$$= \frac{(1100 \text{ ft/s})(5 \text{ ms})}{2}$$
$$= 2.75 \text{ ft}$$

7-12

$$v = 9.8208 \times 10^8 \text{ ft/s}$$
$$t = 5 \text{ ms}$$
$$r = \frac{vt}{2}$$
$$= \frac{(9.8208 \times 10^8 \text{ ft/s})(5 \times 10^{-3} \text{ s})}{2}$$
$$= 2.4552 \times 10^6 \text{ ft}$$

7-13 *Electromagnetic:* Beam cannot be easily concentrated, requires high power for nonmetallic objects, accurate range measurements difficult over short distances.
Light: Accurate range measurements difficult over short distances, except for phase-shift laser system.
Sound: Wide beam width.

7-14 Two types of transducers used in ultrasonic ranging systems are the piezoelectric and electrostatic sound transducers.

7-15 30 to 50 kHz.

7-16

$$\text{Max. count time} = (2^8 - 1) \times \left(\frac{1}{10 \text{ kHz}}\right)$$
$$= 255 \times 0.1 \text{ ms}$$
$$= 25.5 \text{ ms}$$
$$\text{Max. range} = \frac{25.5 \text{ ms}}{1.8 \text{ ms/ft}} = 14.16 \text{ ft}$$

7-17

$$1 \text{ ft} = 1.8 \text{ ms}$$
$$\text{Count} = \frac{1.8 \text{ ms}}{0.1 \text{ ms/count}} = 18 = 0001\ 0010$$

7-18

$$\text{Max. resolution} = \frac{0.1 \text{ ms}}{1.8 \text{ ms/ft}} = 0.055 \text{ ft}$$

7-19 A range image uses different levels of gray to represent distances, while an intensity image uses a gray scale to represent light intensity.

7-20 Noncontact proximity sensors include inductive, capacitive, magnetic, and optical sensors.

7-21 Capacitance is used for proximity sensing by placing the plates of an air dielectric capacitor in an *RC* oscillating circuit. The oscillator output amplitude will change if a nongrounded object passes between the plates.

7-22 Multiple Hall-effect devices are used to sense position by placing a magnet on a moving object. The position of the object is sensed as the sensors are sequentially activated by the magnetic field.

7-23 An optical interrupter has a slot that separates the infrared light source from the phototransistor. Objects are detected as they pass through the slot. In an optical reflector, the infrared light source and phototransistor are mounted side by side. Reflective objects are then detected as they pass across the face of the device.

7-24 A count of 0110 1001 in a 1-kHz counter translates to an elapsed time of 105 ms. Thus, the velocity of the robot manipulator is

$$\frac{1 \text{ ft}}{105 \text{ ms}} = 9.52 \text{ ft/s}$$

Chapter 8

8-1 The three fundamental sensing operations of a tactile system are joint-force sensing, touch sensing, and slip sensing.

8-2 Three ways to measure forces acting on a robot manipulator are (1) measuring arm joint forces, (2) measuring wrist forces, and (3) measuring pedestal forces.

8-3 Arm joint forces in an electric manipulator control system are determined by measuring the armature current of the joint motors.

8-4 The six fundamental forces acting on a robot wrist assembly are lift, sweep, reach, yaw, pitch, and roll.

8-5
$$\begin{aligned}\text{Strain} &= \frac{.2 \text{ mm}}{5 \text{ cm}}\\ &= \frac{0.0002 \text{ m}}{0.05 \text{ m}}\\ &= 0.004\\ &= 4000 \times 10^{-6}\\ &= 4000 \text{ micros}\end{aligned}$$

8-6
$$\begin{aligned}\Delta R &= 2R_{\text{nominal}} \times \text{strain}\\ &= 2(2500\ \Omega) \times (4000 \times 10^{-6})\\ &= 20\ \Omega\end{aligned}$$

$$\text{Strain resistance} = 2500\ \Omega - 20\ \Omega = 2480\ \Omega$$

8-7 Gauge factor (GF) $= \dfrac{\Delta R/R_{\text{nominal}}}{\text{strain}}$

$$= \frac{20\ \Omega/2500\ \Omega}{4000\ \text{micros}}$$

$$= 2$$

8-8 $\text{GF} = \dfrac{\Delta R/R_{\text{nominal}}}{\text{strain}}$

Thus,

$$\Delta R = \text{GF} \times R_{\text{nominal}} \times \text{strain}$$
$$= 100 \times 500\ \Omega \times 4000\ \text{micros}$$
$$= 200\ \Omega$$

8-9 $R_2 = R_3 = 2500\ \Omega + 20\ \Omega = 2520\ \Omega$

$R_1 = R_4 = 2500\ \Omega - 20\ \Omega = 2480\ \Omega$

$$V_2 = \frac{R_2}{R_2 + R_1} V_{\text{REF}}$$
$$= \frac{2520\ \Omega}{2520\ \Omega + 2480\ \Omega}\ 12\ \text{V}$$
$$= 6.048\ \text{V}$$

$$V_4 = \frac{R_4}{R_4 + R_3} V_{\text{REF}}$$
$$= \frac{2480\ \Omega}{2480\ \Omega + 2520\ \Omega}\ 12\ \text{V}$$
$$= 5.952\ \text{V}$$

$$V_{\text{out}} = V_2 - V_4$$
$$= 6.048\ \text{V} - 5.952\ \text{V}$$
$$= 0.096\ \text{V}$$

8-10 $R_2 = R_3 = 500\ \Omega + 200\ \Omega = 700\ \Omega$

$R_1 = R_4 = 500\ \Omega - 200\ \Omega = 300\ \Omega$

$$V_2 = \frac{R_2}{R_2 + R_1} V_{\text{REF}}$$
$$= \frac{700\ \Omega}{700\ \Omega + 300\ \Omega}\ 12\ \text{V}$$
$$= 8.4\ \text{V}$$

$$V_4 = \frac{R_4}{R_4 + R_3} V_{REF}$$

$$= \frac{300\ \Omega}{300\ \Omega + 700\ \Omega}\ 12\ V$$

$$= 3.6\ V$$

$$V_{out} = V_2 - V_4$$

$$= 8.4\ V - 3.6\ V$$

$$= 4.8\ V$$

8-11 The ideal properties of a touch sensor are an array, skin-like, fast response, sensitive, low hysteresis, and smart.

8-12 The two broad categories of touch sensors are binary tactile sensors and analog tactile sensors.

8-13 A tactile sensor that can be used to probe objects without disturbing them is the whisker sensor.

8-14 Slip can be detected using a binary tactile array by comparing several binary tactile images over a given time period.

8-15 Analog tactile sensing elements include conductive rubber, conductive foam, semiconductor pressure sensors, and pressure sensitive paint, among others.

8-16 The conductive sensing elements form a matrix where a separate element exists at each row/column intersection within the matrix. The sensing element forms a variable resistance within a voltage divider along with a fixed resistance. The columns of the matrix are selected one at a time by the column select logic. Then the voltages developed across the row sensors within a given column are multiplexed into an A/D converter. The selected row/column junction must be isolated from the unselected junctions.

8-17 See Figure 8-25.

8-18 Slip sensing is important in a tactile system so that the optimum amount of gripping force can be applied to an object.

8-19 Two general ways to provide slip sensing in a tactile system are by sensing vibration between the object and gripper, or by sensing any movement, or displacement, between the object and gripper.

8-20 10×0.25 inches = 2.5 inches

8-21 2.5 inches/0.5 s = 5 inches/s

8-22 See Figure 8-30.

Index

D

E

F

G

S